THE ESSENCE OF MATHEMATICS
THROUGH ELEMENTARY PROBLEMS

The Essence of Mathematics

Through Elementary Problems

Alexandre Borovik and Tony Gardiner

Knowledge is for sharing

http://www.openbookpublishers.com

© 2019 Alexandre Borovik and Tony Gardiner

This work is licensed under a Creative Commons Attribution 4.0 International license (CC BY 4.0). This license allows you to share, copy, distribute and transmit the work; to adapt the work and to make commercial use of the work providing attribution is made to the author (but not in any way that suggests that they endorse you or your use of the work). Attribution should include the following information:

Alexandre Borovik and Tony Gardiner, *The Essence of Mathematics through Elementary Problems*. Cambridge, UK: Open Book Publishers, 2019. http://dx.doi.org/10.11647/OBP.0168

Further details about CC BY licenses are available at http://creativecommons.org/licenses/by/4.0/

All external links were active on 10/04/2019 and archived via the Internet Archive Wayback Machine: https://archive.org/web/

Every effort has been made to identify and contact copyright holders and any omission or error will be corrected if notification is made to the publisher.

Digital material and resources associated with this volume are available at http://www.openbookpublishers.com/product/979#resources

This is the third volume of the OBP Series in Mathematics:

ISSN 2397-1126 (Print)
ISSN 2397-1134 (Online)

ISBN Paperback 9781783746996
ISBN Hardback: 9781783747009
ISBN Digital (PDF): 9781783747016
DOI: 10.11647/OBP.0168

Cover photo: *Abstract Spiral Pattern* (2015) by Samuel Zeller, https://unsplash.com/photos/j0g8taxHZa0

Cover design by Anna Gatti.

Contents

Preface   vii

About this text   xiii

I. Mental Skills   1
   1.1   Mental arithmetic and algebra   2
       1.1.1   Times tables.   2
       1.1.2   Squares, cubes, and powers of 2.   2
       1.1.3   Primes   5
       1.1.4   Common factors and common multiples   5
       1.1.5   The Euclidean algorithm   6
       1.1.6   Fractions and ratio   7
       1.1.7   Surds   9
   1.2   Direct and inverse procedures   9
       1.2.1   Factorisation   12
   1.3   Structural arithmetic   12
   1.4   Pythagoras' Theorem   13
       1.4.1   Pythagoras' Theorem, trig for special angles, and CAST   14
       1.4.2   Converses and Pythagoras' Theorem   16
       1.4.3   Pythagorean triples   17
       1.4.4   Sums of two squares   19
   1.5   Visualisation   20
   1.6   Trigonometry and radians   22
       1.6.1   Sine Rule   22
       1.6.2   Radians and spherical triangles   23
       1.6.3   Polar form and $\sin(A+B)$ 27
   1.7   Regular polygons and regular polyhedra   28
       1.7.1   Regular polygons are cyclic   28
       1.7.2   Regular polyhedra   28
   1.8   Chapter 1: Comments and solutions   29

II. Arithmetic   51
 2.1 Place value and decimals: basic structure   51
 2.2 Order and factors   53
 2.3 Standard written algorithms   53
 2.4 Divisibility tests   54
 2.5 Sequences   56
      2.5.1 Triangular numbers   56
      2.5.2 Fibonacci numbers   56
 2.6 Commutative, associative and distributive laws   60
 2.7 Infinite decimal expansions   61
 2.8 The binary numeral system   64
 2.9 The Prime Number Theorem   66
 2.10 Chapter 2: Comments and solutions   69

III. Word Problems   91
 3.1 Twenty problems which embody "$3 - 1 = 2$"   93
 3.2 Some classical examples   94
 3.3 Speed and acceleration   95
 3.4 Hidden connections   96
 3.5 Chapter 3: Comments and solutions   97

IV. Algebra   111
 4.1 Simultaneous linear equations and symmetry   112
 4.2 Inequalities and modulus   115
      4.2.1 Geometrical interpretation of modulus, of inequalities,
             and of modulus inequalities   115
      4.2.2 Inequalities   117
 4.3 Factors, roots, polynomials and surds   119
      4.3.1 Standard factorisations   119
      4.3.2 Quadratic equations   123
 4.4 Complex numbers   126
 4.5 Cubic equations   131
 4.6 An extra   133
 4.7 Chapter 4: Comments and solutions   134

V. Geometry   169
 5.1 Comparing geometry and arithmetic   171
 5.2 Euclidean geometry: a brief summary   173
 5.3 Areas, lengths and angles   194
 5.4 Regular and semi-regular tilings in the plane   196
 5.5 Ruler and compasses constructions for regular polygons   199
 5.6 Regular and semi-regular polyhedra   201

5.7	The Sine Rule and the Cosine Rule	206
5.8	Circular arcs and circular sectors	211
5.9	Convexity	217
5.10	Pythagoras' Theorem in three dimensions	217
5.11	Loci and conic sections	220
5.12	Cubes in higher dimensions	227
5.13	Chapter 5: Comments and solutions	230
VI.	**Infinity: recursion, induction, infinite descent**	**283**
6.1	Proof by mathematical induction I	286
6.2	'Mathematical induction' and 'scientific induction'	287
6.3	Proof by mathematical induction II	290
6.4	Infinite geometric series	297
6.5	Some classical inequalities	299
6.6	The harmonic series	304
6.7	Induction in geometry, combinatorics and number theory	311
6.8	Two problems	313
6.9	Infinite descent	314
6.10	Chapter 6: Comments and solutions	317

Simon Phillips Norton
1952–2019

In memoriam

Preface

> *Understanding mathematics cannot be transmitted by painless entertainment ... actual contact with the **content** of living mathematics is necessary. The present book ... is not a concession to the dangerous tendency toward dodging all exertion.*
> Richard Courant (1888–1972) and Herbert Robbins (1915–2001)
> Preface to the first edition of *What is mathematics?*

Interested students of mathematics, who seek insight into the "essence of the discipline", and who read more widely with a view to discovering what the subject is really about, may emerge with the justifiable impression of serious mathematics as an austere, but distant mountain range – accessible only to those who devote their lives to its exploration. And they may conclude that the beginner can only appreciate its rough outline through a haze of unbridgeable distance. The best popularisers sometimes manage to convey more than this – including hints of the human story behind recent developments, and the way different branches and results interact in unexpected ways; but the essence of mathematics still tends to remain elusive, and the picture they paint is inevitably a broad brush substitute for the detail of living mathematics.

This collection takes a different approach. We start out by observing that mathematics is not a fixed entity – as one might unconsciously infer from the metaphor of an "austere mountain range". Mathematics is a *mental universe*, a work-in-progress in our collective imagination, which grows dramatically over time, and whose eventual extent would seem to be unconstrained – without any obvious limits. This boundlessness also works in reverse, when applied to small details: features which we thought we had understood are repeatedly filled in, or reinterpreted, in new ways to reveal finer and finer micro-structures.

Hence whatever the essence of the discipline may be, it is clearly not something which can only be accessed through the complete exploration of some fixed corpus of knowledge. Rather the essential character of mathematics seems to be related to

- the kind of material that counts as mathematical,
- the way this material is addressed,
- the changes in perspective that occur as our understanding grows and deepens, and
- the unexpected connections that regularly emerge between separate strands and layers.

There are a number of books giving excellent *general* advice to prospective students about how university mathematics differs from school mathematics. In contrast, this collection – which we hope will be enjoyed by interested high school students and their teachers, by undergraduates and postgraduates, and by many others is more like a messy workshop than a polished exposition. Here the reader is asked to tackle a sequence of problems, to reflect on what they discover, and mostly to draw their own conclusions (though some key messages are explicitly discussed in the text, or in the solutions at the end of each chapter). This attempt to engage the reader as an active participant along the way is inevitably untidy – and may sometimes prove frustrating. In particular, whereas a polished exposition would break up the text with eye-catching diagrams, an untidy workshop will usually leave the reader to draw their own figures as an essential part of the struggle. This temporary untidiness and frustration is an integral part of "the essence" that we seek to capture – provided it leads to occasional glimpses of the power, and the elegance of mathematics.

Young children and students of all ages regularly experience the power, the economy, the beauty, and the elegance of mathematics and of mathematical thinking *on a small scale*, through struggling with certain elementary results and problems (or groups of problems). For example, one of the problems we have included in Chapter 3 was mentioned explicitly in an interview[1] with the leading Russian mathematician Vladimir Arnold (1937–2010):

> **Interviewer**: *Please tell us a little bit about your early education. Were you already interested in mathematics as a child?*
>
> **Arnold**: [...] *The first real mathematical experience I had was when our schoolteacher I.V. Morotzkin gave us the following problem* [VA then formulated Problem **89** in Chapter 3].
>
> *I spent a whole day thinking on this oldie, and the solution (based on what are now called scaling arguments, dimensional analysis, or toric variety theory, depending on your taste) came as a revelation.*

---

[1] *Notices of the AMS*, vol 44, no. 4.

The feeling of discovery that I had then (1949) was exactly the same as in all the subsequent much more serious problems – be it the discovery of the relation between algebraic geometry of real plane curves and four-dimensional topology (1970), or between singularities of caustics and of wave fronts and simple Lie algebras and Coxeter groups (1972). It is the greed to experience such a wonderful feeling more and more times that was, and still is, my main motivation in mathematics.

This suggests that school mathematics need not be seen solely as an extended apprenticeship, which is somehow different from the craft of mathematics itself. Maybe some aspects of elementary mathematics can be experienced *as if they were a part of mathematics proper*, in which case suitably chosen elementary material, addressed in the appropriate spirit, might serve as a microcosm, or mini-universe, in which many features of the larger mathematical cosmos can be directly, and faithfully experienced by a relative novice (at least to some extent).

This collection of problems (and solutions) is an attempt to embody this idea in a form that might offer students, teachers, and interested readers a glimpse of "the essence of mathematics" – where this insight is experienced, not vicariously through the authors' elegant prose, or broad-brush descriptions, but **through the reader's own engagement with carefully chosen, accessible problems from elementary mathematics**.

Our understanding of the human body and how it works owes much to those (such as the ancient Greeks from 500 BC to Galen in the 2nd century AD, and much later Vesalius in the 16th century AD), who went beyond merely writing *about* such things in high-sounding prose, and who got their hands dirty by procuring cadavers, and cutting them up in order to see things from the inside – while asking themselves all the time how the different parts of the body were connected, and what function they served. In a similar way, the European discovery of the New World in the 15th century, and the confirmation that the Earth can be circumnavigated, depended on those who dared to set sail into uncharted waters and to keep a careful record of what they found.

The process of trying to understand things *from the inside* is not a deterministic procedure: it depends on a mixture of experience and inspiration, intelligence and inference, error and self-criticism. At any given time, the prevailing view may be incomplete, or misguided. But the underlying approach (of checking current ideas against the reality they purport to describe) is the only way we human beings know that allows us to gradually overcome errors and to gain fresh insight.

Our goal in this book is universal (namely to illustrate the idea that a suitably selected elementary microcosm can capture something of the essence of mathematics): hence the problems have all been chosen because we believe they convey something universal in a relatively elementary setting. But the particular set of problems chosen to illustrate the central goal is personal. So we encourage the reader to engage with these problems and results in the same way that old anatomists engaged with cadavers, or old explorers set out on voyages of discovery – getting their hands dirty while asking questions, such as:

> How do the things we see relate to what we know?
> What does this tell us about the subject of mathematics that we want to understand better?

In recent years schools and teachers in many countries have been under increasing political pressure to concentrate on measurable, short term "improvements". Such pressures have often been linked to central testing, with negative consequences for low scores. This has encouraged teachers to play safe, and to focus on *backward-looking* methods that allow students to produce answers to predictable one-step problems. The effect has been to downgrade the more important challenges which every student should face: namely

- of developing a robust mastery of new, *forward-looking* techniques (such as fractions, proportion, and algebra), and

- of integrating the single steps students have at their disposal into larger, systematic schemes, so that they can begin to tackle and solve simple multi-step problems.

Focusing on short-term goals is incompatible with good mathematics teaching. Learning mathematics is a long game; and teachers and students need the freedom to digress, to look ahead, and to build slowly over time. Teachers at each stage must be free to recognise that their primary responsibility is not just to improve their students' performance on the next test, but to establish a firm platform on which subsequent stages can build.

The pressures referred to above will be recognised in many countries, where well-intentioned, but ill-considered, centrally imposed accountability mechanisms have given rise to short-sighted "reforms". A didactical and pedagogical framework that is consistent with the essence, and the educational value of elementary mathematics cannot be rooted in false alternatives to mathematics (such as numeracy, or mathematical literacy). Nor can it be based on tests measuring cheap success on questions that require only one-step routines. We need a framework that encourages a rich combination of childlike curiosity, persistence, fruitful frustration, and the solid satisfaction of structural sense-making.

A problem sequence such as ours should ideally be distilled and refined over decades. However, the best is sometimes the enemy of the good:

Striving to better,
Oft we mar what's well.
(William Shakespeare, *King Lear*)

Hence, as a mild contribution to this process of rediscovering the essence of elementary mathematics, we risk this collection in its present form. And we encourage interested readers to take up pencil and paper, and to join us on this voyage of discovery through elementary mathematics.

Those who enjoy watching professional football (i.e. soccer) must sometimes marvel at the way experienced players seem to be instinctively aware of the movements of other players, and manage to feed the ball into gaps and spaces *that we mere spectators never even noticed were there*. What we overlook is that the best players practise the art of constantly looking around them, and updating their mental record – "viewing the field of play, with their heads up" – so that when the ball arrives and their eyes have to focus on the ball, their ever-changing mental record keeps updating itself to tell them (sometimes apparently miraculously) where the best tactical options lie. Implementing those tactical options depends in part on endless practice of skills; but practice is only one part of the story. What we encourage readers to develop here is the mathematical equivalent of this habit of "viewing the field of play, with one's head up", so that what is noticed can continue to guide the choice of tactical options when one is subsequently immersed in the thick of calculation.

Ours is a unique discipline, which is so much richer than the predictable routines that dominate many contemporary classrooms and assessments. We hope that **all** readers will find that the experience of struggling with, and savouring, this little collection reveals the occasional fresh and memorable insight into "the essence of mathematics".

We should not worry if students don't know everything,
but only if they know everything badly.
Peter Kapitsa, (1894–1984)
Nobel Prize for Physics 1978

To ask larger questions is to risk getting things wrong.
George Steiner (1929–)

Acknowledgements

Our thanks for suggestions, corrections comments and other contributions go to: Jean Bacon, Ayşe Berkman, Anna Borovik, Raul Cordovil, Serkan Dogan, Gwyneth Gardiner, Dick Hudson, Martin Hyland, Hovhannes Khuderverdyan, Ali Nesin, Martin Richards, Simon Singh, Gunnar Traustason, Ozge Uklem, Yusuf Ulum, and numerous students from the 2014 UKMT Summer School in Apperley Bridge.

About this text

> *And as this is done, so all
> similar problems are done.*
> Paolo dell'Abbaco (1282–1374)
> *Trattato d'aritmetica*

> *It is better to solve one problem in five different ways
> than to solve five problems in one way.*
> George Pólya (1887–1985)

> *If you go on hammering away at a problem,
> it seems to get tired, lies down,
> and lets you catch it.*
> Sir William Lawrence Bragg (1890–1971)
> Nobel Prize for Physics 1915

> *Young man,
> in mathematics you don't understand things.
> You just get used to them.*
> John von Neumann (1903–1957)

This is not a random collection of nice problems. Each item or problem, and each group of problems, is included for two reasons:

- they constitute good mathematics – mathematics which repays the effort of engaging with it for the first time, or revisiting it (should it already be familiar);

and

- they embody in a distilled form the quintessential **spirit** of elementary (initially pre-university) *mathematics* in a style which can be actively enjoyed by committed students and teachers in schools and colleges, and by the interested general reader.

Some items exemplify core general methods, which can be used over and over again (as hinted by the dell'Abbaco quotation). Some items require us to take different views of ostensibly the same material (as illustrated by the contrasting Pólya quote). Many items will at first seem elusive; but persistence may sometimes lead to an unexpected reward (in the spirit of the Bragg quote). In other instances, a correct answer may be obtained – yet leave the solver less than fully satisfied (at least in the short term, as illustrated by the von Neumann quote). And some items are of little importance in themselves – except that they force the solver to engage in *a kind of thinking* which is mathematically important.

Almost all of the included items are likely to involve – in some degree – that frustration which characterises all fruitful problem solving (as represented by the Bragg quote, and the William Golding quotation below), where, if we are lucky, a bewildering initial fog of incomprehension is sometimes magically dissipated by the process of struggling intelligently to make sense of things. And since one cannot always expect to succeed, there are bound to be occasions when the fog *fails* to lift. One may then have no choice but to consult the solutions (either because some essential idea or technique is not yet part of one's stock-in-trade, or because one has overlooked some simple connection). The only advice we can give here is: *the longer you can delay looking at the solutions the better*. But these solutions have been included both to help you improve your own efforts, and to show the way when you get truly stuck.

The "essence of mathematics", which we have tried to capture in these problems is mostly implicit, and so is often left for the reader to extract. Occasionally it has seemed appropriate to underline some aspect of a particular problem or its solution. Some comments of this kind have been included in the text that is interspersed between the problems. But in many instances, the comment or observation that needs to be made can only be appreciated **after** readers have struggled to solve a problem for themselves. In such cases, positioning the observation in the main text might risk spilling the beans prematurely. Hence, many important observations are buried away in the solutions, or in the **Notes** which follow many of the solutions. More often still, we have chosen to make no explicit remark, but have simply tried to shape and to group the problems in such a way that the intended message is conveyed silently by the problems themselves.

Roughly speaking, one can distinguish three types of problems: these may be labelled as *Core*, as *Gems*, or as focusing on more general *Cognition*.

1. *Core* problems or ideas encapsulate important mathematical concepts and mathematical knowledge in a relatively mundane way, yet in a manner that is in some way canonical. These have sometimes been included here to emphasise some important aspect, which contemporary treatments may have forgotten.

2. *Gems* constitute some kind of paradigm that all aspiring students of mathematics should encounter at some stage. These are likely to be encountered as fully fresh, or surprising, only once in a lifetime. But they then continue to serve as beacons, or trig points, that help to delineate the mathematical landscape.

3. The third type of problem plays an *auxiliary* role – namely problems which emphasise the importance of basic *cognitive skills* for doing mathematics (for example: instant mental calculation, visualisation of abstract concepts, short-term memory, attention span, etc.)

The items are grouped into chapters – each with a recognisable theme. Later chapters tend to have a higher level of technical demand than earlier chapters; and the sequence is broadly consistent with a rising level of sophistication. However, this is not a didactically organised text. Each problem is listed where it fits most naturally, even if it involves an idea which is not formally introduced until somewhat later. Detailed solutions, together with any comments which would be out of place in the main text, are grouped together at the end of each chapter.

The first few chapters tend to focus on more elementary material – partly to emphasise the hierarchical structure of mathematics, partly as a reminder that the essence of mathematics can be experienced at *all* levels, and partly to offer a gentle introduction to readers who may appreciate something slightly more structured before they tackle selected parts of later chapters. Hence these early chapters include more discursive commentary than later chapters. **Readers who choose to skip these nursery slopes on a first reading may wish to return to them later, and to consider what this relatively elementary material tells us about the essence of mathematics.**

The collection is offered as a *supplement* to the standard school curriculum. Some items could (and perhaps should) be incorporated into any official curriculum. But the collection as a whole is mainly designed for those who have good reason, and the time and inclination, to go beyond the usual institutional constraints, and to begin to explore the broader landscape of elementary mathematics in order to experience real, "free range" mathematics – as opposed to artificially reconstituted, or processed products.

> *It has come to me in a flash! One's intelligence may march about and about a problem, but the solution does not come gradually into view. One moment it is not. The next and it is there.*
>
> William Golding (1911–1993), *Rites of Passage*

I. Mental Skills

> *Even a superficial glance at history shows ...*
> *great innovators ... did vast amounts of computation*
> *and gained much of their insight in this way.*
> *I deplore the fact that contemporary mathematical education*
> *tends to give students the idea that computation is*
> *demeaning drudgery to be avoided at all costs.*
> Harold M. Edwards (1936–)
> *Fermat's Last Theorem*

We start our journey in a way that should be accessible to everyone – with a quick romp through important ideas from secondary school mathematics. The content is at times very elementary; but the problems often hint at something more challenging. The items included in this first chapter also highlight selected facts, techniques and ideas. Some of this early material is included to introduce certain ideas and techniques that later chapters will assume to be "known". A few problems appeal to more advanced ideas (such as *complex numbers*), and are included here to indicate that "mental skills" are not restricted to elementary material.

Pencil and paper will be needed, but the items tend to focus on things which a student of mathematics should know by heart, or should learn to see at a glance, or should be able to calculate inside the head. In later problems (e.g. from Problem **18** onwards) the emphasis on *mental skills* should be interpreted as "ways of thinking", rather than being taken to mean that everything should be done in your head. This is especially true where extended calculations or proofs are required.

Some of the items in this chapter (such as Problems **1** and **2**) should be thoroughly familiar, and are included to underline this fact, rather than because we anticipate that they will need much active attention. Most of the early items in this first chapter are either *core* or *auxiliary*. However, there are also some real *gems*, which may even warrant a place in the the standard *core*.

The chapter is largely devoted to underlining the need for mastery of a repertoire of instantly available techniques, that can be used mentally, quickly, and flexibly to analyse less familiar problems at sight. But it also seeks to emphasise *connections*. Hence readers should be prepared to challenge their previous experience, in case it may have led to methods and results being perceived too narrowly.

We repeat the comment made in the section *About this book*. The "essence of mathematics", which is referred to in the title, is largely implicit in the problems, and is there for the reader to extract. There is some discussion of this essence in the text interspersed between the problems. But, to avoid spilling the beans prematurely, and hence spoiling the problems, many important observations are buried away in the solutions, or in the **Notes** which follow many of the solutions.

1.1. Mental arithmetic and algebra

1.1.1 Times tables.

Problem 1 Using only mental arithmetic:

(a) Compute for yourself, and learn by heart, the times tables up to 9×9.

(b) Calculate instantly:

    (i) 0.004×0.02 (ii) 0.0008×0.07 (iii) 0.007×0.12
    (iv) $1.08 \div 1.2$ (v) $(0.08)^2$ △

Multiplication tables are important for many reasons. They allow us to appreciate directly, at first hand, the efficiency of our miraculous *place value system* – in which representing any number, and implementing any operation, are reduced to a combined mastery of

(i) the arithmetical behaviour of the ten digits 0–9, and

(ii) the index laws for powers of 10.

Fluency in mental and written arithmetic then leaves the mind free to notice, and to appreciate, the deeper patterns and structures which may be lurking just beneath the surface.

1.1.2 Squares, cubes, and powers of 2.

Algebra begins in earnest when we start to calculate with expressions involving *powers*. As one sees in the language we use for *squares* and *cubes* (i.e. 2nd and 3rd powers), these powers were interpreted *geometrically* for hundreds and thousands of years – so that higher powers, beyond the third power, were seen as being somehow unreal (like the 4th dimension). Our uniform algebraic notation covering all powers emerged in the 17th century (with Descartes (1596–1650)). But before one begins to work with *algebraic* powers, one should first aim to achieve complete fluency in working with *numerical* powers.

Problem 2

(a) Compute by mental arithmetic (using pencil only to record results), then learn by heart:

(i) the squares of positive integers: first up to 12^2; then to 31^2

(ii) the cubes of positive integers up to 11^3

(iii) the powers of 2 up to 2^{10}.

(b) How many squares are there: (i) < 1000? (ii) $< 10\,000$? (iii) $< 100\,000$?

(c) How many cubes are there: (i) < 1000? (ii) $< 10\,000$? (iii) $< 1\,000\,000$?

(d) (i) Which powers of 2 are squares? (ii) Which powers of 2 are cubes?

(e) Find the smallest square greater than 1 that is also a cube. Find the next smallest. △

Evaluating powers, and the associated index laws, constitute an example of a *direct* operation. For each *direct* operation, we need to think carefully about the corresponding *inverse* operation – here "extracting roots". In particular, we need to be clear about the distinction between the fact that the equation $x^2 = 4$ has *two* different solutions, while $\sqrt{4}$ has *just one* value (namely 2).

Problem 3

(a) The operation of "squaring" is a **function**: it takes a single real number x as *input*, and delivers a definite real number x^2 as *output*.

– Every positive number arises as an output ("is the square of something").

– Since $x^2 = (-x)^2$, each output (other than 0) arises from **at least two** different inputs.

– If $a^2 = b^2$, then $0 = a^2 - b^2 = (a-b)(a+b)$, so either $a = b$, or $a = -b$. Hence no two positive inputs have the same square, so each output (other than 0) arises from **exactly two** inputs (one positive and one negative).

– Hence each positive output y corresponds to **just one** positive input, called $\sqrt{y}$.

Find:

4   Mental Skills

(i) $\sqrt{49}$ (ii) $\sqrt{144}$ (iii) $\sqrt{441}$ (iv) $\sqrt{169}$
(v) $\sqrt{196}$ (vi) $\sqrt{961}$ (vii) $\sqrt{96\,100}$

(b) Let $a > 0$ and $b > 0$. Then $\sqrt{ab} > 0$, and $\sqrt{a} \times \sqrt{b} > 0$, so both expressions are positive.

Moreover, they *have the same square*, since
$$(\sqrt{ab})^2 = ab = (\sqrt{a})^2 \cdot (\sqrt{b})^2 = (\sqrt{a} \times \sqrt{b})^2.$$

$\therefore \sqrt{a \times b} = \sqrt{a} \times \sqrt{b}.$

Use this fact to simplify the following:

(i) $\sqrt{8}$ (ii) $\sqrt{12}$ (iii) $\sqrt{50}$
(iv) $\sqrt{147}$ (v) $\sqrt{288}$ (vi) $\sqrt{882}$

(c) [This part requires some written calculation.] Exact expressions involving square roots occur in many parts of elementary mathematics. We focus here on just one example – namely the regular pentagon.

Suppose that a regular pentagon $ABCDE$ has sides of length 1.

(i) Prove that the diagonal AC is parallel to the side ED.
(ii) If AC and BD meet at X, explain why $AXDE$ is a rhombus.
(iii) Prove that triangles ADX and CBX are similar.
(iv) If AC has length x, set up an equation and find the exact value of x.
(v) Find the exact length of BX.
(vi) Prove that triangles ABD and BXA are similar.
(vii) Find the exact values of $\cos 36°$, $\cos 72°$.
(viii) Find the exact values of $\sin 36°$, $\sin 72°$.   △

Every calculation with square roots depends on the fact that "$\sqrt{}$ is a function". That is: given $y > 0$,

$\sqrt{y}$ denotes a **single value** – the **positive** number whose square is y.

The equation $x^2 = y$ has **two** roots, namely $x = \pm\sqrt{y}$; however, $\sqrt{y}$ has **just one** value (which is positive).

The mathematics of the regular pentagon is important – and generally neglected. It is included here to underline the way exact expressions involving square roots arise naturally.

In Problem **3**(c), parts (iii) and (vi) require one to identify similar triangles using angles. The fact that "corresponding sides are then proportional" leads to a quadratic equation – and hence to square roots.

Parts (vii) and (viii) illustrate the fact that basic tools, such as

- the trigonometric identity $\cos^2 \theta + \sin^2 \theta = 1$,
- the *Cosine Rule*, and
- the *Sine Rule*

should be part of one's stock-in-trade. Notice that the exact values for

$$\cos 36°, \ \cos 72°, \ \sin 36°, \ \text{and} \ \sin 72°$$

also determine the exact values of

$\sin 54° = \cos 36°, \ \sin 18° = \cos 72°, \ \cos 54° = \sin 36°, \ \text{and} \ \cos 18° = \sin 72°.$

1.1.3 Primes

Problem 4

(a) Factorise 12 345 as a product of primes.

(b) Using only mental arithmetic, make a list of all prime numbers up to 100.

(c)(i) Find a prime number which is one less than a square.

(ii) Find another such prime. △

There are 4 prime numbers less than 10; 25 prime numbers less than 100; and 168 prime numbers less than 1000.

Problem **4(c)** is included to emphasise a frequently neglected message:

> Words and images are part of the way we communicate.
> But most of us cannot *calculate* with words and images.

To make use of mathematics, we must routinely translate *words* into *symbols*. For example, unknown numbers need to be represented by symbols, and points in a geometric diagram need to be properly labelled, before we can begin to calculate, and to reason, effectively.

1.1.4 Common factors and common multiples

To add two fractions we need to find a common multiple, or the LCM, of the two given denominators. To cancel fractions, or to simplify ratios, we need to be able to spot common factors and to find HCFs. Two positive integers a, b which have no (positive) common factors other than 1 (that is, with $HCF(a, b) = 1$) are said to be *relatively prime*, or *coprime*.

6 Mental Skills

Problem 5 [This problem requires a mixture of serious thought and written proof.]

(a) I choose six integers between 10 and 19 (inclusive).

  (i) Prove that some pair of integers among my chosen six must be relatively prime.

  (ii) Is it also true that some pair must have a common factor?

(b) I choose six integers in the nineties (from 90–99 inclusive).

  (i) Prove that some pair among my chosen integers must be relatively prime.

  (ii) Is it also true that some pair must have a common factor?

(c) I choose $n+1$ integers from a run of $2n$ consecutive integers.

  (i) Prove that some pair among the chosen integers must be relatively prime.

  (ii) Is it also true that some pair must have a common factor? △

1.1.5 The Euclidean algorithm

School mathematics gives the impression that to find the HCF of two integers m and n, one must first obtain the *prime power factorisations* of m and of n, and can then extract the HCF from these two expressions. This is fine for beginners. But arithmetic involves unexpected subtleties. It turns out that, as the numbers get larger, factorising integers quickly becomes extremely difficult – a difficulty that is exploited in modern encryption systems. (The limitations of any method that depends on first finding the prime power factorisation of an integer should have become clear in Problem **4(b)**, where it is all too easy to imagine that 91 is prime, and in Problem **4(c)(ii)**, where students regularly think that 143, or that 323 are prime.)

Hence we would like to have a simple way of finding the HCF of two integers without having to *factorise* each of them first. That is what the Euclidean algorithm provides. We will look at this in more detail later. Meanwhile here is a first taste.

Problem 6

(a)(i) Explain why any integer that is a factor (or a divisor) of both m and n must also be a factor of their difference $m - n$, and of their sum $m + n$.

  (ii) Prove that
$$HCF(m, n) = HCF(m - n, n).$$

(iii) Use this to calculate in your head $HCF(1001, 91)$ without factorising either number.

(b)(i) Prove that: $HCF(m, m+1) = 1$.

(ii) Find $HCF(m, 2m+1)$.

(iii) Find $HCF(m^2 + 1, m - 1)$. △

1.1.6 Fractions and ratio

Problem 7 Which is bigger: 17% of nineteen million, or 19% of seventeen million? △

Problem 8

(a) Evaluate
$$\left(1 + \frac{1}{2}\right)\left(1 + \frac{1}{3}\right)\left(1 + \frac{1}{4}\right)\left(1 + \frac{1}{5}\right).$$

(b) Evaluate
$$\sqrt{1 + \frac{1}{2}} \times \sqrt{1 + \frac{1}{3}} \times \sqrt{1 + \frac{1}{4}} \times \sqrt{1 + \frac{1}{5}} \times \sqrt{1 + \frac{1}{6}} \times \sqrt{1 + \frac{1}{7}}.$$

(c) We write the product "$4 \times 3 \times 2 \times 1$" as "4!" (and we read this as "4 factorial"). Using only pencil and paper, how quickly can you work out the number of weeks in 10! seconds? △

Problem 9 The "DIN A" series of paper sizes is determined by two conditions. The basic requirement is that all the DIN A rectangles are *similar*; the second condition is that when we fold a given size exactly in half, we get the next smaller size. Hence

- a sheet of paper of size A3 folds in half to give a sheet of size A4 – which is *similar* to A3; and
- a sheet of size A4 folds in half to give a sheet of size A5; etc..

(a) Find the constant ratio
$$r = \text{"(longer side length) : (shorter side length)"}$$
for all DIN A paper sizes.

(b)(i) To enlarge A4 size to A3 size (e.g. on a photocopier), each length is enlarged by a factor of r. What is the "enlargement factor" to get from A3 size back to A4 size?

(ii) To "enlarge" A4 size to A5 size (e.g. on a photocopier), each length is "enlarged" by a factor of $\frac{1}{r}$. What is the enlargement factor to get from A5 size back to A4 size? △

Problem 10

(a) In a sale which offers "15% discount on all marked prices" I buy three articles: a pair of trainers priced at £57.74, a T-shirt priced at £17.28, and a yo-yo priced at £4.98. Using only mental arithmetic, work out how much I should expect to pay altogether.

(b) Some retailers display prices without adding VAT – or "sales tax" – at 20% (because their main customers need to know the pre-VAT price). Suppose the prices in part (a) are the prices before adding VAT. Each price then needs to be adjusted in two ways – adding VAT and subtracting the discount. Should I add the VAT first and then work out the discount? Or should I apply the discount first and then add the VAT?

(c) Suppose the discount in part (b) is no longer 15%. What level of discount would exactly cancel out the addition of VAT at 20%? △

Problem 11

(a) Using only mental arithmetic:

(i) Determine which is bigger:

$$\frac{1}{2} + \frac{1}{5} \quad \text{or} \quad \frac{1}{3} + \frac{1}{4}?$$

(ii) How is this question related to the observation that $10 < 12$?

(b) [This part will require some written calculation and analysis.]

(i) For positive real numbers x, compare

$$\frac{1}{x+2} + \frac{1}{x+5} \quad \text{and} \quad \frac{1}{x+3} + \frac{1}{x+4}.$$

(ii) What happens in part (i) if x is negative? △

1.1.7 Surds

Problem 12

(a) Expand and simplify in your head:

(i) $(\sqrt{2}+1)^2$ (ii) $(\sqrt{2}-1)^2$ (iii) $(1+\sqrt{2})^3$

(b) Simplify:

(i) $\sqrt{10+4\sqrt{6}}$ (ii) $\sqrt{5+2\sqrt{6}}$

(iii) $\sqrt{\frac{3+\sqrt{5}}{2}}$ (iv) $\sqrt{10-2\sqrt{5}}$ △

The expressions which occur in exercises to develop fluency in working with surds often appear arbitrary. But they may not be. The arithmetic of surd arises naturally: for example, some of the expressions in the previous problem have already featured in Problem **3**(c). In particular, surds will feature whenever Pythagoras' Theorem is used to calculate lengths in geometry, or when a proportion arising from similar triangles requires us to solve a quadratic equation. So surd arithmetic is important. For example:

- A regular octagon with side length 1 can be surrounded by a square of side $\sqrt{2}+1$ (which is also the diameter of its incircle); so the area of the regular octagon equals $(\sqrt{2}+1)^2 - 1$ (the square minus the four corners).

- $\sqrt{2}-1$ features repeatedly in the attempt to apply the Euclidean algorithm, or *anthyphairesis*, to express $\sqrt{2}$ as a "continued fraction".

- $\sqrt{10-2\sqrt{5}}$ may look like an arbitrary, uninteresting repeated surd, but is in fact very interesting, and has already featured as $4\sin 36°$ in Problem **3**(c).

- One of the simplest ruler and compasses constructions for a regular pentagon $ABCDE$ (see Problem **185**) starts with a circle of radius 2, centre O, and a point A on the circle, and in three steps constructs the next point B on the circle, where $\underline{AB}$ is an edge of the inscribed regular pentagon, and
$$\underline{AB} = \sqrt{10-2\sqrt{5}}.$$

1.2. Direct and inverse procedures

We all learn to calculate – with numbers, with symbols, with functions, etc. But we may not notice that most calculating procedures come *in pairs*.

- First we learn a *direct*, deterministic, handle-turning technique, where answers are easy to churn out (as with addition, or multiplication, or working out powers, or multiplying brackets in algebra, or differentiating).

- Then we try to work backwards, or to "undo" this direct operation (as with subtraction, or division, or finding roots, or factorising, or integrating). This *inverse* procedure requires one to be completely fluent in the corresponding *direct* procedure; but it is much more demanding, in that one has to juggle possibilities as one goes, in order to home in on the required answer.

To master *inverse* procedures requires a surprising amount of time and effort. And because they are harder to master, they can easily get neglected. Even where they receive a lot of time, there are aspects of *inverse* procedures which tend to go unnoticed.

Problem 13 In how many different ways can the missing digits in this short multiplication be completed?

$$\begin{array}{r} \square\,6 \\ \times\quad \square \\ \hline \square\,2\,8 \end{array}$$

△

One would like students not only to master the *direct* operation of multiplying digits effectively, but also to notice that the *inverse* procedure of

"identifying the multiples of a given integer that give rise to a specified output"

depends on

the HCF of the *multiplier* and the *base* (10) of the numeral system.

- Multiplying by 1, 3, 7, or 9 induces a *one-to-one mapping* on the set of ten digits 0–9; so an inverse problem such as "7 × □ ends in 6" has just one digit-solution.

- Multiplying by 2, 4, 6, or 8 induces a *two-to-one mapping* onto the set of even digits (multiples of 2); so an inverse problem such as "6 × □ ends in 4" has two digit-solutions, and an inverse problem such as "6 × □ ends in 3" has no digit-solutions.

- Multiplying by 5 induces a *five-to-one mapping* onto the multiples (0 and 5) of 5, so an inverse problem such as "5 × □ ends in 0" has five digit-solutions and an inverse problem such as "5 × □ ends in 3" has no digit-solutions at all.

- Multiplying by 0 induces a *ten-to-one mapping* onto the multiples of 0 (namely 0); so an inverse problem such as "0 × □ ends in 0" has ten digit-solutions and an inverse problem such as "0 × □ ends in 3 (or any digit other than 0)" has no digit-solutions at all.

The next problem shows – in a very simple setting – how elusive inverse problems can be. Here, instead of being asked to perform a *direct* calculation, the rules and the answer are given, and we are simply asked to invent a calculation that gives the specified output.

Problem 14

(a) In the "24 game" you are given four numbers. Your job is to use each number once, and to combine the four numbers using any three of the four basic arithmetical operations – using the same operation more than once if you wish, and as many brackets as you like (but never concatenating different numbers, such as "3" and "4" to make "34"). If the given numbers are 3, 3, 4, 4, then one immediately sees $3 \times 4 + 3 \times 4 = 24$. With 3, 3, 5, 5 it may take a little longer, but is still fairly straightforward. However, you may find it more challenging to make 24 in this way:

  (i) using the four numbers 3, 3, 6, 6
  (ii) using the four numbers 3, 3, 7, 7
  (iii) using the four numbers 3, 3, 8, 8.

(b) Suppose we restrict the numbers to be used each time to "four 4s" (4, 4, 4, 4), and change the goal from "make 24", to "make each answer from 0–10 using exactly four 4s".

  (i) Which of the numbers 0–10 cannot be made?
  (ii) What if one is allowed to use squaring and square roots as well as the four basic operations? What is the first inaccessible integer? △

Calculating by turning the handle deterministically (as with addition, or multiplication, or multiplying out brackets, or differentiating) is a valuable skill. But such *direct* procedures are usually only the beginning. Using mathematics and solving problems generally depend on the corresponding *inverse* procedures – where a certain amount of juggling and insight is needed in order to work backwards (as with subtraction, or division, or factorisation, or integration). For example, in applications of calculus, the main challenge is to solve *differential equations* (an *inverse* problem) rather than to differentiate known functions.

Problem **14** captures the spirit of this idea in the simplest possible context of arithmetic: the required answer is given, and we have to find how (or whether) that answer can be generated. We will meet more interesting examples of this kind throughout the rest of the collection.

1.2.1 Factorisation

Problem 15

(a)(i) Expand $(a+b)^2$ and $(a+b)^3$.

 (ii) Without doing any more work, write out the expanded forms of $(a-b)^2$ and $(a-b)^3$.

(b) Factorise (i) $x^2 + 2x + 1$ (ii) $x^4 - 2x^2 + 1$ (iii) $x^6 - 3x^4 + 3x^2 - 1$.

(c)(i) Expand $(a-b)(a+b)$.

 (ii) Use (c)(i) and (a)(i) to write down (with no extra work) the expanded form of
$$(a-b-c)(a+b+c)$$
and of
$$(a-b+c)(a+b-c).$$

(d) Factorise $3x^2 + 2x - 1$. △

1.3. Structural arithmetic

Whenever the answer to a question turns out to be unexpectedly nice, one should ask oneself whether this is an accident, or whether there is some explanation which should perhaps have led one to expect such a result. For example:

- Exactly 25 of the integers up to 100 are prime numbers – and 25 is exactly **one quarter** of 100. This is certainly a beautifully memorable fact. But it is a numerical fluke, with no hidden mathematical explanation.

- 11 and 101 are prime numbers. Is this perhaps a way of generating lots of prime numbers:
$$11, 101, 1001, 10\,001, 100\,001, \ldots ?$$
It may at first be tempting to think so – until, that is, you remember what you found in Problem **6**(a)(iii).

Problem 16 Write out the first 12 or so powers of 4:

$$4,\ 16,\ 64,\ 256,\ 1024,\ 4096,\ 16\,384,\ 65\,536,\ \ldots$$

Now create two sequences:

the sequence of **final** digits: 4, 6, 4, 6, 4, 6, ...
the sequence of **leading** digits: 4, 1, 6, 2, 1, 4, 1, 6, ...

Both sequences seem to consist of a single "block", which repeats over and over for ever.

(a) How long is the apparent repeating block for the first sequence? How long is the apparent repeating block for the second sequence?

(b) It may not be immediately clear whether either of these sequences really repeats forever. Nor may it be clear whether the two sequences are alike, or whether one is quite different from the other. Can you give a simple proof that one of these sequences *recurs*, that is, repeats forever?

(c) Can you explain why the other sequence seems to recur, and decide whether it really does recur forever? △

Problem 17 The 4 by 4 "multiplication table" below is completely familiar.

$$\begin{array}{cccc}
1 & 2 & 3 & 4 \\
2 & 4 & 6 & 8 \\
3 & 6 & 9 & 12 \\
4 & 8 & 12 & 16
\end{array}$$

What is the total of all the numbers in the 4 by 4 square? How should one write this answer in a way that makes the total obvious? △

1.4. Pythagoras' Theorem

From here on the idea of "mental skills" tends to refer to *ways of thinking* rather than to doing everything in your head.

Pythagoras' Theorem is one of the first truly surprising results in school mathematics: it is hard to see why anyone would think of "adding the squares of the two shorter sides". Despite the apparent attribution to a named person (Pythagoras), the origin of the theorem, and its proof, are unclear. There certainly was someone called Pythagoras (around 500 BC). But the main ancient references to him were written many hundreds of years after he died, and are not very reliable. The truth is that we know very little about him, or his theorem. The proof in Problem **18** below appeared in Book I of Euclid's thirteen books of *Elements* (written around 300 BC – two hundred years after Pythagoras). Much that is said (wrongly) to stem from Pythagoras is attributed in some sources to the *Pythagoreans* – a loose term which refers to any philosopher in what is seen as a tradition going back to Pythagoras.

(This is a bit like interpreting anything called Christian in the last 2000 years as stemming directly from Christ himself.)

Clay tablets from around 1700 BC suggest that some Babylonians must have known "Pythagoras' Theorem"; and it is hard to see how one could know the result without having some kind of justification. But we have no evidence of either a clear statement, or a proof, at that time. There are also Chinese texts that refer to Pythagoras' Theorem (or as they call it, "Gougu"), which are thought to have originated BC – though the earliest surviving edition is from the 13th century AD. There is even an interesting little book by Frank Swetz, with the tongue-in-cheek title *Was Pythagoras Chinese?*.

The history may be confused, but the result – and its Euclidean proof – embodies something of the surprise and elegance of the very best mathematics. The Euclidean proof is included here partly because it is one that can, and should, be remembered (or rather, reconstructed – once one realises that there is really only one possible way to split the "square on the hypotenuse" in the required way). But, as we shall see, the result also links to *exact mental calculation* with surds, to trigonometry, to the familiar mnemonic "CAST", to the idea of a "converse", to sums of two squares, and to Pythagorean triples.

1.4.1 Pythagoras' Theorem, trig for special angles, and CAST

Problem 18 (Pythagoras' Theorem) Let $\triangle ABC$ be a right angled triangle, with a right angle at C. Draw the squares $ACQP$, $CBSR$, and $BAUT$ on the three sides, external to $\triangle ABC$. Use the resulting diagram to prove *in your head* that the square $BAUT$ on BA is equal to the sum of the other two squares by:

- drawing the line through C perpendicular to AB, to meet AB at X and UT at Y

- observing that PA is parallel to QCB, so that $\triangle ACP$ (half of the square $ACQP$, with base AP and perpendicular height AC) is equal in area to $\triangle ABP$ (with base AP and the same perpendicular height)

- noting that $\triangle ABP$ is SAS-congruent to $\triangle AUC$, and that $\triangle AUC$ is equal in area to $\triangle AUX$ (half of rectangle $AUYX$, with base AU and height AX).

- whence $ACQP$ is equal in area to rectangle $AUYX$

- similarly $BCRS$ is equal in area to $BTYX$. $\triangle$

The proof in Problem **18** is the proof to be found in Euclid's *Elements* Book 1, Proposition 47. Unlike many proofs,

- it is clear what the proof depends on (namely SAS triangle congruence, and the area of a triangle), and

- it reveals exactly **how** the square on the hypotenuse AB divides into two summands – one equal to the square on AC and one equal to the square on BC.

Problem 19

(a) Use Pythagoras' Theorem in a square $ABCD$ of side 1 to show that the diagonal AC has length $\sqrt{2}$. Use this to work out *in your head* the exact values of $\sin 45°$, $\cos 45°$, $\tan 45°$.

(b) In an equilateral triangle $\triangle ABC$ with sides of length 2, join A to the midpoint M of the base BC. Apply Pythagoras' Theorem to find AM. Hence work out *in your head* the exact values of $\sin 30°$, $\cos 30°$, $\tan 30°$, $\sin 60°$, $\cos 60°$, $\tan 60°$.

(c)(i) On the unit circle with centre at the origin $O:(0,0)$, mark the point P so that P lies in the first quadrant, and so that OP makes an angle θ with the positive x-axis (measured anticlockwise from the positive x-axis). Explain why P has coordinates $(\cos\theta, \sin\theta)$.

(ii) Extend the definitions of $\cos\theta$ and $\sin\theta$ to apply to angles beyond the first quadrant, so that for any point P on the unit circle, where OP makes an angle θ measured anticlockwise from the positive x-axis, the coordinates of P are $(\cos\theta, \sin\theta)$. Check that the resulting functions sin and cos satisfy:

* sin and cos are both positive in the first quadrant,
* sin is positive and cos is negative in the second quadrant,
* sin and cos are both negative in the third quadrant, and
* sin is negative and cos is positive in the fourth quadrant.

(iii) Use (a), (b) to calculate the exact values of $\cos 315°$, $\sin 225°$, $\tan 210°$, $\cos 120°$, $\sin 960°$, $\tan(-135°)$.

(d) Given a circle of radius 1, work out the exact area of a regular n-gon **inscribed in** the circle:

(i) when $n = 3$ (ii) when $n = 4$ (iii) when $n = 6$
(iv) when $n = 8$ (v) when $n = 12$.

(e) Given a circle of radius 1, work out the area of a regular n-gon **circumscribed around** the circle:

(i) when $n = 3$ (ii) when $n = 4$ (iii) when $n = 6$
(iv) when $n = 8$ (v) when $n = 12$.   △

16 Mental Skills

Knowing the exact values of sin, cos and tan for the special angles $0°$, $30°$, $45°$, $60°$, $90°$ is like knowing one's tables. In particular, it allows one to evaluate trigonometric functions mentally for related angles in all four quadrants (using the CAST mnemonic – C being in the SE of the unit circle, A in the NE quadrant, S in the NW quadrant, and T in the SW quadrant – to remind us which functions are positive in each quadrant). These special angles arise over and over again in connection with equilateral triangles, squares, regular hexagons, regular octagons, regular dodecagons, etc., where one can use what one knows to calculate *exactly* in geometry.

Problem 20

(a) Use Pythagoras' Theorem to calculate the exact length of the diagonal AC in a square $ABCD$ of side length 2.

(b) Let X be the centre of the square $ABCD$ in part (a). Draw lines through X parallel to the sides of $ABCD$ and so divide the large square into four smaller squares, each of side 1. Find the length of the diagonals AX and XC.

(c) Compare your answers to parts (a) and (b) and your answer to Problem **3**(b)(i). △

Pythagoras' Theorem holds the key to calculating exact distances in the plane. To calculate distances on the Earth's surface one needs a version of Pythagoras for "right angled triangles" on the sphere. We address this in Chapter 5.

1.4.2 Converses and Pythagoras' Theorem

Each mathematical statement of the form

> "if ... (Hypothesis H),
> then ... (Consequence C)"

has a *converse* statement – namely

> "if C,
> then H".

If the first statement is true, there is no *a priori* reason to expect its converse to be true. For example, part (c) of Problem **25** below proves that

> "if an integer has the form $4k + 3$,
> then it cannot be written as the sum of two squares".

However, the converse of this statement

"if an integer cannot be written as a sum of two squares, then it has the form $4k + 3$"

is false – since 6 cannot be written as the sum of two squares.

Despite this counterexample, whenever we prove a standard result, it makes sense to ask whether the converse is also true. For example,

"if $PQRS$ is a parallelogram, then opposite angles are equal: $\angle P = \angle R$, and $\angle Q = \angle S$" (see Problem **157**(ii)).

However you may not have considered the truth (or otherwise) of the converse statement:

If $ABCD$ is a quadrilateral in which opposite angles are equal ($\angle A = \angle C$ and $\angle B = \angle D$), is it true that $ABCD$ has to be a parallelogram?

The next problem invites you to prove the converse of Pythagoras' Theorem. You should not use the Cosine Rule, since this is a generalisation of both Pythagoras' Theorem and its converse.

Problem 21 Let ABC be a triangle. We use the standard labelling convention, whereby the side BC opposite A has length a, the side CA opposite B has length b, and the side AB opposite C has length c.

Prove that, if $c^2 = a^2 + b^2$, then $\angle BCA$ is a right angle. △

1.4.3 Pythagorean triples

The simplest example of a right angled triangle with integer length sides is given by the familiar triple 3, 4, 5:

$$3^2 + 4^2 = 5^2.$$

Any such integer triple is called a *Pythagorean triple*.

The classification of *all* Pythagorean triples is a delightful piece of elementary number theory, which is included in this chapter both because the result deserves to be memorised, and because (like Pythagoras' Theorem itself) the proof only requires one to juggle a few simple ideas that should be part of one's armoury.

Pythagorean triples arise in many contexts (e.g. see the text after Problem **180**). The classification given here shows that Pythagorean triples form a family depending on *three* parameters p, q, s (in which s is simply a "scaling" parameter, so the most important parameters are p, q). As a warm-up we consider two "one-parameter subfamilies" related to the triple 3, 4, 5.

Problem 22 Suppose $a^2 + b^2 = c^2$ and that b, c are consecutive integers.

(a) Prove that a must be odd – so we can write it as $a = 2m + 1$ for some integer m.

(b) Prove that c must be odd – so we can write it as $c = 2n + 1$ for some integer n. Find an expression for n in terms of m. △

Problem **22** reveals the triple $(3, 4, 5)$ as the first instance $(m = 1)$ of a one-parameter infinite family of triples, which continues

$$(5, 12, 13)\ (m = 2),\ (7, 24, 25)\ (m = 3),\ (9, 40, 41)\ (m = 4), \ldots,$$

whose general term is

$$(2m + 1,\ 2m(m + 1),\ 2m(m + 1) + 1).$$

The triple $(3, 4, 5)$ is also the first member of a quite different "one-parameter infinite family" of triples, which continues

$$(6, 8, 10),\ (9, 12, 15),\ \ldots.$$

Here the triples are scaled-up versions of the first triple $(3, 4, 5)$.

In general, common factors simply get in the way:

If $a^2 + b^2 = c^2$ and $HCF(a, b) = s$, then s^2 divides $a^2 + b^2$, and $a^2 + b^2 = c^2$; so s divides c.
And if $a^2 + b^2 = c^2$ and $HCF(b, c) = s$, then s^2 divides $c^2 - b^2 = a^2$, so s divides a.

Hence a typical Pythagorean triple has the form (sa, sb, sc) for some scale factor s, where (a, b, c) is a triple of integers, no two of which have a common factor: any such triple is said to be *primitive* (that is, basic – like prime numbers). Every Pythagorean triple is an integer multiple of some *primitive Pythagorean triple*. The next problem invites you to find a simple formula for all primitive Pythagorean triples.

Problem 23 Let (a, b, c) be a primitive Pythagorean triple.

(a) Show that a and b have opposite parity (i.e. one is odd, the other even) – so we may assume that a is odd and b is even.

(b) Show that

$$\left(\frac{b}{2}\right)^2 = \left(\frac{c-a}{2}\right)\left(\frac{c+a}{2}\right),$$

where
$$HCF\left(\frac{c-a}{2}, \frac{c+a}{2}\right) = 1$$
and $\frac{c-a}{2}, \frac{c+a}{2}$ have opposite parity.

(c) Conclude that
$$\frac{c+a}{2} = p^2 \text{ and } \frac{c-a}{2} = q^2,$$
where $HCF(p,q) = 1$ and p and q have opposite parity, so that $c = p^2 + q^2$, $a = p^2 - q^2$, $b = 2pq$.

(d) Check that any pair p, q having opposite parity and with $HCF(p,q) = 1$ gives rise to a primitive Pythagorean triple
$$c = p^2 + q^2, \quad a = p^2 - q^2, \quad b = 2pq$$
satisfying $a^2 + b^2 = c^2$. △

Problem 24 The three integers $a = 3$, $b = 4$, $c = 5$ in the Pythagorean triple $(3, 4, 5)$ form an *arithmetic progression*: that is, $c - b = b - a$. Find all Pythagorean triples (a, b, c) which form an arithmetic progression – that is, for which $c - b = b - a$. △

1.4.4 Sums of two squares

The classification of Pythagorean triples tells us precisely which **squares** can be written as the sum of two squares. We now turn to the wider question: "Which *integers* are equal to the sum of two squares?"

Problem 25

(a) Which of the prime numbers < 100 can be written as the sum of two squares?

(b) Find an easy way to immediately write $(a^2 + b^2)(c^2 + d^2)$ in the form $(x^2 + y^2)$. (This shows that the set of integers which can be written as the sum of two squares is "closed" under multiplication.)

(c) Prove that no integer (and hence no prime number) of the form $4k + 3$ can be written as the sum of two squares.

(d) The only *even* prime number can clearly be written as a sum of two squares: $2 = 1^2 + 1^2$. Euler (1707–1783) proved that every *odd* prime number of the form $4k + 1$ can be written as the sum of two squares in exactly one way. Find all integers < 100 that can be written as a sum of two squares.

(e) For which integers $N < 100$ is it possible to construct a square of area N, with vertices having integer coordinates? △

In Problem **25** parts (a) and (d) you had to decide which integers < 100 can be written as a sum of two squares as an exercise in mental arithmetic. In part (b) the fact that this set of integers is closed under multiplication turned out to be an application of the arithmetic of *norms* for complex numbers. Part (e) then interpreted sums of two squares geometrically by using Pythagoras' Theorem on the square lattice. These exercises are worth engaging in for their own sake. But it may also be of interest to know that writing an integer as a sum of two squares is a serious mathematical question – and in more than one sense.

Gauss (1777–1855), in his book *Disquisitiones arithmeticae* (1801) gave a complete analysis of when an integer can be represented by a 'quadratic form', such as $x^2 + y^2$ (as in Problem **25**) or $x^2 - 2y^2$ (as in Problem **54**(c) in Chapter 2).

A completely separate question (often attributed to Edward Waring (1736–1798)) concerns which integers can be expressed as a kth power, or as a sum of n such powers. If we restrict to the case $k = 2$ (i.e. squares), then:

- When $n = 2$, Euler (1707–1783) proved that the integers that can be written as a sum of *two squares* are precisely those of the form

$$m^2 \times p_0 \times p_1 \times p_2 \times \cdots \times p_s,$$

where $p_0 = 1$ or 2, and $p_1 < p_2 < \cdots < p_s$ are odd primes of the form $4l + 1$.

- When $n = 3$, Legendre (1752–1833) and Gauss proved between them that the integers which can be written as a sum of *three squares* are precisely those that are **not** of the form $4^m \times (8l + 7)$.

- When $n = 4$, Lagrange (1736–1813) had previously proved that **every** positive integer can be written as a sum of *four squares*.

1.5. Visualisation

Problem 26 (Pages of a newspaper) I found a (double) sheet from an old newspaper, with pages 14 and 27 next to each other. How many pages were there in the original newspaper? △

Problem 27 (Overlapping squares) A square $ABCD$ of side 2 sits on top of a square $PQRS$ of side 1, with vertex A at the centre O of the small square, side AB cutting the side PQ at the point X, and $\angle AXQ = \theta$.

(a) Calculate the area of the overlapping region.

(b) Replace the two squares in part (a) with two equilateral triangles. Can you find the area of overlap in that case? What if we replace the squares (i.e. regular 4-gons) in part (a) with regular $2n$-gons? △

Problem 28 (A folded triangle) The equilateral triangle $\triangle ABC$ has sides of length 1 cm. D and E are points on the sides AB and AC respectively, such that folding $\triangle ADE$ along DE folds the point A onto A' which lies outside $\triangle ABC$.

What is the total perimeter of the region formed by the three single layered parts of the folded triangle (i.e. excluding the quadrilateral with a folded layer on top)? △

Problem 29 ($A + B = C$) The 3 by 1 rectangle $ADEH$ consists of three adjacent unit squares: $ABGH$, $BCFG$, $CDEF$ left to right, with A in the top left corner. Prove that

$$\angle DAE + \angle DBE = \angle DCE.$$ △

Problem 30 (Dissections)

(a) Joining the midpoints of the edges of an equilateral triangle ABC cuts the triangle into four identical smaller equilateral triangles. Removing one of the three outer small triangles (say AMN, with M on AC) leaves three-quarters of the original shape in the form of an isosceles trapezium $MNBC$. Show how to cut this isosceles trapezium into four congruent pieces.

(b) Joining the midpoints of opposite sides of a square cuts the square into four congruent smaller squares. If we remove one of these squares, we are left with three-quarters of the original square in the form of an L-shape. Show how to cut this L-shape into four congruent pieces. △

Problem 31 (Yin and Yang) The shaded region in Figure 1, shaped like a large comma, is bounded by three semicircles – two of radius 1 and one of radius 2.

Cut each region (the shaded region and the unshaded one) into two 'halves', so that all four parts are congruent (i.e. of identical size and shape, but with possibly different orientations). △

Figure 1: Yin and Yang

In Problem **31** your first thought may have been that this is impossible. However, since the wording indicated that you are expected to succeed, it was clear that you must be missing something – so you tried again. The problem then tests both flexibility of thinking, and powers of visualisation.

1.6. Trigonometry and radians

1.6.1 Sine Rule

School textbooks tend to state the *Sine Rule* for a triangle ABC without worrying why it is true. So they often fail to give the result in its full form:

Theorem If R is the radius of the circumcircle of the triangle ABC, then

$$\frac{a}{\sin A} = \frac{b}{\sin B} = \frac{c}{\sin C} = 2R.$$

This full form explains that the three ratios

$$\frac{a}{\sin A}, \quad \frac{b}{\sin B}, \quad \frac{c}{\sin C}$$

are all equal *because they are all equal to the diameter $2R$ of the circumcircle* of $\triangle ABC$ – an additional observation which may well suggest how to prove the result (see Problem **32**).

Problem 32 Given any triangle ABC, construct the perpendicular bisectors of the two sides AB and BC. Let these two perpendicular bisectors meet at O.

(a) Explain why $OA = OB = OC$.

(b) Draw the circle with centre O and with radius OA. There are three possibilities:

  (i) The centre O lies on one of the sides of triangle ABC.
  (ii) The centre O lies inside triangle ABC.
  (iii) The centre O lies outside triangle ABC.

Case (i) leads directly to the *Sine Rule* for a right angled triangle ABC (remembering that $\sin 90° = 1$). We address case (ii), and leave case (iii) to the reader.

(ii) Extend the line BO to meet the circle again at the point A'. Explain why $\angle BA'C = \angle BAC = \angle A$, and why $\angle A'CB$ is a right angle. Conclude that

$$\sin A = \frac{BC}{A'B} = \frac{a}{2R},$$

and hence that

$$\frac{a}{\sin A} = 2R \quad \left(= \frac{b}{\sin B} = \frac{c}{\sin C}\right). \qquad \triangle$$

Problem 33 Let $\Delta = \text{area}(\triangle ABC)$.

(a) Prove that
$$\Delta = \frac{1}{2} \cdot ab \cdot \sin C.$$

(b) Prove that $4R\Delta = abc$. $\qquad \triangle$

1.6.2 Radians and spherical triangles

There is no God-given unit for measuring distance; different choices of unit give rise to answers that are related by *scaling*. However the situation is different for *angles*. In primary and secondary school we measure *turn* in *degrees* – where a half turn is 180°, a right angle is 90°, and a complete turn is 360°. This angle unit dates from the ancient Babylonians ($\sim$ 2000 BC). We are not sure why they chose 360 units in a full turn, but it seems to be related to the approximate number of days in a year (the time required for the heavens to make a complete rotation in the night sky), and to the fact that they wrote their numbers in "base 60". However the choice is no more objectively mathematical than measuring distance in inches or in centimetres.

After growing up with the idea that angles are measured in degrees, we discover towards the end of secondary school that:

there is another unit of measure for *angles* – namely **radians**.

It may not at first be clear that this is an entirely natural, God-given unit. The size of, or amount of turn in, an angle at the point A can be captured in an absolute way by drawing a circle of radius r centred at the point A, and measuring the arc length which the angle cuts off on this circle. The angle size (in *radians*) of the angle at A is then defined to be the ratio

$$\frac{\text{arc length}}{\text{radius}}.$$

That is,

size of angle at the point A = arc length cut off on a circle of radius 1, centred at the apex A.

Hence a right angle is of size $\frac{\pi}{2}$ radians; a half turn is equal to π (radians); a full turn is equal to 2π (radians); each angle in an equilateral triangle is equal to $\frac{\pi}{3}$ (radians); the three angles of a triangle have sum π; and the angles of a polygon with n sides have sum $(n-2)\pi$ (see Problem **230** in Chapter 6).

For a while after the introduction of radians we continue to emphasise the word *radians* each time we give the measure of an angle in order to stress that we are no longer using degrees. But this is not really a switch to a new unit: this new way of measuring angles is in some sense objective – so we soon drop all mention of the word "radians" and simply refer to the size of an angle (in radians) as if it were a pure number.

This switch affects the meaning of the familiar *trigonometric functions*. And though we continue to use the same names (sin, cos, tan, etc.), they become slightly different *as functions*, since the inputs are now always assumed to be in radians.

The real payoff for making this change stems from the way it recognizes the connection between *angles* and *circles*. This certainly makes calculating circular arc lengths and areas of sectors easy (an arc with angle θ on a circle of radius r now has length θr; and a circular sector with angle 2θ now has area θr^2). But the main benefit – which one hopes all students appreciate eventually – is that this change of perspective highlights the fundamental link between $\sin x$, $\cos x$, and e^x:

- "$\cos x$" becomes the derivative of $\sin x$
- "$-\sin x$" becomes the derivative of $\cos x$, and
- the three functions are related by the totally unexpected identity

$$e^{i\theta} = \cos\theta + i\sin\theta.$$

The next problem draws attention to a beautiful result which reveals, *in a pre-calculus, pre-complex number setting*, a beautiful consequence of thinking about angles in terms of radians. The goal is to discover a formula for the area of a *spherical triangle* in terms of its angles and π, which links the formula for the circumference of a circle with that for the surface area of a sphere.

Suppose we wish to do geometry on the sphere. There is no problem deciding how to make sense of *points*. But it is less clear what we mean by (*straight*) *lines*, or line segments.

Before making due allowance for the winds and the tides, an airline pilot and a ship's Captain both need to know how to find the *shortest path* joining two given points A, B on a sphere. If the two points both lie on the equator, it is plausible (and correct) that the shortest route is to travel from A to B *along the equator*. If we think of the equator as being in a *horizontal* plane through the centre O of the sphere, then we may notice that we can change the equator into a circle of longitude by rotating the sphere so that the "horizontal" plane (through O) becomes a "vertical" plane (through O). So we may view two points A and B which both lie on the same circle of longitude as lying on a "vertical equator" passing through A, B and the North and South poles: the shortest distance from A to B must therefore lie along that circle of longitude.

If we now rotate the sphere through some other angle, we get a "tilted equator" passing through the images of the (suitably tilted) points A and B: these "tilted equators" are called *great circles*. Each great circle is the intersection of the sphere with a plane through the centre O of the sphere. So

to find the shortest path from A to B:

- take the plane determined by the points A, B and the centre of the sphere O;

- find the great circle where this plane cuts the sphere;

- then follow the arc from A to B along this great circle.

Once we have points and line segments (i.e. arcs of great circles) on the sphere, we can think about *triangles*, and about the *angles* in such a triangle. In a triangle ABC on the sphere, the sides AB and AC are arcs of great circles meeting at A. By rotating the sphere we can imagine A as being at the North pole; so the two sides AB and AC behave just like arcs of two circles of longitude emanating from the North pole. In particular, we can measure the angle between them (this is exactly how we measure *longitude*): the two arcs AB, AC of circles of longitude set off from the North pole A in different horizontal directions before curving southwards, and the angle between them is the angle between these two initial horizontal directions

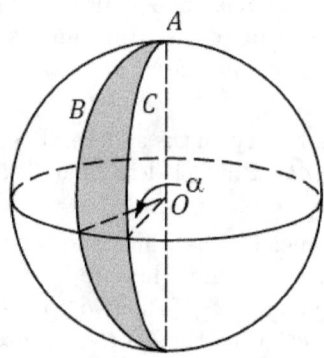

Figure 2: Angles on a sphere

(that is, the angle between the plane determined by O, A, B and the plane determined by O, A, C).

Problem 34 Imagine a triangle ABC on the unit sphere (with radius $r = 1$), with angle α between AB and AC, angle β between BC and BA, and angle γ between CA and CB. You are now in a position to derive the remarkable formula for the area of such a spherical triangle.

(a) Let the two great circles containing the sides AB and AC meet again at A'. If we imagine A as being at the North pole, then A' will be at the South pole, and the angle between the two great circles at A' will also be α. The slice contained between these two great circles is called a *lune* with angle α.

  (i) What fraction of the surface area of the whole sphere is contained in this lune of angle α? Write an expression for the actual area of this lune.

  (ii) If the sides AB and AC are extended backwards through A, these backward extensions define another lune with the same angle α, and the same surface area. Write down the total area of these two lunes with angle α.

(b)(i) Repeat part (a) for the two sides BA, BC meeting at the vertex B, to find the total area of the two lunes meeting at B and B' with angle β.

  (ii) Do the same for the two sides CA, CB meeting at the vertex C, to find the total area of the two lunes meeting at C and C' with angle γ.

(c)(i) Add up the areas of these six lunes (two with angle α, two with angle β, and two with angle γ). Check that this total includes every part of the sphere at least once.

(ii) Which parts of the sphere have been covered more than once? How many times have you covered the area of the original triangle ABC? And how many times have you covered the area of its sister triangle $A'B'C'$?

(iii) Hence find a formula for the area of the triangle ABC in terms of its angles – α at A, β at B, and γ at C. $\triangle$

1.6.3 Polar form and $\sin(A+B)$

The next problem is less elementary than most of Chapter 1, but is included here to draw attention to the ease with which the addition formulae in trigonometry can be reconstructed once one knows about the *polar form* representation of a complex number. Those who are as yet unfamiliar with this material may skip the problem – but should perhaps remember the underlying message (namely that, once one *is* familiar with this material, there is no need ever again to get confused about the trig addition formulae).

Problem 35

(a) You may know that any complex number $z = \cos\theta + i\sin\theta$ of modulus 1 (that is, which lies on the unit circle centred at the origin) can be written in the modulus form $z = e^{i\theta}$. Use this fact to reconstruct in your head the trigonometric identities for $\sin(A+B)$ and for $\cos(A+B)$. Use these to derive the identity for $\tan(A+B)$.

(b) By choosing X, Y so that $A = \frac{X+Y}{2}$, and $B = \frac{X-Y}{2}$, use part (a) to reconstruct the standard trigonometric identities for

$$\sin X + \sin Y, \quad \sin X - \sin Y, \quad \cos X + \cos Y, \quad \cos X - \cos Y.$$

(c)(i) Check your answer to (a) for $\sin(A+B)$ by substituting $A = 30°$, and $B = 60°$.

(ii) Check your answer to (b) for $\cos X - \cos Y$ by substituting $X = 60°$, and $Y = 0°$.

(d)(i) If $A + B + C + D = \pi$, prove that

$$\sin A \sin B + \sin C \sin D = \sin(B+C)\sin(B+D).$$

(ii) Given a cyclic quadrilateral $WXYZ$, with $\angle XWY = A$, $\angle WXZ = B$, $\angle YXZ = C$, $\angle WYX = D$, deduce *Ptolemy's Theorem*:

$$WX \times YZ + WZ \times XY = WY \times XZ.$$ $\triangle$

1.7. Regular polygons and regular polyhedra

Regular polygons have already featured rather often (e.g. in Problems **3, 12, 19, 27, 28, 29**). This is a general feature of elementary mathematics; so the neglect of the geometry of regular polygons, and their 3D companions, the regular polyhedra, is all the more unfortunate. We end this first chapter with a first brief look at polygons and polyhedra.

1.7.1 Regular polygons are cyclic

Problem 36 A polygon $ABCDE \cdots$ consists of n vertices A, B, C, D, E, ..., and n sides AB, BC, CD, DE ... which are disjoint except that successive pairs meet at their common endpoint (as when AB, BC meet at B). A polygon is *regular* if any two sides are congruent (or equal), and any two angles are congruent (or equal). Can a regular n-gon $ABCDE \cdots$ always be inscribed in a circle? In other words, does a regular polygon automatically have a "centre", which is equidistant from all n vertices? △

1.7.2 Regular polyhedra

Problem 37 (Wrapping)

(a) You are given a regular tetrahedron with edges of length 2. Is it possible to choose positive real numbers a and b so that an a by b rectangular sheet of paper can be used to "wrap", or cover, the regular tetrahedron without leaving any gaps or overlaps?

(b) Given a cube with edges of length 2, what is the smallest sized rectangle that can be used to wrap the cube in the same way without cutting the paper? (In other words, if we want to completely cover the cube, what is the smallest area of overlap needed? How small a fraction of the paper do we have to waste?) △

Problem 38 (Cross-sections) Can a cross-section of a cube be:

(i) an equilateral triangle?

(ii) a square?

(iii) a polygon with more than six sides?

(iv) a regular hexagon?

(v) a regular pentagon? △

Problem 39 (Shadows) Can one use the Sun's rays to produce a plane shadow of a cube:

(i) in the form of an equilateral triangle?

(ii) in the form of a square?

(iii) in the form of a pentagon?

(iv) in the form of a regular hexagon?

(v) in the form of a polygon with more than six sides? △

> *The imparting of factual knowledge is for us a secondary consideration. Above all we aim to promote in the reader a correct attitude, a certain discipline of thought, which would appear to be of even more essential importance in mathematics than in other scientific disciplines. ...*
>
> *General rules which could prescribe in detail the most useful discipline of thought are not known to us. Even if such rules could be formulated, they would not be very useful. Rather than knowing the correct rules of thought theoretically, one must have assimilated them into one's flesh and blood ready for instant and instinctive use. Therefore for the schooling of one's powers of thought only the practice of thinking is really useful.*
>
> G. Pólya (1887–1985) and G. Szegő (1895–1985)

1.8. Chapter 1: Comments and solutions

1.

(a) Assuming that the 2×, 3×, 4×, and 5× tables are known, and that one has understood that the order of the factors does not matter, all that remains to be learned is 6×6, 6×7, 6×8, 6×9; 7×7, 7×8, 7×9; 8×8, 8×9; and 9×9.

(b)(i) $(4 \times 10^{-3}) \times (2 \times 10^{-2}) = 8 \times 10^{-5} = 0.00008$

(ii) $(8 \times 10^{-4}) \times (7 \times 10^{-2}) = 56 \times 10^{-6} = 0.000056$

(iii) $(7 \times 10^{-3}) \times (12 \times 10^{-2}) = 84 \times 10^{-5} = 0.00084$

(iv) $1.08 \div 1.2 = 10.8 \div 12 = 108 \div (12 \times 10) = 0.9$

(v) $(8 \times 10^{-2})^2 = 64 \times 10^{-4} = 0.0064$

2.

(a)(i) 1, 4, 9, 16, 25, 36, 49, 64, 81, 100, 121, 144;
169, 196, 225, 256, 289, 324, 361, 400, 441, 484, 529, 576, 625, 676, 729, 784, 841, 900, 961

30   Mental Skills

(ii) 1, 8, 27, 64, 125, 216, 343, 512, 729, 1000, 1331

(iii) $1 = 2^0$, 2, 4, 8, 16, 32, 64, 128, 256, 512, 1024

(b)(i) 31 ($31^2 = 961$, $32^2 = 2^{10} = 1024$)

(ii) 99 ($100^2 = 10^4 = 10\,000$)

(iii) 316 ($310^2 = 96\,100 < 100\,000 < 320^2$; so look more carefully between 310 and 320)

(c)(i) 9 ($9^3 = 729 < 10^3 = 1000$)

(ii) 21 ($20^3 = 8000$, $22^3 = 10\,648$)

(iii) 99 ($100^3 = 10^6 = 1\,000\,000$)

(d)(i) Those powers 2^e of the form 2^{2n} for which the exponent e is a multiple of 2: i.e. $e \equiv 0 \pmod{2}$.

(ii) Those powers 2^e of the form 2^{3n} for which the exponent e is a multiple of 3: i.e. $e \equiv 0 \pmod{3}$.

(e) $64 = 2^6 = 4^3 = 8^2$. $729 = 3^6 = 9^3 = 27^2$.

3.

(a) (i) 7; (ii) 12; (iii) 21; (iv) 13; (v) 14; (vi) 31; (vii) $10 \times 31 = 310$

(b) (i) $2\sqrt{2}$; (ii) $2\sqrt{3}$; (iii) $5\sqrt{2}$; (iv) $7\sqrt{3}$; (v) $12\sqrt{2}$; (vi) $21\sqrt{2}$

(c)(i) $\angle ABC = 108°$. $\triangle BAC$ is isosceles ($BA = BC$), so $\angle BAC = \angle BCA = 36°$.
$\therefore \angle CAE = \angle BAE - \angle BAC = 72° = 180° - \angle AED$.
So AC is parallel to ED (since corresponding angles add to 180°).

(ii) AX is parallel to ED; similarly DX is parallel to EA. Hence $AXDE$ is a parallelogram, with $EA = ED$.

(iii) The two triangles are both isosceles and $\angle AXD = \angle CXB = 108°$ (vertically opposite angles).
Hence $\angle XAD = \angle XCB = 36°$, and $\angle XDA = \angle XBC = 36°$.

(iv) $AD : CB = DX : BX$, so $x : 1 = 1 : (x-1)$; hence $x^2 - x - 1 = 0$ and $x = \frac{1+\sqrt{5}}{2}$ – the *Golden Ratio*, usually denoted by the Greek letter τ (*tau*), with approximate value $1.6180339887\ldots$.

(v) $BX = x - 1 = \frac{\sqrt{5}-1}{2}$ ($= \frac{1}{\tau} = \tau - 1$), with approximate value $0.6180339887\ldots$.

(vi) We may either check that corresponding angles are equal in pairs $(36°, 72°, 72°)$, or that corresponding sides are in the same ratio $x : 1 = 1 : (x-1)$.

(vii) $\cos 36° = \frac{\sqrt{5}+1}{4}$; $\cos 72° = \frac{\sqrt{5}-1}{4}$ (drop perpendiculars from D to AB and from X to BC; or use the Cosine Rule).

(viii) Use $\sin^2 36° + \cos^2 36° = 1$: $\sin 36° = \frac{\sqrt{10-2\sqrt{5}}}{4}$; $\sin 72° = \frac{\sqrt{10+2\sqrt{5}}}{4}$.

Note: The *Golden Ratio* crops up in many unexpected places (including the regular pentagon, and the *Fibonacci* numbers). Unfortunately much that is written about its ubiquity is pure invention. One of the better popular treatments, that highlights the number's significance, while taking a sober view of spurious claims, is the book *The Golden Ratio* by Mario Livio.

4.

(a) $12\,345 = 5 \times 2469 = 3 \times 5 \times 823$. But is 823 a prime number?

It is easy to check that 823 is not divisible by 2, or 3, or 5, or 7, or 11. The *Square Root Test* (displayed below) tells us that it is only necessary to check four more potential prime factors.

Square Root Test: If $N = a \times b$ with $a \leqslant b$, then $a \times a \leqslant a \times b = N$, so the smaller factor $a \leqslant \sqrt{N}$.

Hence, if N (= 823 say) is not prime, its smallest factor > 1 is at most equal to $\sqrt{N}$ $(= \sqrt{823} < 29)$. Checking $a = 13, 17, 19, 23$ shows that the required prime factorisation is $12\,345 = 3 \times 5 \times 823$.

(b) There are 25 (2, 3, 5, 7, 11, 13, 17, 19, 23, 29, 31, 37, 41, 43, 47, 53, 59, 61, 67, 71, 73, 79, 83, 89, 97).

Notes:

(i) For small primes, mental arithmetic should suffice. But one should also be aware of the general *Sieve of Eratosthenes* (a Greek polymath from the 3rd century BC). Start with the integers 1–100 arranged in ten columns, and proceed as follows:

1	2	3	4	5	6	7	8	9	10
11	12	13	14	15	16	17	18	19	20
21	22	23	24	25	26	27	28	29	30
31	32	33	34	35	36	37	38	39	40
41	42	43	44	45	46	47	48	49	50
61	62	63	64	65	66	67	68	69	70
71	72	73	74	75	76	77	78	79	80
81	82	83	84	85	86	87	88	89	90
91	92	93	94	95	96	97	98	99	100

Delete 1 (which is not a prime: see (ii) below).
Circle the first undeleted integer; remove all other multiples of 2.
Circle the first undeleted integer; remove all other multiples of 3.
Circle the first undeleted integer; remove all other multiples of 5.
Circle the first undeleted integer; remove all other multiples of 7.
All remaining undeleted integers < 100 must be prime (by the *Square Root Test* (see part (a)).

(ii) The multiplicative structure of integers is surprisingly subtle. The first thing to notice is that 1 has a special role, in that it is the *multiplicative identity*: for each integer n, we have $1 \times n = n$. Hence 1 is "multiplicatively neutral" – it has no effect.

The "multiplicative building blocks" for integers are the *prime numbers*: every integer > 1 can be broken down, or *factorised* as a product of prime numbers, in exactly one way. The integer 1 has no proper factors, and has no role to play in breaking down larger integers by *factorisation*. So 1 is **not** a prime. (If we made the mistake of counting 1 as a prime number, then we would have to make all sorts of silly exceptions – for example, to allow for the fact that $2 = 2 \times 1 = 2 \times 1 \times 1 = \ldots$, so 2 could then be factorised in infinitely many ways.)

(iii) Notice that $91 = 7 \times 13$ is not a prime; so there is exactly one prime in the 90s – namely 97.

How many primes are there in the next run of 10 (from 100–109)?

How many primes are there from 190–199? How many from 200–210?

(c) (i) $3 = 2^2 - 1$.

(ii) Many students struggle with this, and may suggest 143, or 323, or even 63. The problem conceals a (very thinly) disguised message:

One cannot calculate with words.

To make use of mathematics, we must routinely translate *words* into *symbols*. As soon as "one less than a square" is translated into symbols, bells should begin to ring. For you know that $x^2 - 1 = (x - 1)(x + 1)$, so $x^2 - 1$ can only be prime if the smaller factor $(x - 1)$ is equal to 1.

5.

(a) (i) If we try to avoid such a "relatively prime pair", then we must **not** choose any of 11, 13, 17, 19 (since they are prime, and have no multiples in the given range). So we are forced to choose the other six integers: 10, 12, 14, 15, 16, 18 – and there are then exactly two pairs which are relatively prime, namely 14, 15 and 15, 16.

(ii) If we try to avoid such a pair, then we can choose at most one even integer. So we are then forced to choose **all five** available odd integers, and our list will be: "unknown even, 11, 13, 15, 17, 19". If the even integer is chosen to be 14, or 16, then every pair in my list has LCM = 1. So the answer is "No".

(b) (i) If we try to avoid such a pair, then we must **not** choose 97 (the only prime number in the nineties). And we must not choose $95 = 5 \times 19$ (which is relatively prime to all other integers in the given range – except for 90); and we must not choose $91 = 7 \times 13$ (which is relatively prime to all other integers in the given range – except for 98). So we are forced to choose six integers from 90, 92, 93, 94, 96, 98, 99. Whichever integer we then omit leaves a pair which is relatively prime.

(ii) If we try to avoid such a pair, then we can choose at most one even integer. So we are then forced to choose all five available odd integers, and our list will be: "unknown even, 91, 93, 95, 97, 99", and so must include the pair 93, 99 – with common factor 3. So the answer is "Yes".

(c) In parts (a) and (b), the possible integers are limited (in (a) to the "teens", and in (b) to the "nineties"); so it is natural to reach for *ad hoc* arguments as we did above. But in part (c) you know nothing about the numbers chosen.

Note: The question says that "*I* choose", and asks whether "*you*" can be sure. So you have to find **either** a *general* argument that works for any n, **or** a counterexample. And the theme of this chapter indicates that it should not require any extended calculation.

The relevant "general idea" is the *Pigeon Hole Principle* which we may meet in the second part of this collection. So this problem may be viewed as a gentle introduction.

(i) Group the $2n$ consecutive integers

$$a+1, a+2, \ldots, a+2n$$

into n pairs of consecutive integers

$$\{a+1, a+2\}, \{a+3, a+4\}, \ldots, \{a+(2n-1), a+2n\}.$$

- If we choose at most one integer from each pair, then we never get more than n integers.
- So as soon as we choose $n+1$ integers from $2n$ consecutive integers, we are forced to choose **both** integers in some pair k, $k+1$, and this pair of consecutive integers is always relatively prime (see Problem **6**(b)(i)).

(ii) We saw in part (a)(ii) that, if $n = 5$ and the $2n$ integers start at 10, then we can choose six integers (either 11, 13, 14, 15, 17, 19, or 11, 13, 15, 16, 17, 19), and in each case every pair has LCM $= 1$. So for $n = 5$ the answer is "No" (because there is *at least one case* where one cannot be sure).

However, as soon as n is at least 6, we show that the argument in part (a)(ii) breaks down. As before, if we try to choose a subset in which no pair has a common factor, then we can choose at most one even integer. So we are forced to choose **all** the odd integers. But any run of **at least six** consecutive odd integers includes two multiples of 3. So for $n \geqslant 6$, the answer is "Yes".

6.

(a)(i) Suppose k is a factor of m and n. Then we can write $m = kp$ and $n = kq$ for some integers p, q. Hence $m - n = k(p - q)$, so k is a factor of $m - n$. Also $m + n = k(p + q)$, so k is a factor of $m + n$.

(ii) Any factor of m and n is also a factor of their difference $m - n$; so the set of common factors of m and n is a **subset** of the set of common factors of $m - n$ and n.

And any factor of $m - n$ and n is also a factor of their sum m; so the set of common factors of $m - n$ and n is a **subset** of the set of common factors of m and n.

Hence the two sets of common factors are identical. In particular, the two "highest common factors" are equal.

(iii) Subtract 91 from 1001 ten times to see that

$$HCF(1001, 91) = HCF(1001 - 910, 91) = 91.$$

34   *Mental Skills*

(b)(i) Subtract m from $m+1$ once to see that
$$HCF(m+1,m) = HCF(1,m) = 1.$$

(ii) Subtract m from $2m+1$ twice to see that
$$HCF(2m+1,m) = HCF(m+1,m) = HCF(1,m) = 1.$$

(iii) Subtract $m-1$ from m^2+1 "$m+1$ times" to see that
$$HCF(m^2+1, m-1) = HCF((m^2+1) - (m^2-1), m-1) = HCF(2, m-1).$$

Hence, if m is odd, the HCF $= 2$; if m is even, the HCF $= 1$.

7. They are equal. (The first is
$$\frac{17}{100} \times 19\,000\,000,$$
the second is
$$\frac{19}{100} \times 17\,000\,000,$$
which are equal since multiplication is commutative and associative.)

8.

(a)
$$\frac{3}{2} \times \frac{4}{3} \times \frac{5}{4} \times \frac{6}{5} = \frac{6}{2} = 3$$

(b)
$$\sqrt{\frac{3}{2} \times \frac{4}{3} \times \frac{5}{4} \times \frac{6}{5} \times \frac{7}{6} \times \frac{8}{7}} = \sqrt{\frac{8}{2}} = 2$$

(c)
$$10 \times 9 \times 8 \times 7 \times \times 5 \times 4 \times 3 \times 2 \times 1 \text{ seconds}$$
$$= \frac{10 \times 9 \times 8 \times 7 \times 6 \times 5 \times 4 \times 3 \times 2 \times 1}{60} \text{ minutes}$$
$$= \frac{10 \times 9 \times 8 \times 7 \times 6 \times 5 \times 4 \times 3 \times 2 \times 1}{60 \times 60} \text{ hours}$$
$$= \frac{10 \times 9 \times 8 \times 7 \times 6 \times 5 \times 4 \times 3 \times 2 \times 1}{60 \times 60 \times 24} \text{ days}$$
$$= \frac{10 \times 9 \times 8 \times 7 \times 6 \times 5 \times 4 \times 3 \times 2 \times 1}{60 \times 60 \times 24 \times 7} \text{ weeks}$$
$$= 6 \text{ weeks (after cancelling).}$$

Note: These three questions underline what we mean by *structural arithmetic*. Fractions should **never** be handled by *evaluating* numerators and denominators. Instead one should always be on the lookout for structural features which simplify calculation – such as *cancellation*.

9.

(a) Suppose a rectangle in the "DIN A" series has dimensions a by b, with $a < b$. Folding in half produces a rectangle of size $\frac{b}{2}$ by a. Hence $b : a = a : \frac{b}{2}$, so $b^2 = 2a^2$, and $b : a = \sqrt{2} : 1$.

(b) (i) $\frac{1}{r}$. (ii) r.

10.

(a) "15% discount" means the price actually charged is "85% of the marked price". Hence each marked price needs to be **multiplied by 0.85**.

The distributive law says we may add the marked prices first and then multiply the total (exactly £80) by 0.85 to get

$$£\left(\frac{85}{100} \times 80\right) = £(17 \times 4) = £68.$$

Note: The context (shopping, sales tax, and discount) is mathematically uninteresting. What matters here is the underlying *multiplicative structure* of the solution, which arises in many different contexts.

(b) "Add 20% VAT" means multiplying the discounted pre-VAT total (£68) by 1.2, or $\frac{6}{5}$. Hence the final price, with VAT added, is £$(1.2 \times 0.85 \times 80)$.

If the VAT were added first, the price before discount would be £(1.2×80), and the final price after allowing for the discount would be £$(0.85 \times 1.2 \times 80)$.

Since multiplication is commutative, the two calculations have the same result, so the order does not matter (just as the final result in Problem **9** is the same whether one first enlarges A4 to A3 and then reduces A3 to A4, or first reduces A4 to A5 and then enlarges A5 to A4).

Note: Notice that we did not evaluate the two answers to see that they gave the same output £81.60. If we had, then the equality might have been a fluke due to the particular numbers chosen. Instead we left the answer unevaluated, in *structured form*, which showed that the equality would hold for any input.

(c) To cancel out multiplying by $\frac{6}{5}$ we need to multiply by $\frac{5}{6}$ – a discount of $\frac{1}{6}$, or $16\frac{2}{3}\%$.

Note: This question has nothing to do with financial applications. It is included to underline the fact that although percentage change questions use the language of *addition* and *subtraction* ("increase", or "decrease"), the mathematics suggests they should be handled **multiplicatively**.

11.

(a)(i) $2 \times 5 < 3 \times 4$, so

$$\frac{7}{2 \times 5} > \frac{7}{3 \times 4}.$$

Hence

$$\frac{1}{2} + \frac{1}{5} > \frac{1}{3} + \frac{1}{4}.$$

(ii) At first sight, "$10 < 12$" may not seem related to "$\frac{1}{2} + \frac{1}{5} > \frac{1}{3} + \frac{1}{4}$". Yet the crucial fact we started from in part (i) was "$2 \times 5 = 10 < 12 = 3 \times 4$".

(b) $10 < 12$, so
$$(x+2)(x+5) = x^2 + 7x + 10 < x^2 + 7x + 12 = (x+3)(x+4).$$

(i) If all four brackets are positive (i.e. if $x > -2$), then we also have $2x + 7 > 0$, and it follows that
$$\frac{1}{x+2} + \frac{1}{x+5} = \frac{2x+7}{(x+2)(x+5)}$$
$$> \frac{2x+7}{(x+3)(x+4)}$$
$$= \frac{1}{x+3} + \frac{1}{x+4}.$$

(ii) When calculating with the given algebraic expression, the values
$$x = -2, -3, -4, -5$$
are "forbidden values".

If $x > -2$, then (as in part (i)) we have
$$\frac{1}{x+2} + \frac{1}{x+5} = \frac{2x+7}{(x+2)(x+5)}$$
$$> \frac{2x+7}{(x+3)(x+4)}$$
$$= \frac{1}{x+3} + \frac{1}{x+4}.$$

For permitted values of $x < -2$, one or more of the brackets $(x+2)$, $(x+5)$, $(x+3)$, $(x+4)$ will be negative. However, one can still carry out the algebra to simplify
$$\frac{1}{x+2} + \frac{1}{x+5} = \frac{2x+7}{(x+2)(x+5)} \quad \text{and} \quad \frac{1}{x+3} + \frac{1}{x+4} = \frac{2x+7}{(x+3)(x+4)}$$

When $x = -\frac{7}{2}$ both expressions are **equal**, and equal to 0. The simplified numerators are both positive if $x > -\frac{7}{2}$, and both negative if $x < -\frac{7}{2}$; and the sign of the denominators changes as one moves through the four intervals $-3 < x < -2$, $-4 < x < -3$, $-5 < x < -4$, $x < -5$, with the inequality switching

from " $>$ " (for $x > -2$) to " $<$ " (for $-3 < x < -2$),
to " $>$ " (for $-3.5 < x < -3$),
to " $<$ " (for $-4 < x < -3.5$),
to " $>$ " (for $-5 < x < -4$),
to " $<$ " (for $x < -5$).

12.

(a) (i) $3 + 2\sqrt{2}$; (ii) $3 - 2\sqrt{2}$; (iii) $7 + 5\sqrt{2}$.

Note: Notice that you can write down the answer to (ii) as soon as you have finished (i), without doing any further calculation.

(b) (i) $2 + \sqrt{6}$; (ii) $\sqrt{2} + \sqrt{3}$; (iii) $\frac{1+\sqrt{5}}{2}$ (the *Golden Ratio* $\frac{1+\sqrt{5}}{2} = \tau$ is the larger root of the quadratic equation $x^2 - x - 1 = 0$. Hence $\frac{3+\sqrt{5}}{2} = \tau + 1 = \tau^2$);
(iv) $\sqrt{10 - 2\sqrt{5}}$: this does not simplify further.

Note: Some readers may think an apology is in order for part (iv). The lesson here is that, while one should always try to simplify, there is no way of knowing in advance whether a simplification is possible. And there is no way out of this dilemma. So one is reduced to thinking: any simplification would involve $\sqrt{5}$, and if one tries to solve $(a + b\sqrt{5})^2 = 10 - 2\sqrt{5}$, then the solutions for a and b do not lead to anything "simpler". (This repeated surd should perhaps have rung bells, as it was equal to the exact expression for $4\sin 36°$ in Problem 3(c). It was included here partly because the question of its simplification should already have arisen when it featured in that context.)

13. In reconstructing the missing digits the number of possible solutions is determined by *the highest common factor of the multiplier and* 10. At the first step (in the units column):

because $HCF(6, 10) = 2$, $\square \times 6 = 8$ (mod 10) has **two** solutions which differ by 5 – namely 3 and 8.

The first possibility then requires us to solve $(\square \times 3) + 1 = 2$ (mod 10): because $HCF(3, 10) = 1$, this has just **one** solution – namely 7. This gives rise to the solution **76 × 3 = 228**.

The second possibility requires us to solve $(\square \times 8) + 4 = 2$ (mod 10): because $HCF(8, 10) = 2$, this has **two** solutions which differ by 5 – namely 1 and 6. This gives rise to two further solutions: **16 × 8 = 128**, and **66 × 8 = 528**.

14.

(a) The solutions are entirely elementary, with no trickery. But they can be surprisingly elusive. And since this elusiveness is the only reason for including the problem, we hesitate to relieve any frustration by giving the solution.
The whole thrust of the "24 game" is to underline the scope for "getting to know" the many faces of a number like 24: for example, as $24 = 12 + 12 \, (= 3 \times 4 + 3 \times 4$ for 3, 3, 4, 4); as $24 = 25 - 1 \, (= 5 \times 5 - 3 \div 3$ for 3, 3, 5, 5); and as $24 = 27 - 3$ $(= 3 \times 3 \times 3 - 3$ for 3, 3, 3, 3). So one should be looking for ways of exploiting other important arithmetical aspects of 24 – in particular, as 4×6 and as 3×8.

(b)(i) $0 = (4-4)+(4-4)$; $1 = (4 \div 4) \times (4 \div 4)$; $2 = (4 \div 4)+(4 \div 4)$; $3 = (4+4+4) \div 4$; $4 = ((4-4) \times 4)+4$; $5 = ((4 \times 4)+4) \div 4$; $6 = 4+((4+4) \div 4)$; $7 = 4+4-(4 \div 4)$; $8 = (4+4) \times (4 \div 4)$; $9 = (4+4)+(4 \div 4)$. The output 10 seems to be impossible with the given restrictions.

(ii) With squaring and $\sqrt{}$ allowed we can manage $10 = 4+4+4-\sqrt{4}$. Indeed, one can make everything up to 40 except (perhaps) 39.

15.

(a)(i) $a^2 + 2ab + b^2$; $a^3 + 3a^2b + 3ab^2 + b^3$
  (ii) Replace b by $(-b)$: $a^2 - 2ab + b^2$; $a^3 - 3a^2b + 3ab^2 - b^3$
(b) (i) $(x+1)^2$; (ii) $(x^2-1)^2$; (iii) $(x^2-1)^3$
(c) (i) $a^2 - b^2$
  (ii) Replace "b" by "$b+c$": $a^2 - (b+c)^2 = a^2 - b^2 - c^2 - 2bc$
  Replace "b" by "$b-c$": $a^2 - (b-c)^2 = a^2 - b^2 - c^2 + 2bc$
(d) One way is to rewrite this expression as a difference of two squares:

$$\begin{aligned} (2x)^2 - (x^2 - 2x + 1) &= (2x)^2 - (x-1)^2 \\ &= (2x - (x-1))(2x + (x-1)) \\ &= (x+1)(3x-1) \end{aligned}$$

Note: As so often, the messages here are largely implicit. In part (a)(ii) we explicitly highlight the intention to use what you already know (by simply substituting "$-b$" in place of "b". In part (b), you are expected to recognise (i), and then to see (ii) and (iii) as mild variations on the expansions of $(a-b)^2$ and $(a-b)^3$ in part (a). Part (c) repeats (in silence) the message of (a)(ii): **think** – don't slog it out. And part (d) encourages you to keep an eye out for thinly disguised instances of "a difference of two squares".

16.

(a) **Final digits**: 'block' 4, 6 of length 2;
  leading digits: "block" 4, 1, 6, 2, 1 of length 5.
(b) **Claim** The sequence of "units digits" really does recur.
  Proof Given a power of 4 that has units digit 4, the usual multiplication algorithm for multiplying by 4 produces a number with units digit 6.
  Given this new power of 4 with units digit 6, the usual multiplication algorithm for multiplying by 4 produces a number with units digit 4.
  At this stage the sequence of units digits begins a new cycle.
  [Alternatively: The units digit is simply equal to the relevant power of 4 (mod 10). Multiplying by 4 changes 4 to 6 (mod 10); multiplying by 4 changes 6 to 4 (mod 10); – and the cycle repeats.]
(c) The sequence of **leading** digits *seems* to recur every 5 terms, because $4^5 = 2^{10} = 1024$ is almost exactly equal to 1000. Each time we move on 5 steps in the sequence, we multiply by $4^5 = 1024$. As far as the leading digit is concerned, this has the same effect as multiplying the initial term (4) by slightly more than 1.024 (then adding any 'carries'), which is very like multiplying by 1 – and so does not change the leading digit (yet).

However, each time we move on 10 steps in the sequence, we multiply by $4^{10} = 1024^2 = 1\,048\,576$. As far as the leading digit is concerned, this has the same effect as multiplying by slightly more than 1.048576.

When we move on 25 steps, we multiply by $4^{25} = 1\,125\,899\,906\,842\,624$. And as far as the leading digit is concerned, this has the same effect as multiplying by slightly more than 1.12599906842624. And so on.

Eventually, the multiplier becomes large enough to change one of the leading digits.

17. The total is 100.

Having found this by *direct* calculation, we should think *indirectly* and notice that $100 = 10^2$.

And we should then ask: "Why 10? What has **10** got to do with the 4× multiplication table?"

A quick check of the 1× multiplication table (total = 1), the 2× multiplication table (total = 9), etc. may suggest what we should have seen immediately.

The first row has sum:	$(1+2+3+4)$.
The second row has total	$2 \times (1+2+3+4)$.
The third row has total	$3 \times (1+2+3+4)$.
The fourth row has total	$4 \times (1+2+3+4)$.
∴ The total is	$(1+2+3+4) \times (1+2+3+4) = (1+2+3+4)^2$.

19.

(a) $\sin 45° = \frac{1}{\sqrt{2}} = \frac{\sqrt{2}}{2} = \cos 45°$; $\tan 45° = 1$

(b) $AM = \sqrt{3}$; $\sin 30° = \frac{1}{2}$, $\cos 30° = \frac{\sqrt{3}}{2}$, $\tan 30° = \frac{1}{\sqrt{3}} = \frac{\sqrt{3}}{3}$; $\sin 60° = \frac{\sqrt{3}}{2}$, $\cos 60° = \frac{1}{2}$, $\tan 60° = \sqrt{3}$.

(c) (iii) $\cos 315° = \cos 45° = \frac{\sqrt{2}}{2}$; $\sin 225° = -\sin 45° = -\frac{\sqrt{2}}{2}$; $\tan 210° = \tan 30° = \frac{\sqrt{3}}{3}$; $\cos 120° = -\cos 60° = -\frac{1}{2}$; $\sin 960° = \sin 240° = -\sin 60° = -\frac{\sqrt{3}}{2}$; $\tan(-135°) = \tan 45° = 1$.

(d) Cut the n-gon into n "cake slices", and use the formula "$\frac{1}{2}ab\sin C$" for each slice.

(i) $\frac{3\sqrt{3}}{4}$; (ii) 2; (iii) $\frac{3\sqrt{3}}{2}$; (iv) $2\sqrt{2}$; (v) 3

(e) Work out the side length of the n-gon, then cut the n-gon into n "slices", and use the formula "$\frac{1}{2}$(base × height)" for each slice.

(i) $3\sqrt{3}$; (ii) 4; (iii) $2\sqrt{3}$; (iv) $8(\sqrt{2}-1)$; (v) $12(2-\sqrt{3})$

Note: There is no hidden trig here: all you need is Pythagoras' Theorem. For example, in part (e)(iv) we can extend alternate sides of the regular octagon to form the circumscribed 2 by 2 square. The four corner triangles are isosceles right angled triangles with hypotenuse of length s (the side of the octagon). Hence each side of the square is equal to $s + 2 \cdot \frac{s}{\sqrt{2}} = 2$, whence $s = 2(\sqrt{2}-1)$.

20. (a) $\sqrt{8}$; (b) $\sqrt{2}, \sqrt{2}$; (c) $\sqrt{8} = \sqrt{4 \times 2} = 2\sqrt{2}$

21. Construct the perpendicular from A to BC (possibly extended); let this meet the line BC at X. There are four possibilities:

(i) either $X = C$, in which case $\angle BCA$ is a right angle as required; or $X = B$, in which case $b^2 = a^2 + c^2$, contradicting $a^2 + b^2 = c^2$;

(ii) $X \neq B, C$, and C lies between B and X;

(iii) $X \neq B, C$, and X lies between B and C;

(iv) $X \neq B, C$, and B lies between X and C.

We analyse case (ii) and leave cases (iii) and (iv) to the reader.

(ii) $\triangle AXC$ and $\triangle AXB$ are both right angled triangles; so by Pythagoras' Theorem we know that

$$\begin{aligned} AC^2 &= AX^2 + XC^2, \text{ and} \\ AB^2 &= AX^2 + XB^2 \\ &= AX^2 + (XC + CB)^2 \\ &= AX^2 + XC^2 + CB^2 + 2XC \cdot CB \\ &= AC^2 + CB^2 + 2XC \cdot CB. \end{aligned}$$

Since we are told that $AC^2 + CB^2 = AB^2$, it follows that $2XC \cdot CB = 0$, contrary to $X \neq C$.

Note: Notice that the proof of the converse of Pythagoras' Theorem makes use of Pythagoras' Theorem itself.

22.

(a) $c = b+1$, so $a^2 = c^2 - b^2 = 2b+1$. Hence a is odd, and we can write $a = 2m+1$.

(b) Suppose $b = 2n - 1$ is also odd. Then $c^2 = 4n^2$ is divisible by 4 – which contradicts the fact that $b^2 = 4(n^2 - n) + 1$, and $a^2 = 4(m^2 + m) + 1$, so $a^2 + b^2$ leaves remainder 2 on division by 4.

Hence $b = 2n$ is even and $c = 2n + 1$ is odd. But then

$$(2m+1)^2 + (2n)^2 = a^2 + b^2 = c^2 = (2n+1)^2,$$

so $4(m^2 + m) = 4n$, and $n = m(m+1)$.

23.

(a) If a and b are both even, then $HCF(a, b) \neq 1$, so the triple would not be primitive.

If a and b are both odd, we use the idea from Problem **22**(b). Suppose $a = 2m+1$, $b = 2n+1$; then $a^2 = 4(m^2+m)+1$, and $b^2 = 4(n^2+n)+1$, so $a^2+b^2 = 2\times(2(m^2+m+n^2+n)+1)$. But this is "twice an odd number", so cannot be equal to c^2 (since c would have to be even, and any even square must be a multiple of 4).

Hence we may assume that a is odd and b is even: so c is is odd.

(b) Then $a^2+b^2 = c^2$ yields $b^2 = c^2 - a^2 = (c-a)(c+a)$, so
$$\left(\frac{b}{2}\right)^2 = \left(\frac{c-a}{2}\right)\left(\frac{c+a}{2}\right).$$

Any common factor of $\frac{c+a}{2}$ and $\frac{c-a}{2}$ divides their sum c and their difference a, so $HCF(\frac{c-a}{2}, \frac{c+a}{2}) = 1$. Since the difference of these two factors is a, which is odd, they have opposite parity.

(c) If two integers are relatively prime, and their product is a square, then each of the factors has to be a square (consider their prime factorisations). Hence $\frac{c+a}{2} = p^2$ and $\frac{c-a}{2} = q^2$, where $HCF(p,q) = 1$ and p and q have opposite parity. Therefore
$$c = p^2+q^2, \quad a = p^2-q^2, \quad b = 2pq.$$

(d) It is easy to check that any triple of the given form is (i) primitive, and (ii) satisfies $a^2+b^2 = c^2$.

24. Claim The only such triples are those of the form $(3s, 4s, 5s)$.

Proof We show that the only *primitive* Pythagorean triple which forms an arithmetic progression is the familiar triple $(3,4,5)$.

By Problem **23**, one of the numbers in any *primitive* Pythagorean triple is *even* (namely $2pq$) and two are *odd* (p and q are of opposite parity, so p^2-q^2 and p^2+q^2 are both odd).

$\therefore$ $2pq$ is the "middle term", and the smallest and largest terms differ by $2q^2$.

$\therefore$ the common difference $d = c-b = b-a$ is equal to q^2.

$\therefore$ $2pq = p^2$, so $p = 2q$.

Finally, since p and q are relatively prime, we must have $q = 1$, $p = 2$. QED

Note: Alternatively, let (a,b,c) be any Pythagorean triple (not necessarily primitive), which forms an arithmetic progression. Then
$$a^2 = c^2 - b^2 = (c-b)(c+b) = (b-a)(c+b).$$
So $b(c+b) = a(a+b+c)$. Hence $a \cdot 3b = a(a+b+c) = b(c+b)$. It then follows that $3b^2 = b(a+b+c) = 4ba$, so $3b = 4a$ and $a:b = 3:4$.

25.

(a) $2 = 1^2 + 1^2$, $5 = 2^2 + 1^2$, $13 = 3^2 + 2^2$, $17 = 4^2 + 1^2$, $29 = 5^2 + 2^2$, $37 = 6^2 + 1^2$, $41 = 5^2 + 4^2$, $53 = 7^2 + 2^2$, $61 = 6^2 + 5^2$, $73 = 8^2 + 3^2$, $89 = 8^2 + 5^2$, $97 = 9^2 + 4^2$.

(b) $(a^2 + b^2)(c^2 + d^2) = (ac - bd)^2 + (ad + bc)^2$.

Note: It is easy to check this identity once it is given, but most of us are not so fluent in algebra as to spot this handy identity without help! However, Chapter 1 is about "Mental skills", and one such technique (once you have mastered it) arises from the arithmetic of *complex numbers*. If you have met complex numbers, then this identity can be written down immediately. Let us explain briefly how.

Every complex number $w = a + bi$ (where $i^2 = -1$) can be represented as a point in the complex plane with coordinates (a, b). The "size", or *modulus*, of w is its length $|w|$ (the distance of (a, b) from the origin $(0, 0)$); and the square of this length $a^2 + b^2$ is referred to as the *norm* of the complex number $w = a + bi$. The required identity is an immediate consequence of the two facts:

- the modulus of a product is equal to the product of the two moduli: $|wz| = |w| \cdot |z|$, and
- the norm $a^2 + b^2$ can be expressed algebraically as $a^2 + b^2 = (a + bi)(a - bi)$.

Once we know these facts:

- $a^2 + b^2$ can be interpreted as the norm of $w = a + bi$, and
- $c^2 + d^2$ as the norm of $z = c + di$;

the product of the two norms $(a^2 + b^2)(c^2 + d^2)$ is then equal to the norm of the product $w \cdot z = (ac - bd) + (ad + bc)i$.

Note: If we choose $z = c - di$, then $wz = (ac + bd) + (bc - ad)i$, and we get a second identity: $(a^2 + b^2)(c^2 + d^2) = (ac + bd)^2 + (bc - ad)^2$.

(c) The square $(2n)^2$ of any even number $2n$ is a multiple of $2^2 = 4$.
Any odd number has the form $2n + 1$; its square $(2n + 1)^2 = 4n^2 + 4n + 1$ is 1 more than a multiple of 4. So in the sum of two squares, we have

(i) both squares are even and their sum is a multiple of 4, or

(ii) one square is even and one is odd and their sum is of the form $4k + 1$, or

(iii) both squares are odd and their sum is of the form $4k + 2$.

Hence no number of the form $4k + 3$ can be written as a sum of two squares.

(d) We are told that $2 = 1^2 + 1^2$, and that Euler showed every prime of the form $4k + 1$ can be written as the sum of two squares. Part (b) then shows that any product of such primes can be written as the sum of two squares. And if we multiply a sum of two squares by a square, the result can again be written as the sum of two squares. This allows us to construct the list of forty six integers $N < 100$ which can be so written. These are precisely the integers of the form "(a square) × (a product of distinct primes p, where $p = 2$ or $p = 4k+1$)":

0, 1, 2, 4, 5, 8, 9, 10, 13, 16, 17, 18, 20, 25, 26, 29, 32, 34, 36, 37, 40, 41, 45, 49, 50, 52, 53, 58, 60, 61, 64, 65, 68, 72, 73, 74, 80, 81, 82, 83, 85, 87, 89, 90, 97, 98.

(e) The side of such a square is the hypotenuse of a right angled triangle whose legs run in the x- and y- directions, and have integer lengths (because their vertices are at points with integer coordinates). Hence the answer is exactly the same as for part (d) (provided one does not quibble about the idea of a square with side 0 and area 0).

26. Most sheets in a newspaper are double sheets with four pages. If we assume that all sheets are double sheets, then the 13 pages before page 14 match up with the 13 pages after page 27, so there are $27 + 13 = 40$ pages in all. (If the paper included inserted 'single sheets' – with just two pages, then there is no solution.)

27.

(a) If $\theta = 90°$, then the overlap is clearly one quarter of the small square. In general, the continuations of the sides BA and DA cut the small square into four congruent quadrilaterals, one of which is the area of overlap. So the overlap is always **one quarter** of the lower square.

(b) The area of overlap for "a large equilateral $\triangle ABC$ on top of a small equilateral $\triangle PQR$" is not constant, but depends on the angle of orientation of the large triangle. However, if viewed in the right way, something similar works for a large regular $2n$-gon on top of a small regular $2n$-gon with one corner of the large polygon at the centre of the small one.

The key is to realise how the fraction "one quarter" arises for a regular 4-gon. There $2n = 4$, so $n = 2$, and each vertex angle is equal to $\left(\frac{360°}{2n}\right)(n-1) = 90°$, which is exactly $\frac{n-1}{2n} = \frac{1}{4}$ of $360°$. For a regular $2n$-gon, the large polygon always covers a fraction equal to exactly $\frac{n-1}{2n}$ of the small polygon.

28. 3 cm, the same as the perimeter of triangle ABC.[2]

What if A were folded to some point A'' on BC?

29. *In extremis* one may reach for trigonometry: if we denote the three angles by α (at A), β (at B), and γ (at C), then the arrangement of squares implies that $\tan\alpha = \frac{1}{3}$, $\tan\beta = \frac{1}{2}$, and $\tan\gamma = 1$, so we can use the standard identity

$$\tan(\alpha + \beta) = \frac{\tan\alpha + \tan\beta}{1 - \tan\alpha \tan\beta}$$

to see that $\tan(\alpha + \beta) = 1 = \tan\gamma$.

---

[2] From: Y. Wu, The examination system in China: the case of zhongkao mathematics. 12th International Congress on Mathematical Education. 8 July – 15 July, 2012, COEX, Seoul, Korea

However, it is worth looking for a more elementary explanation than 'brute force calculation'. If we embed the horizontal 3 by 1 rectangle $ADEH$ in the top right hand corner of a 4 by 4 square $ZDXY$, (with Z labelling the top left hand corner), then we can complete the square $AEPQ$, which has AE as one side, with P on side XY and Q on side YZ.

Then $\angle AEH = \angle DAE$, and $\angle AEQ = \angle DCE$. So all we need to explain is why $\angle HEQ = \angle DBE$ – and this follows from the fact that EQ passes through the centre of the 4 by 4 square $ZDXY$.

30.

(a) Construct points P and Q inside the trapezium so that $MNPQ$ is similar to $BCMN$. If the line through P parallel to AB meets BC at X, and the line through Q parallel to AC meets BC at Y, then $MNPQ$, $NBXP$, $XYQP$, $MCYQ$ are the required pieces.

(b) Each of the four pieces must be *three-quarters* of one of the small squares. So we have to lose one quarter of each small square. There are various ways to do this, but most create non-congruent parts. Cut each of the smaller squares into quarters as for the original square. If O is the centre of the original square, lump together the three mini-squares which have O as a vertex to form an L-shape. Each of the three remaining small squares has lost a quarter and forms an identical L-shape.

31. To divide the shaded region in 2 congruent parts, rotate the lower small semicircle through the angle $\frac{\pi}{2}$ anticlockwise about the centre of the large circle.

Note: The same idea allows one to divide the shaded region into n congruent parts: rotate the lower small semicircle successively through the angle $\frac{\pi}{n}$ anticlockwise about the centre of the large circle.[3]

32.

(a) The point O lies on the perpendicular bisector of AB, and so is equidistant from the two endpoints A and B, so $OA = OB$. The point O also lies on the perpendicular bisector of BC, and so is equidistant from the two endpoints B and C, so $OB = OC$.

(b) (ii) $\angle BA'C$ and $\angle BAC$ are angles subtended in the same segment of the circle by the same chord BC, so are equal ("angles in the same segment").

$\angle A'CB$ is the angle subtended on the circumference (at C) by the diameter $A'B$, and so must be a right angle. In $\triangle A'BC$ we then see that

$$\sin A = \sin A' = \frac{a}{2R}.$$

If we now switch attention from the angle at A to the angle at B, and then to the angle at C, we can show that $\sin B = \frac{b}{2R}$, and that $\sin C = \frac{c}{2R}$.

---

[3] From: Introductory Assignment, Gelfand Correspondence Program in Mathematics.

33.

(a) Drop a perpendicular from A to meet the line BC at X. Then $AX = b \cdot \sin C$, so
$$\Delta = \frac{1}{2} \cdot (a \times b \sin C).$$

(b) Substitute "$\sin C = \frac{c}{2R}$" (from the Sine Rule) into the formula in part (a).

34.

(a) (i) $\frac{\alpha}{2\pi}$; $\frac{\alpha}{2\pi} \times$ (surface area of unit sphere $= 4\pi$) $= 2\alpha$.

  (ii) 4α

(b) (i) 4β; (ii) 4γ

(c) (i) $4(\alpha + \beta + \gamma)$

  (ii) Triangle ABC and its sister triangle $A'B'C'$ are congruent, and each is covered 3 times.

  (iii)
$$\begin{aligned} 4(\alpha + \beta + \gamma) &= \text{(total surface area of the unit sphere)} \\ &\quad + (4 \times \text{(area of the spherical triangle } ABC)) \\ \therefore \text{area}(\triangle ABC) &= (\alpha + \beta + \gamma) - \pi. \end{aligned}$$

Note: In particular, the formula for the area of a spherical triangle implies:

- the angle sum $\alpha + \beta + \gamma$ in any spherical triangle is **always greater than** π, and
- the larger the triangle ABC, the more its angle sum must exceed π.

35.

(a)
$$\begin{aligned} \cos(A+B) + i\sin(A+B) &= e^{i(A+B)} \\ &= e^{iA} \cdot e^{iB} \\ &= (\cos A + i \sin A)(\cos B + i \sin B). \end{aligned}$$

Hence
$$\sin(A+B) = \sin A \cos B + \cos A \sin B,$$
and
$$\cos(A+B) = \cos A \cos B - \sin A \sin B.$$

To reconstruct $\tan(A+B)$, divide these two expressions, and then divide numerator and denominator by "$\cos A \cos B$" to get
$$\tan(A+B) = \frac{\sin(A+B)}{\cos(A+B)} = \frac{\tan A + \tan B}{1 - \tan A \tan B}.$$

(b)
$$\begin{aligned}\sin X &= \sin(A+B) \\ &= \sin A \cos B + \cos A \sin B \\ &= \sin\left(\frac{X+Y}{2}\right)\cos\left(\frac{X-Y}{2}\right) + \cos\left(\frac{X+Y}{2}\right)\sin\left(\frac{X-Y}{2}\right),\end{aligned}$$

and

$$\begin{aligned}\sin Y &= \sin(A-B) \\ &= \sin A \cos B - \cos A \sin B \\ &= \sin\left(\frac{X+Y}{2}\right)\cos\left(\frac{X-Y}{2}\right) - \cos\left(\frac{X+Y}{2}\right)\sin\left(\frac{X-Y}{2}\right),\end{aligned}$$

$$\therefore \sin X + \sin Y = 2\sin\left(\frac{X+Y}{2}\right)\cos\left(\frac{X-Y}{2}\right).$$

For $\sin X - \sin Y$, substitute "$-Y$" in place of Y to get:

$$\sin X - \sin Y = 2\sin\left(\frac{X-Y}{2}\right)\cos\left(\frac{X+Y}{2}\right).$$

Similarly

$$\begin{aligned}\cos X + \cos Y &= \cos(A+B) + \cos(A-B) \\ &= (\cos A \cos B - \sin A \sin B) + (\cos A \cos B + \sin A \sin B) \\ &= 2\cos A \cos B \\ &= 2\cos\left(\frac{X+Y}{2}\right)\cos\left(\frac{X-Y}{2}\right)\end{aligned}$$

$$\begin{aligned}\cos X - \cos Y &= \cos(A+B) - \cos(A-B) \\ &= (\cos A \cos B - \sin A \sin B) - (\cos A \cos B + \sin A \sin B) \\ &= -2\sin A \sin B \\ &= -2\sin\left(\frac{X+Y}{2}\right)\sin\left(\frac{X-Y}{2}\right).\end{aligned}$$

(c)(i) $\sin(A+B) = \sin 90° = 1$;

$$\sin A \cos B + \cos A \sin B = \left(\frac{1}{2}\right)^2 + \left(\frac{\sqrt{3}}{2}\right)^2 = 1.$$

(ii) $\cos X - \cos Y = \frac{1}{2} - 1 = -\frac{1}{2}$;

$$-2\sin\left(\frac{X+Y}{2}\right)\sin\left(\frac{X-Y}{2}\right) = -2\sin^2 30° = -\frac{1}{2}.$$

(d) (i) $2\sin A \sin B = \cos(A-B) - \cos(A+B)$.

$\therefore 2\sin A \sin B + 2\sin C \sin D$
$= [\cos(A-B) - \cos(A+B)]$
$+ [\cos(C-D) - \cos(C+D)]$
$= \cos(A-B) + \cos(C-D)$
 (since $C+D = \pi - (A+B)$)
$= 2\cos\left(\dfrac{A+C-(B+D)}{2}\right) \cos\left(\dfrac{A+D-(B+C)}{2}\right)$
$= 2\cos\left(\dfrac{\pi}{2} - (B+D)\right) \cos\left(\dfrac{\pi}{2} - (B+C)\right)$
$= 2\sin(B+D)\sin(B+C)$.

Note: We can swap A and B without changing the expression "$\sin A \sin B + \sin C \sin D$". Hence the same should be true of the RHS "$\sin(B+C)\sin(B+D)$". Fortunately, since $A+B+C+D = \pi$, we know that $\sin(A+C) = \sin(B+D)$, and $\sin(A+D) = \sin(B+C)$.

(ii) In triangle WXY we see that $A+B+C+D = \pi$. Hence

$$\sin A \sin B + \sin C \sin D = \sin(A+D)\sin(B+D).$$

Now let R be the radius of the circle. Use "equality of angles in the same segment" and the Sine Rule (in its full form: see Problem **32**) to write:

$2R\sin A = XY$, $2R\sin B = WZ$, $2R\sin C = YZ$, $2R\sin D = WX$,

$2R\sin(A+D) = WY$, $2R\sin(B+D) = XZ$.

$\therefore WX \times YZ + WZ \times XY = WY \times XZ$.

36. Yes.

Let the perpendicular bisectors of AB and BC meet at the point O.

Then $OA = OB$ and $OB = OC$, so the circle with centre O passing through A also goes through B and C.

We have to prove that this circle also passes through D, E, etc..

To do this we prove that $\triangle OBC \equiv \triangle OCD$.

We know that $\triangle OAB \equiv \triangle OBC$ (by SSS-congruence: $OA = OB$ and $OB = OC$ by the construction of O; and $AB = BC$ since both are sides of a regular n-gon). Moreover

$\angle OAB = \angle OBA$ (base angles of the isosceles triangle $\triangle OAB$)
$ = \angle OCB$ (since $\triangle OAB \equiv \triangle OBC$)
$ = \angle OBC$ (base angles of the isosceles triangle $\triangle OBC$).

And $\angle ABC = \angle BCD$ (angles of the same regular n-gon).

$\therefore \angle OCD = \angle BCD - \angle OCB = \angle ABC - \angle OBA = \angle OBC$.
$\therefore \triangle OBC \equiv \triangle OCD$ (by the SAS-congruence criterion).
Hence $OC = OD$.
Continuing in this way we can prove that $OA = OD = OE$, etc..

37.

(a) Yes. There are two nets for a regular tetrahedron. One of these consists of four equilateral triangles in a row (alternately right side up and right side down). In making the tetrahedron, the two sloping ends are glued together. So if we cut half a triangle from one end and stick it on the other end, we get a 4 by $\sqrt{3}$ rectangle which folds round the tetrahedron exactly without any gaps or overlaps.

(b) The usual way to wrap a cube with edges of length 2 is to take a 4 by 8 rectangular piece of wrapping paper, to position the cube centrally on an edge of length 4 (1 unit from each edge), and to wrap the paper to cover a circuit of four faces. The overlapping residue can then be folded down to cover each side face, with overlaps. Hence the ratio

"area of paper" : "surface area of cube" $= 8 : 6$.

The same 'wastage rate' can be achieved with a square $4\sqrt{2}$ by $4\sqrt{2}$ piece of paper. Position the cube centrally on the paper, but turned through an angle of 45°. Then fold the four corners of the paper up each of the four side faces (with folds to tuck in four 'wasted' isosceles right angled triangles – one in the middle of each edge of the paper, with total wasted area 8). Finally, the four isosceles right angled triangles at the four corners of the paper can be folded in to exactly cover the top face without further overlaps.

However, one can do significantly better if the paper can be folded back on itself. Take a 2 by 14 rectangle, and think of this as being marked into seven 2 by 2 squares. Place the cube to cover the central 2 by 2 square – leaving three 2 by 2 squares sticking out each side. Fold one 2 by 6 strip up to cover the top square, before folding back along a diagonal of the top square to reveal the inside of the paper and to cover half of the top square *twice* before folding down to cover one side square. Do the same with the other 2 by 6 strip, with the reverse fold along the diagonal of the top square resulting in the other half of the top square being covered twice, with the tail folding down to cover the other side square. Hence the ratio

"area of paper" : "surface area of cube" $= 7 : 6$.

(This lovely solution was provided by Julia Gog. We do not know whether one can do better.)

38.

(i) Yes. (Cut off a corner A say with a plane passing through the three neighbours of A.)

(ii) Yes. (Cut the cube parallel to a face.)

(iii) No. (Any cross-section of the cube is a polygon. Each edge of this polygonal cross-section is the line segment formed by the intersection of the cutting plane with one of the faces of the cube. Since the cube has just six faces, the cross-section can have at most six sides.)

(iv) Yes. (Let A and G be two opposite corners – so that AG passes through the centre O of the cube. Then the plane through O which is perpendicular to AG cuts the surface of the cube in a regular hexagon.)

(v) No. (It may not be clear how to prove this easily. It is perfectly possible to obtain a pentagonal cross-section by cutting with a plane that misses exactly one face. But if the cutting plane misses exactly one face, we can be sure that it must cut both faces belonging to some "opposite pair"; and these two faces are **parallel**, so the resulting edges of the cross-sectional pentagon are **parallel**. Hence the pentagon can never be regular.)

39.

(i) No; (ii) Yes; (iii) No; (iv) Yes; (v) No.

The 12 edges of the cube come in three groups of four – namely the four parallel edges in each of three directions.

Consider the four edges in one of these parallel groups. If the Sun's rays are parallel to these four edges, then each of these edges projects to a single vertex of the outline of the shadow – which is a projection of a square.

In all other cases two of the four parallel edges in the group give rise to shadows that form part of the *boundary* of the shadow polygon, while the other two edges project to the inside of the shadow (and so do not feature in the boundary of the shadow). Hence each of the three groups provides two edges to the boundary of the shadow polygon, and we obtain a hexagon.

To obtain a *regular* hexagon, align the Sun's rays parallel to the line AG joining two opposite corners A and G of the cube, and position the shadow plane perpendicular to this direction. The three edges at these two corners A and G then project to the inside of the shadow, while the six remaining edges project to a regular hexagon. (Since there are four body diagonals like AG, there are four ways to make such a projection. In each case, the six edges of the cube that project to the regular hexagon form a non-planar hexagon on the surface of the cube, that zig-zags its way round the polyhedron like a 'wobbly equator', turning alternately left and right each time it reaches a vertex. Such a closed circuit on a regular polyhedron is called a *Petrie polygon* – named after John Flinders Petrie (1907–1972), son of the famous Egyptologist Flinders Petrie).

II. Arithmetic

> *A child of the new generation*
> *Refused to learn multiplication*
> *He said, "Don't conclude*
> *That I'm stupid, or rude.*
> *I am simply without motivation."*
> Joel Henry Hildebrand (1881–1983)

Many important aspects of serious mathematics have their roots in the world of arithmetic. This is a world everyone can enjoy and master. In this chapter we re-visit, or maybe meet for the first time, key aspects of arithmetic that are often overlooked – ending with an introduction to the basic result on the distribution of primes.

The place of arithmetic in elementary mathematics can only be understood if one realises that, from upper primary school onwards, mathematics should no longer focus on more and more complicated calculations. Rather it moves beyond a set of procedures for grinding out answers, and should become a *structural laboratory*, where we gain insight into simple phenomena, and where we begin to appreciate how calculation can be managed, or tamed. The focus on structure leads in the main to matters which can be best expressed *algebraically*. This chapter concentrates mainly on structural aspects of number that are strictly arithmetical (e.g. related to numerals and place value), or where the relevant structural approach is "pre-algebraic" – with occasional forays into the world of algebra.

We repeat the observation that the "essence of mathematics" in the title is mostly left implicit in the problems. And while there is some discussion of this "essence" in the text between the problems, most of the relevant observations that we make are to be found in the solutions, or in the **Notes** which follow many of the solutions.

2.1. Place value and decimals: basic structure

Problem 40 Without using a calculator:

(a) Work out

(i) $12\,345\,679 \times 9$

(ii) $7 \times 9 \times 11 \times 13$.

(b) Divide

(i) $123\,123\,123$ by 123

(ii) $111\,111\,111$ by 111

(iii) $111\,111\,111$ by 37. △

Problem 41 Work out in your head (i) 11^2 (ii) 11^3 (iii) 101^2. △

Problem 42 Try to answer the following questions using only mental arithmetic:

(a)(i) What is the largest and the smallest possible number of digits in the answer when you multiply a 3-digit integer by a 5-digit integer?

(ii) What if we multiply an m-digit integer by an n-digit integer?

(b)(i) How many (base 10) digits are there in the evaluated form of 2^{20}?

(ii) Estimate $\left(\frac{1}{2}\right)^{20}$ to 6 decimal places.

(c) Can a natural number (i.e. a positive integer) be smaller than the product of its (base 10) digits?

(d) Work out how many zeros there are on the end, and work out the last non-zero digit of (i) $2^{15} \times 5^3$ (ii) $20!$. △

Problem 43 Imagine the sequence of positive integers from 1 to 60 written in a single row as the digits of a very large integer:

$$12345678910111213141516171819202122 23 \cdots 5960.$$

You have to cross out 100 of these digits.

(a) Suppose you want to make the remaining number as *small* as possible. What number is left?

(b) Now suppose that you want to make the remaining number as *large* as possible. What number is left? △

2.2. Order and factors

Problem 44 Find the remainder when we divide

$$1111\cdots 1111 \text{ (with 1111 digits 1)}$$

by 1111. △

Problem 45 Which of the numbers

$$\frac{100\,001}{100\,002} \text{ and } \frac{10\,000\,001}{10\,000\,002}$$

is bigger? △

Problem 46 Show that the integer

$$100\,000\,000\,003\,000\,000\,000\,000\,700\,000\,000\,021$$

is not prime. △

Problem 47 How many prime numbers are there in each of these sequences? (Can you identify infinitely many primes in either sequence? Can you identify infinitely many non-primes?)

(a) 1, 11, 111, 1111, 11 111, 111 111, 1 111 111, ...

(b) 11, 1001, 100 001, 10 000 001, ... △

2.3. Standard written algorithms

Problem 48 Use standard column arithmetic (i.e. long multiplication) to evaluate 9009×37. Why should you have foreseen the outcome? △

Problem 49 In the long division shown here, all the digits are missing.

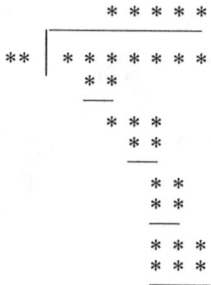

But the "shape" of the constituent numbers is clear.

Can you work out all possibilities for the two-digit divisor?   △

Problem 50 (*For those readers who can write simple computer code.*) In these problems you may choose your favourite programming language, and a device of your choice.

(a) Two non-negative integers m and n are to be entered in base 10, digit by digit, via a keyboard. Write computer code to implement the standard algorithms of column arithmetic in order to output to the screen (in the same format):

   (i) $m + n$
   (ii) $m - n$
   (iii) $m \times n$
   (iv) (if n is a divisor of m) $m \div n$
   (v) (if n is not a divisor of m) the integer part q of the quotient $m \div n$ and the remainder r.

(b) Repeat the challenge of part (a), but this time try to write shorter code by using recursion (or other programming tricks).

(c) Repeat the challenge of parts (a) and (b), but this time with inputs and outputs in the binary numeral system (see Section 2.8).   △

2.4. Divisibility tests

An integer written in base 10:

   is divisible by 10 precisely when the units digit is 0.

2.5. Sequences

Because $10 = 2 \times 5$, it follows that an integer (in base 10):

is divisible by 5 precisely when the units digits is 0 or 5 (i.e. a multiple of 5); and

is divisible by 2 precisely when the units digit is 0, 2, 4, 6, or 8 (i.e. a multiple of 2).

Because $100 = 4 \times 25$, it follows that an integer:

is divisible by 4 precisely when the integer formed by its last two digits is a multiple of 4; and

is divisible by 25 precisely when its last two digits are 00, 25, 50, or 75 (that is, a multiple of 25).

Because $1000 = 8 \times 125$, it follows that an integer:

is divisible by 8 precisely when the integer formed by its last three digits is a multiple of 8.

Hence simple tests for divisibility by 2, by 4, by 5, by 8, and by 10 all follow easily from the way we write numbers in base 10.

Problem 51

(a) Prove that, when an integer is written in base 10, the *remainder* when it is divided by 9 is equal to the *remainder* when its "digit-sum" is divided by 9. Conclude that the remainder when an integer is divided by 3 is equal to the remainder when its "digit-sum" is divided by 3.

(b) Explain why an integer is divisible by 6 precisely when it is divisible both by 2 and by 3. △

Problem 52

(a) What can you say about an integer N which is divisible by three times the sum of its base 10 digits?

(b) Find all integers which are *equal* to three times the sum of their base 10 digits.

(c) Find the smallest positive multiple of 9 with no odd digits. △

Problem 53 Prove than an integer written in base 11 is divisible by ten precisely when its digit-sum is divisible by ten. △

2.5. Sequences

We have already met

- the sequence of natural numbers (1, 2, 3, 4, 5, ...),
- the sequence of squares (1, 4, 9, 16, 25, ...),
- the sequence of cubes (1, 8, 27, 64, 125, ...),
- the sequence of prime numbers (2, 3, 5, 7, 11, 13, 17, ...),
- the sequence of powers of 2 (1, 2, 4, 8, 16, 32, ...), and the sequence of powers of 4 (1, 4, 16, 64, 256, ...).

We have also considered

- the sequence of *units* digits of the powers of 4 (1, 4, 6, 4, 6, 4, 6, ...),
- the sequence of *leading* digits of the powers of 4 (1, 4, 1, 6, 2, 1, 4, ...).

2.5.1 Triangular numbers

Problem 54

(a) Evaluate the first twelve terms of the sequence of *triangular* numbers:
$$1, 1+2, 1+2+3, 1+2+3+4, \ldots, 1+2+3+\cdots+10+11+12.$$

(b) Find and prove a formula for the nth triangular number
$$T_n = 1 + 2 + 3 + \cdots + n.$$

(c) Which triangular numbers are also (i) powers of 2? (ii) prime? (iii) squares? (iv) cubes? △

2.5.2 Fibonacci numbers

The Hindu-Arabic numeral system emerged in the Middle East in the 10th and 11th centuries. Fibonacci, also known as *Leonardo of Pisa*, is generally credited with introducing this system to Europe around 1200 – especially through his book *Liber Abaci* (1202). One of the problems in that book introduced the sequence that now bears his name.

The sequence of *Fibonacci numbers* begins with the terms $F_0 = 0$, $F_1 = 1$, and continues via the Fibonacci recurrence relation:
$$F_{n+1} = F_n + F_{n-1}.$$

The sequence was introduced through a curious problem about breeding rabbits; but to this day it continues to feature in many unexpected corners of mathematics and its applications.

Problem 55

(a)(i) Generate the first twelve terms of the Fibonacci sequence:
$$F_0, F_1, \ldots, F_{11}.$$

(ii) Use this to generate the first eleven terms of the sequence of "differences" between successive Fibonacci numbers. Then generate the first ten terms of the sequence of "differences between successive differences".

(iii) Find an expression for the mth term of the kth sequence of differences.

(b)(i) Generate the first twelve terms of the sequence of powers of 2:
$$2^0, 2^1, 2^2, \ldots, 2^{11}.$$

(ii) Use this to generate the first eleven terms of the sequence of "differences" between successive powers of 2. Then generate the first ten terms of the sequence of "differences between successive differences".

(iii) Find an expression for the mth term of the kth sequence of differences.
△

The sequence of differences between successive terms in the sequence of triangular numbers is just the sequence of natural numbers (starting with 2):
$$2, 3, 4, 5, 6, \ldots;$$
and the sequence of "second differences" is then *constant*:
$$1, 1, 1, 1, 1, \ldots.$$

The sequences of powers of 2 and the Fibonacci numbers behave very differently from this, in that taking differences reproduces something very like the initial sequence. In particular, taking differences can never lead to a *constant* sequence.

Logically the next four problems should wait until Chapter 6, where we address the delicate matter of "proof by mathematical induction". However, that would deprive us of the chance to sample the kind of surprises that lie just beneath the surface of the Fibonacci sequence, and to experience the process of fumbling our way towards a structural understanding of the apparent patterns that emerge. Of course, each time we think we have

58   Arithmetic

managed to *guess* what seems to be true, we face the challenge of *proof.* Those who have not yet mastered "proof by induction" are encouraged to get what they can from the solutions, and to view this as an informal introduction to ideas that will be squarely addressed in Chapter 6.

Problem 56

(a)(i) Generate the sequence of *partial sums* of the sequence of powers of 2:
$$2^0, \; 2^0 + 2^1, \; 2^0 + 2^1 + 2^2, \; 2^0 + 2^1 + 2^2 + 2^3, \ldots$$

(ii) Prove that each partial sum is 1 less than the *next* power of 2.

(b)(i) Generate the sequence of partial sums of the Fibonacci sequence:
$$F_0, \; F_0 + F_1, \; F_0 + F_1 + F_2, \; F_0 + F_1 + F_2 + F_3, \ldots$$

(ii) Prove that each partial sum is 1 less than the *next but one* Fibonacci number. △

Problem **56**(b) starts out with the observation that
$$F_0 + F_1 = F_3 - 1$$
which is a consequence of the first two instances of the fundamental recurrence relation
$$F_{n-1} + F_n = F_{n+1}$$
and derives a surprising value for the nth partial sum:
$$F_0 + F_1 + F_2 + \cdots + F_{n-1}.$$

Fibonacci numbers make their mathematical presence felt in a quiet way – partly through the almost spooky range of unexpected internal relations which they satisfy, as illustrated in Problem **56**(b) and in the next few problems.

Problem 57

(a) Note that
$$F_n^2 = F_{n-0}F_{n+0} = F_n^2 + (-1)^{n-1}F_0.$$

(i) Evaluate the succession of terms:
$$F_{1-1}F_{1+1}, \; F_{2-1}F_{2+1}, \; F_{3-1}F_{3+1}, \; F_{4-1}F_{4+1}, \; \ldots.$$

(ii) Guess a simpler expression for the product $F_{n-1}F_{n+1}$. Prove your guess is correct.

(b) Let $a, b, c, d \geqslant 0$.

   (i) Show that the parallelogram $OABC$ spanned by the origin O, and the points $A = (a, b)$, $C = (c, d)$ and their sum $B = (a + c, b + d)$ has area $|ad - bc|$.

   (ii) Find the area of the first parallelogram in the sequence of "Fibonacci parallelograms", spanned by the origin O, and the points $A = (F_0, F_1) = (0, 1)$, $C = (F_1, F_2) = (1, 1)$.

   (iii) Show that the nth parallelogram $OACB$ in this sequence, spanned by the origin O, and the points $A = (F_{n-1}, F_n)$ and $B = (F_n, F_{n+1})$, and the $(n+1)^{\text{th}}$ parallelogram $OBDC$ spanned by the origin O, and the points $B = (F_n, F_{n+1})$ and $C = (F_{n+1}, F_{n+2})$ overlap in the triangle OBC, which is exactly half of each parallelogram.

   Conclude that every such parallelogram has area 1. Relate this to the conclusion of (a)(ii). △

The basic recurrence relation for Fibonacci numbers specifies the next term as the sum of two successive terms. We now consider what this implies about the sum of the *squares* of two successive terms.

Problem 58

(a) Evaluate the first few terms of the sequence

$$F_0^2 + F_1^2, \; F_1^2 + F_2^2, \; F_2^2 + F_3^2, \; \ldots.$$

(b) Guess a simpler expression for the sum $F_{n-1}^2 + F_n^2$. Prove your guess is correct. △

Problem 59

(a) Note that
$$F_0 F_4 = 0 = F_2^2 - 1, \quad F_1 F_5 = 5 = F_3^2 + 1.$$

   (i) Evaluate the succession of terms:
   $$F_2 - 2F_{2+2}, \; F_3 - 2F_{3+2}, \; F_4 - 2F_{4+2}, \; F_5 - 2F_{5+2}, \; F_6 - 2F_{6+2}, \; \ldots.$$

   (ii) Guess a simpler expression for the product $F_{n-2}F_{n+2}$. Prove your guess is correct.

(b)(i) Evaluate the succession of terms:
$$F_{3-3}F_{3+3},\ F_{4-3}F_{4+3},\ F_{5-3}F_{5+3},\ F_{6-3}F_{6+3}, \ldots.$$

(ii) Guess a simpler expression for the product $F_{n-3}F_{n+3}$. Prove your guess is correct. △

2.6. Commutative, associative and distributive laws

In this short section we re-emphasise the shift away from blind calculation, and towards consideration of the *structure* of arithmetic, which was already implicit in Problems **7–10**, and Problems **16–17** in Chapter 1.

Problem 60 Each of two positive numbers a and b is increased by 10%.

(i) What is the change of their sum $a + b$?

(ii) What is the percentage change of their product $a \times b$?

(iii) What is the percentage change in their quotient $\frac{a}{b}$? △

Problem 61 The numbers a, b, c, d, e, f are positive. How will the value of the expression
$$a \div (b \div (c \div (d \div (e \div f))))$$
change if the value of f is doubled? △

Problem 62 In Problem **17** we saw that it is no accident that the sum of entries in the 4 by 4 'multiplication table' is equal to 100.

1	2	3	4
2	4	6	8
3	6	9	12
4	8	12	16

(a) Go back to the proof that the total is equal to $(1 + 2 + 3 + 4)^2$ and see how this depends on the distributive law.

(b) The total of all entries in the multiplication square can be broken down into a succession of "reverse L-shapes", such as the one formed by the bottom row and right hand column (shown above in **bold**).

(i) Work out the subtotal in each of the four reverse L-shapes in the 4 by 4 multiplication table. What do you notice about these four subtotals?

(ii) Use the formulae for the kth and $(k-1)^{\text{th}}$ triangular numbers T_k and T_{k-1} to prove that, in the n by n multiplication table, the kth reverse L-shape always gives rise to a subtotal k^3.
Conclude that
$$T_n^2 = 1^3 + 2^3 + 3^3 + \cdots + n^3.$$
Hence find a simple formula for the sum C_n of the first n cubes. △

Now that we have a compact formula

- for the sum T_n of the first n positive integers, and
- for the sum C_n of the first n positive cubes,

we would naturally like to find a similar formula

- for the sum S_n of the first n squares:

$$S_n = 1^2 + 2^2 + 3^2 + \cdots + n^2$$

(that is, the sum of the entries on the leading diagonal of the n by n multiplication square).

This can be surprisingly elusive. But one way of obtaining it is to look instead for the sum of the entries in the sloping diagonal 2, 6, 12, 20, ... *just above* the main diagonal in the n by n multiplication square.

Problem 63 Consider the n by n multiplication square.

(a) Express the rth term in the sloping diagonal just above the main diagonal in terms of r. Hence show that the sum of entries in this sloping diagonal is equal to $S_{n-1} + T_{n-1}$.

(b) **Multiply by 3** each of the terms in the sloping diagonal just above the main diagonal.

   (i) Guess a formula for the successive sums of these terms (6, 6 + 18, 6 + 18 + 36, ...), and prove that your formula is correct.

   (ii) Hence derive a formula for the sum S_n of the first n squares. △

2.7. Infinite decimal expansions

The standard written algorithms for calculating with integers extend naturally to *terminating* decimals. But how is one supposed to calculate *exactly* with decimals that go on for ever?

Arithmetic

Problem 64 The decimals listed here all continue forever, recurring in the expected way. Calculate:

(a) $0.55555\cdots + 0.66666\cdots =$

(b) $0.99999\cdots + 0.11111\cdots =$

(c) $1.11111\cdots - 0.22222\cdots =$

(d) $0.33333\cdots \times 0.66666\cdots =$

(e) $1.22222\cdots \times 0.818181\cdots =$ △

Problem 65

(a) Show that any decimal $b_n b_{n-1} \cdots b_0.b_{-1}b_{-2}\cdots b_{-k}$ that terminates can be written as a fraction with denominator a power of 10.

(b) Show that any fraction that is equivalent to a fraction with denominator a power of 10 has a decimal that terminates.

(c) Conclude that a fraction $\frac{p}{q}$, for which $HCF(p,q) = 1$, has a decimal that terminates precisely when q divides some power of 10 (that is, when $q = 2^a \times 5^b$ for some non-negative integers a, b).

(d) Prove that any fraction $\frac{p}{q}$, for which $HCF(p,q) = 1$, and where q is not of the form $q = 2^a \times 5^b$, has a decimal which recurs, with a recurring block of length at most $q - 1$.

(e) Prove that any decimal which recurs is the decimal of some fraction. △

Problem 66

(a) Find the fraction equivalent to each of these recurring decimals:

   (i) $0.037037037\cdots$
   (ii) $0.370370370\cdots$
   (iii) $0.703703703\cdots$

(b) Let a, b, c be digits ($0 \leqslant a, b, c \leqslant 9$).

   (i) Write the recurring decimal "$0.aaaaa\cdots$" as a fraction.
   (ii) Write the recurring decimal "$0.ababababab\cdots$" as a fraction.
   (iii) Write the recurring decimal "$0.abcabcabcabcabc\cdots$" as a fraction. △

2.7. Infinite decimal expansions

Problem 67 Find the lengths of the recurring blocks for:

(a) $\dfrac{1}{6}, \dfrac{5}{6}$

(b) $\dfrac{1}{7}, \dfrac{2}{7}, \dfrac{3}{7}, \dfrac{4}{7}, \dfrac{5}{7}, \dfrac{6}{7}$

(c) $\dfrac{1}{11}, \dfrac{2}{11}, \dfrac{3}{11}, \dfrac{4}{11}, \dfrac{5}{11}, \dfrac{6}{11}, \dfrac{7}{11}, \dfrac{8}{11}, \dfrac{9}{11}, \dfrac{10}{11}$

(d) $\dfrac{1}{13}, \dfrac{2}{13}, \dfrac{3}{13}, \dfrac{4}{13}, \dfrac{5}{13}, \dfrac{6}{13}, \dfrac{7}{13}, \dfrac{8}{13}, \dfrac{9}{13}, \dfrac{10}{13}, \dfrac{11}{13}, \dfrac{12}{13}$ △

Problem 68 Decide whether each of these numbers has a decimal that recurs. Prove each claim.

(a) 0.12345678910111213141516171819202122232425262728293031 $\cdots$

(b) 0.10010001000010000010000001000000010000000010000000010 $\cdots$

(c) $\sqrt{2}$ △

Problem 69 For which real numbers x is the decimal representation of x unique? △

Problem 68 raises the question as to whether one person, who has total control, can specify the digits of a decimal so as to be sure that it neither terminates nor recurs: that is, so that it represents an *irrational* number. The next problem asks whether one person can achieve the same outcome with less control over the choice of digits.

Problem 70 Players A and B specify a real number between 0 and 1. The first player A tries to make sure that the resulting number is *rational*; the second player B tries to make sure that the resulting number is *irrational*. In each of the following scenarios, decide whether either player has a strategy that guarantees success.

(a) Can either player guarantee a "win" if the two players take turns to specify successive digits: first A chooses the entry in the first decimal place, then B chooses the entry in the second decimal place, then A chooses the entry in the third decimal place, and so on?

(b) Can either player guarantee a win if A chooses the digits to go in the odd-numbered places, and (entirely separately) B chooses the digits to go in the even-numbered places?

(c) What if A chooses the digits that go in almost all the places, but allows B to choose the digits that are to go in a sparse infinite collection of decimal places (e.g. the prime-numbered positions; or the positions numbered by the powers of 2; or ...)?

(d) What if A controls the choice of all but a finite number of decimal digits?

$\triangle$

2.8. The binary numeral system

There are all sorts of reasons why one should give thought to numeral systems using bases different from the familiar base 10. This is especially true of base 2, which is the simplest system of all, and is also (in some sense) the most widely used. What follows is only intended to offer a restricted glimpse into this alternative universe.

Problem 71 The numbers in this item are all written in base 2.

(a) Carry out the addition

$$\begin{array}{r} 11100 \\ +\ \ 1110 \\ \hline \end{array}$$

without changing the numbers into their base 10 equivalents – simply by applying the rules for base 2 column addition and "carrying".

(b) Carry out these long multiplications without changing the numbers into their base 10 equivalents – simply by applying the rules for base 2 column multiplication.

(i) $\begin{array}{r} 10110 \\ \times\ \ \ \ 10 \\ \hline \end{array}$
(ii) $\begin{array}{r} 1110 \\ \times\ \ \ 11 \\ \hline \end{array}$
(iii) $\begin{array}{r} 110 \\ \times 111 \\ \hline \end{array}$

(c) Try to add these fractions (where the numerators and denominators are numerals written in base 2) without changing the fractions into more familiar base 10 form.

$$\frac{110}{1111} + \frac{1}{10} + \frac{1001}{1110}$$

$\triangle$

The next problem invites you to devise divisibility tests for integers written in base 2 like those for base 10 (that is, tests which implement some check involving the base 2 digits in place of carrying out the actual division).

2.8. The binary numeral system

Problem 72 Let N be a positive integer written in base 2. Describe and justify a simple test, based on the digits of N_{base2}:

(i) for N to be divisible by 2

(ii) for N to be divisible by 3

(iii) for N to be divisible by 4

(iv) for N to be divisible by 5.   △

Problem 73 A mathematical merchant has a pair of scales and an infinite set of calibrated integral weights with values $w_0, w_1, w_2, \ldots$ (where $w_0 < w_1 < w_2 < \ldots$), but with only one weight of each value.

(a) Suppose that, for each object of positive integer weight w whose weight is to be determined, when the object is placed in one scale pan, the merchant is able to select some combination of his weights $w_0, w_1, w_2, \ldots$ to put in the other scale pan to balance, and hence to determine the weight of, the object to be weighed.

   (i) If for each weight w there is a *unique* choice of weights w_i that balance w, prove that the collection of weights must consist of all the powers of 2.

   (ii) If every object of unknown integral weight w can be balanced by some collection of the weights w_i, but some weights w can be balanced, or "represented", in more than one way, is it true that the merchant's collection of weights has to *include* all the powers of 2?

(b) What can you prove if the merchant's set of weights allow him to balance every unknown integer weight w in exactly one way by varying his weighing procedure, so that he can place his "known weights" *in either scale pan* (either in the same scale pan as the unknown weight to add to its weight, or in the opposite scale pan to balance it)?   △

Problem 74 Explain how to express any fraction
$$\frac{m}{2^n}$$
where $0 < m < 2^n$ as a sum of distinct *unit* fractions with denominator a power of 2.   △

You may have heard of an algorithm (a bit like long division) which allows one to compute by hand the *square root* of any number N given in base 10. The algorithm starts by grouping the digits of N in pairs, starting from the decimal point. It then extracts the square root, digit by digit, with the square root having one digit for each successive pair of digits of N, starting with the left-most pair (which may be a single digit).

We all know how to start the process. For example, if the left most pair of digits in N is "12", then we know that the square root starts with a "3". Successive digits are then identified using the algebraic identity

$$N = (x+y)^2 = x^2 + 2xy + y^2,$$

where x is the sequence of leading digits in the "partial square root" extracted so far (followed by an appropriate string of 0s), and y stands for the residual part of the required square root.

The key is to concentrate each time on the leading digit Y of the residue "$N - x^2$", and at each stage to choose the leading digit Y of y so that $2xy + y^2$ does not exceed $N - x^2$. This sequence of steps is traditionally (and helpfully) laid out in much the same way as long division, where at each stage we subtract the square of the current approximate square root x, from the original number N, and "bring down" the next pair of digits, and then choose the next digit Y in the square root (the leading digit of y) so that "$2xy + y^2$" does not exceed the residue $N - x^2$.

In base 10 each stage requires one to juggle possibilities to decide on the next digit in the partial square root. However, in base 2 the process should be simpler, since at each stage we only have to decide whether the next digit is a 1 or a 0.

Problem 75 Work out how to calculate the square root of any square given in *base* 2. △

2.9. The Prime Number Theorem

We have already observed that there are 4 primes less than 10, 25 primes less than 100, and 168 primes less than 1000. And there are 78 498 primes less than 10^6. So

- 40% of integers < 10 are prime;
- 25% of integers < 100 are prime;
- 16.8% of integers < 1000 are prime; and
- 7.8498% of integers < 10^6 are prime.

In other words, the fraction of integers which are prime numbers diminishes as we go up.

2.9. The Prime Number Theorem

The first question to ask is whether prime numbers themselves "run out" at some stage, or whether they go on for ever. The answer is very like that for the counting numbers, or positive integers $1, 2, 3, 4, 5, \ldots$:

> the counting process certainly gets started (with 1); and
> no matter how far we go, we can always "add 1" to get a larger counting number.

Hence we conclude that the counting numbers "go on for ever".

Problem 76

(a)(i) Start the process of generating prime numbers by choosing your favourite small prime number and call it p_1.

(ii) Then define $n_1 = p_1 + 1$.

(b)(i) Since $n_1 > 1$, n_1 must be divisible by some prime. Explain why p_1 is not a factor of n_1. (What is the *remainder* when we divide n_1 by p_1?)

(ii) Let p_2 be the *smallest* prime factor of n_1.

(iii) Define $n_2 = p_1 \times p_2 + 1$

(c)(i) Since $n_2 > 1$, n_2 must be divisible by some prime. Explain why p_1 and p_2 are not factors of n_2. (What is the remainder when we divide n_1 by p_1, or by p_2?)

(ii) Let p_3 be the *smallest* prime factor of n_2.

(iii) Define $n_3 = p_1 \times p_2 \times p_3 + 1$

(d) Suppose we have constructed k distinct prime numbers $p_1, p_2, p_3, \ldots, p_k$. Explain how we can always construct a prime number p_{k+1} different from $p_1, p_2, \ldots, p_k$.

(e) Apply the above process with $p_1 = 2$ to find p_2, p_3, p_4, p_5. △

Once we know that the prime numbers go on for ever, we would like to have a clearer idea as to the *frequency* with which prime numbers occur among the positive integers. We have already noted that

- there are 4 primes between 1 and 10,
- and again 4 primes between 10 and 20;
- but there is only 1 prime in the 90s;
- and then 4 primes between 100 and 110.
- And there are *no primes at all* between 200 and 210.

In other words, the distribution of prime numbers seems to be fairly chaotic. Our understanding of the full picture remains fragmentary, but we are about to see that the apparent chaos in the distribution of prime numbers conceals a remarkable pattern just below the surface.

The next item is only an experiment; but it is a very suggestive experiment. It is artificial, in that what you are invited to count has been carefully chosen to point you in the right direction. The resulting observation is generally referred to as the *Prime Number Theorem*. The result was conjectured by Legendre (1752–1833) and by Gauss (1777–1855) in the late 1790s – and was proved 100 years later (independently and almost simultaneously) in 1896 by the French mathematician Hadamard (1865–1963) and by the Belgian mathematician de la Vallée Poussin (1866–1962). You will need to access a list of prime numbers up to 5000 say.

Problem 77 Let $\pi(x)$ denote the number of prime numbers $\leq x$: so $\pi(1) = 0$, $\pi(2) = 1$, $\pi(3) = \pi(4) = 2$, $\pi(100) = 25$. You are invited to count the number of primes up to certain carefully chosen numbers, and then to study the results. The pattern you should notice works just as well for other numbers – but is considerably harder to spot.

The special values we choose for "x" are

the next integer above successive powers of the special number e,

where e is an important constant in mathematics – an irrational number whose decimal begins $2.7182818\cdots$, and which has its own button on most calculators (see Problem **248**).

(a) Complete the following table.

n	e^n	next integer N	$\pi(N)$
1	2.718···	3	2
2	7.389···		
3	20.08···		
4	54.59···		
5	148.41···		
6	403.42···		
7	1096.63···		
8	2980.95···		
9	8103.08···		1019

(b) Find an expression that seems to specify $\pi(N)$ as a function of n. Hence conjecture an expression for $\pi(x)$ in terms of x. △

> *Durch planmässiges Tattonieren.*
> [Through systematic fumbling.]
> Carl Friedrich Gauss (1777–1855),
> when asked how he came to make so many
> profound discoveries in mathematics.

2.10. Chapter 2: Comments and solutions

40.

(a)(i) 111 111 111

(ii) 9009 ($1001 = 7 \times 11 \times 13$ is a factorisation that is worth remembering for all sorts of reasons: for example, it incorporates $91 = 7 \times 13$; and it lies behind certain tests for divisibility by 7).

(b) (i) 1 001 001; (ii) 1 001 001; (iii) 3 003 003 (since $111 = 3 \times 37$)

41.

(i) $(10+1)^2 = 10^2 + 2 \times 10 + 1^2 = 121$;

(ii) $(10+1)^3 = 10^3 + 3 \times 10^2 + 3 \times 10 + 1^3 = 1331$;

(iii) $(100+1)^2 = 100^2 + 2 \times 100 + 1^2 = 10\,000 + 200 + 1 = 10\,201$

42.

(a)(i) Largest 8, smallest 7. (The smallest 3-digit number is 100 and the smallest 5-digit number is 10 000, so the smallest possible product is $10^2 \times 10^4 = 10^6$ – and so has 7 digits. The largest 3-digit number is just less than 1000 and the largest 5-digit number is just less than 100 000, so the largest possible product is just less than $10^3 \times 10^5 = 10^8$ – and so has 8 digits.)

(ii) Largest $m+n$, smallest $m+n-1$. (The smallest m-digit number is 10^{m-1} and the smallest n-digit number is 10^{n-1}, so the smallest possible product is 10^{m+n-2} – and so has $m+n-1$ digits. The largest m-digit number is just less than 10^m and the largest n-digit number is just less than 10^n, so the largest possible product is just less than $10^m \times 10^n = 10^{m+n}$ – and so has $m+n$ digits.)

(b)(i) $2^{10} = 1024$ is very slightly larger than 10^3. Hence $2^{20} = 1024^2$ is very slightly larger than 10^6, so has 7 digits.

(ii) 2^{20} is very slightly larger than 10^6. In fact

$$(10^3 + 24)^2 = 10^6 + 2 \times 10^3 + 24^2 = 10^6 + 2 \times 10^3 + 576 = 1\,002\,576.$$

$\left(\frac{1}{2}\right)^{20}$ is its reciprocal, so is slightly smaller than $10^{-6} = 0.000001$, so it starts with six 0s after the decimal point and rounds up to 0.000001 (to 6 d.p.).

(c) No. (It can be **equal** to the product of its digits if it has just 1 digit. If a number N has k digits, with leading digit $= m$, then $N \geq m \times 10^{k-1}$, but the product of its digits is at most $m \times 9^{k-1}$.)

(d)(i) 3, 6. ($2^{15} \times 5^3 = 2^{12} \times 10^3 = 4096 \times 10^3$)

(ii) 4, 4. (Most of us will need some rough work to supplement mental arithmetic here.

$$\begin{aligned} 20! &= 20 \times 19 \times 18 \times \cdots \times 2 \times 1 \\ &= 2^{18} \times 3^8 \times 5^4 \times 7^2 \times 11 \times 13 \times 17 \times 19 \\ &= 10^4 \times 2^{14} \times 3^8 \times 7^2 \times 11 \times 13 \times 17 \times 19. \end{aligned}$$

So 20! ends in 4 zeros, and its last non-zero digit is equal to the units digit of $2^{14} \times 3^8 \times 7^2 \times 11 \times 13 \times 17 \times 19$. If we work "mod 10" this is equal to the units digit of $4 \times 1 \times 9 \times 1 \times 3 \times 7 \times 9$.)

Note: The reader may notice that we have used "congruences", or "modular arithmetic" (mod 10) here and at several points in Chapter 1 (e.g. in the solutions to Problem **2**(d), Problem **13**, Problem **16**(b)).

In all these contexts one only needs to know that, if we fix the divisor n, then the *remainders* on division by n can be added and multiplied like ordinary numbers, since

$$(an + r) + (bn + s) = (a + b)n + (r + s),$$

and

$$(an + r)(bn + s) = (abn + as + br)n + rs.$$

Division is more delicate. We leave the reader to look up the details in any book on elementary number theory.

43. (a) 00 000 123 450 (b) 99 999 785 960

The initial number ($12 \cdots 9\,10\,11 \cdots 59\,60$) has $9 + 50 \times 2 + 2 = 111$ digits. Hence we are left with a number having exactly 11 digits.

For the smallest integer, we delete digits to leave the smallest initial digits (preferably 0s).

For the largest integer, we delete digits to leave as many 9s at the front as possible (and then sort out the tail).

44.
$$11\,111\,111 = 11\,110\,000 + 1111 = 1111 \times 10\,001.$$
In much the same way
$$1111 \cdots 1111000$$
(with 1108 1s and three 0s) is exactly divisible by 1111. So the remainder is **111**.

45. Compare $(10^5 + 1)(10^7 + 2)$ and $(10^5 + 2)(10^7 + 1)$.

The second is $10^7 - 10^5$ bigger than the first, so the second fraction is bigger than the first.

46. The fact that $3 \times 7 = 21$, and the position of the zeros, suggests that we express the integer as:

$$10^{35} + 3 \times 10^{24} + 7 \times 10^{11} + 3 \times 7 = (10^{11} + 3)(10^{24} + 7).$$

Note: If you feel you should have been "given a hint", then pause for a moment. There is nothing misleading here. We have no standard techniques for analysing such large numbers. The very size of the number forces you to think whether there is anything familiar about it that you might use. And the number is so simple that the only thing that can possibly stand out is the 3, 7, and 21. The rest is up to you.

47.

(a) 11 is prime. And 111 is a multiple of 3: $111 = 3 \times 37$. You should also be able to see that 1111 is a multiple of 11: $1111 = 11 \times 101$.

It is unclear whether 11 111 is prime or not. The *Square Root Test* says that to decide, we only need to check possible prime factors up to $\sqrt{11\,111} < 107$. We can eliminate 2, 3, 5, 7, 11 mentally, with very little effort. And with a calculator, it is easy to check 13, 17, 19, 23, 29, 31, 37, 41, ... and to discover that $11\,111 = 41 \times 271$.

Clearly $111\,111 = 11 \times 10\,101 = 111 \times 1001$.

So the sequence does not look too promising. All the even-numbered terms are divisible by 11; every third term is divisible by 111 (and of course, by 3); every fourth term is divisible by 1111 (and hence by 101); and so on. So the only possible candidates for primes are the second, third, fifth, seventh, eleventh, ... terms: that is the terms in **prime** positions.

Each of these terms is equal to the second bracket in the factorisation:

$$10^p - 1 = (10-1)(10^{p-1} + 10^{p-2} + \cdots + 10 + 1),$$

where p is a prime number.

We have seen that $111 = 3 \times 37$, and that $11\,111 = 41 \times 271$, which is not very encouraging. The 7th, 11th, 13th, and 17th terms are also not prime. But the 19th term and the 23rd terms are prime.

So primes seem scarce, but 11 is **not** the only prime in the sequence.

Note: Again, if you feel the problem was misleading, then pause for a moment. Part of "the essence of mathematics" is learning that some problems have a tidy solution, while others open up a rather different agenda. The only obvious way to begin to recognise this distinction is occasionally to be left to struggle to solve something that is presented as if it were a *closed* problem (with a tidy solution), only to discover that it is messier than one expected.

(b) We have already seen that $1001 = 7 \times 11 \times 13$.

Another reason for remembering this is that it is a simple instance of the standard factorisation:

$$10^3 + 1 = (10+1)(10^2 - 10 + 1)$$

Because the signs in the second bracket are alternately "+" and "−", this factorisation extends to all **odd** powers of 10: for example,

$$100\,001 = 10^5 + 1 = (10+1)(10^4 - 10^3 + 10^2 - 10 + 1)$$

So this time, 11 **is** the only prime in the list.

Note: The missing "odd" terms

$$101, 10\,001, 1\,000\,001, 100\,000\,001, \ldots\ldots$$

are slightly different – each being of the form $x^2 + 1$.
The fact that there is an algebraic factorisation of

$$x^3 + 1 = (x+1)(x^2 - x + 1)$$

implies that $1001 = 10^3 + 1$ has to factorise. But the lack of an algebraic factorisation of $x^2 + 1$ does not *prevent* any particular number of the form $x^2 + 1$ from factorising: for example, $3^2 + 1 = 2 \times 5$, and $5^2 + 1 = 2 \times 13$ both factorise; but $4^2 + 1$, $6^2 + 1$, and $10^2 + 1$ do not.

One may be forgiven for not knowing that $10^4 + 1 = 10\,001 = 73 \times 137$. But one should realize that

$$10^6 + 1 = 100^3 + 1 = (100 + 1)(100^2 - 100 + 1).$$

48. The prime factorisation $111 = 3 \times 37$ is worth remembering. If this is second nature, then one can do better in this problem than merely grind out the answer using long multiplication, by seeing how the output to the calculation 1001×333 simply positions "333 thousands" and "333 units" next to each other:

$3 \times 37 = 111$, so $9 \times 37 = 333$.

Hence $9009 \times 37 = 1001 \times 333 = 333\,333$.

Note: The prime factorisation of 1001 is not needed here. But it is important elsewhere.

49. The very first step shows that the leading digit of the dividend must be 1; and since "three-digit minus two-digit leaves one-digit (d say)" the divisor has a multiple in the 90s.

The very next stage again gives "three-digit minus two-digit leaves one-digit", and the remainder from the first division is now the hundreds digit, so $d = 1$. Hence the two-digit divisor has 99 as a multiple (at the first step of the long division) – so the divisor must be 11, 33, or 99.

The next division shows that the divisor has a two-digit multiple, which when subtracted from a two-digit number leaves a two-digit remainder, so the divisor cannot be 99.

The final stage shows that the divisor has a three-digit multiple, so it cannot be 11.

Hence the divisor must be 33.

50. Your solution will depend on the programming language used. We use this problem to attract the reader's attention to some not so frequently discussed issues:

- The "formal written algorithms" of arithmetic are not entirely obvious.
- Their practical use is not really "formal", it uses a number of unstated conventions. For example, it requires from the user an intuitive feel for the "data

structures" involved and starts by writing one base 10 integer under another keeping digits in the same decimal position aligned in a column (a computer scientist would call it "parsing the input").

- Base 10 integers contain different numbers of digits and shorter ones may need to be padded with zeroes (mentally, in calculations on paper, or explicitly, as may be necessary when writing code), that is, 1234 + 56 has to be treated as 1234 + 0056.
- Digits in the number are read and used from right to left, the opposite way to reading text. (This may be a piece of fossilised history: our decimals are Arabic, and Arabs write from right to left.)

51.

(a) This exploits the fact that

$$(10^k - 1) = (10 - 1)(10^{k-1} + 10^{k-2} + \cdots + 10 + 1),$$

and so is divisible by $(10-1)$ (a fact which is obvious when we write $10 - 1 = 9$, $10^2 - 1 = 99$, $10^3 - 1 = 999$, etc.). For example:

$$\begin{aligned}
\mathbf{12\,345} &= \mathbf{1 \times 10^4 + 2 \times 10^3 + 3 \times 10^2 + 4 \times 10 + 5} \\
&= [\mathbf{1 \times (10^4 - 1) + 2 \times (10^3 - 1) + 3 \times (10^2 - 1) + 4 \times (10 - 1)}] \\
&\quad + [\mathbf{1 + 2 + 3 + 4 + 5}] \\
&= [\text{a sum of terms, each of which is a multiple of 9}] \\
&\quad + [\text{the sum of the digits of "}\mathbf{12\,345}"]
\end{aligned}$$

If the LHS is divided by 9, the remainder from the first bracket on the RHS is zero, so the overall remainder is the same as the remainder from dividing the digit sum by 9.

Since 9 is a multiple of 3, the first bracket is exactly divisible by 3. Hence if the LHS is divided by 3, the remainder from the first bracket on the RHS is zero, and the overall remainder is the same as the remainder from dividing the digit sum by 3.

Note: If we were only interested in "divisibility by 9", then we could have managed without appealing to the algebraic factorisation

$$(10^k - 1) = (10 - 1)(10^{k-1} + 10^{k-2} + \cdots + 10 + 1),$$

since

$$10 - 1 = 9, \quad 10^2 - 1 = 99, \quad 10^3 - 1 = 999, \ldots$$

are all visibly "multiples of 9". However, the structure of the above solution extends naturally to prove that, when an integer is written in base b, the remainder on division by $b-1$ is the same as the remainder on dividing the base b "digit sum" by $b-1$.

(b) If an integer N is divisible by 6, then we can write $N = 6m$ for some integer m.

Hence $N = (2 \times 3)m = 2 \times (3m)$, so N is a multiple of 2; and $N = 3 \times (2m)$, so N is a multiple of 3.

If an integer N is divisible by 2, then we can write $N = 2k$ for some integer k. If N is also divisible by 3, then 3 divides exactly into $2k$. But $HCF(2,3) = 1$, so the 3 must go exactly into the second factor k, so $k = 3m$ for some integer m, and $N = 6m$ is divisible by 6.

Note: It is crucial that $HCF(2,3) = 1$. (E.g. 12 is divisible by 6 and by 4; but 12 is **not** divisible by 6×4.)

52.

(a) N is divisible by 3. Hence its digit-sum is divisible by 3.
But then "three times the sum of its digits" is a multiple of 9: hence the integer is divisible by 9, and so the sum of its digits is divisible by 9.
But then it is divisible by "three time a multiple of 9" – that is divisible by 27. So $N = 27$, or 54, or 81, or 108, or .... (However, you soon come to the first multiple of 27 that is **not** "divisible by 3 times the some of its digits".)

(b) 27. (Suppose the integer N has k digits. Then $N \geqslant 10^{k-1}$, and its digit-sum is at most $9k$. If N is equal to "three times the sum of its digits", then $10^{k-1} \leqslant N \leqslant 27k$ which means $k \leqslant 2$. And from part (a) we know that N is a multiple of 27.)

(c) 288. (If the digit sum is equal to 9 (or any odd multiple of 9), then at least one digit must be odd; so we only need to worry about integers with digit-sum equal to 18, or 36, or .... The only such multiple of 9 up to 100 is 99. All multiples of 9 between 100 and 200 have an odd hundreds digit. In the 200s, the first integer with digit-sum 18 is 279 – with two odd digits. The next is 288.)

53. (a) This exploits the fact that

$$(11^k - 1) = (11 - 1)(11^{k-1} + 11^{k-2} + \cdots + 11 + 1),$$

and so is divisible by $(11 - 1)$ – a fact which is obvious if we introduce a new digit X in base 11 to stand for "ten", and then notice that

$$11 - 1 = X_{\text{base 11}}, \quad 11^2 - 1 = XX_{\text{base 11}}, \quad 11^3 - 1 = XXX_{\text{base 11}}, \text{ etc.}$$

For example:

$$\begin{aligned}
\mathbf{12\,345}_{\text{base 11}} &= \mathbf{1 \times 11^4 + 2 \times 11^3 + 3 \times 11^2 + 4 \times 11 + 5} \\
&= [\mathbf{1 \times (11^4 - 1) + 2 \times (11^3 - 1) + 3 \times (11^2 - 1) + 4 \times (11 - 1)}] \\
&\quad + [\mathbf{1 + 2 + 3 + 4 + 5}] \\
&= [\text{a sum of terms, each of which is a multiple of ten}] \\
&\quad + [\text{the sum of the digits of "}\mathbf{12\,345}\text{"}]
\end{aligned}$$

If the LHS is divided by ten, the remainder from the first bracket on the RHS is zero, so the overall remainder is the same as the remainder from dividing the digit sum by ten.

54.

(a) 1, 3, 6, 10, 15, 21, 28, 36, 45, 55, 66, 78.

(b) Combine two copies of the required sum. If we do this algebraically, we get

$$\begin{array}{ccccccccc} 1 & + & 2 & + & 3 & + & \cdots & + & n \\ n & + & n-1 & + & n-2 & + & \cdots & + & 1 \end{array}$$

and observe that each of the n vertically aligned columns adds to $n+1$. Hence

$$T_n = 1 + 2 + 3 + \cdots + n = \frac{n(n+1)}{2}.$$

If we do the same geometrically, then we can combine two "staircases"

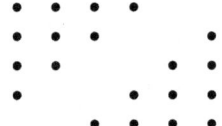

of dots (one of which is inverted) into an n by $n+1$ array of dots (either with n columns and $n+1$ dots in each column, or with $n+1$ columns and n dots in each column).

Note: The nth triangular number is defined by the "formula"

$$T_n = 1 + 2 + 3 + \cdots + n.$$

But this "formula" has serious limitations: in particular, there is no way to calculate T_{100} without first calculating T_1, then T_2, then T_3, ... all the way up to T_{99}. Hence it is just a "recurrence relation", which tells us how to find T_n once we know T_{n-1} (just "add n").

The formula

$$T_n = \frac{n(n+1)}{2}$$

derived in part (b) is much more useful, in that it allows us to work out the value of T_n as soon as we know the value of n. This is what we call a "**closed formula**". (The language may seem strange, but it refers to the fact that the calculation is direct, and that the formula involves a small, fixed number of operations – whereas using the recurrence requires more and more work as n gets larger.)

(c) **Note:** There are two reasons why these questions are worth asking. The first is that whenever we focus attention on certain special classes of objects, it is always good practice to consider whether the notions we have defined are completely separate, and to try to identify any overlaps. The second reason is less obvious, but can be surprisingly fruitful: sometimes two ideas may be interesting, yet have nothing to do with each other; but at other times, the two ideas may not only be interesting in their own right, but may "combine" in a way that gives rise to surprising subtleties. Here two of the combinations are routine and uninteresting; but two combinations generate more interesting mathematics than we have a right to expect.

(i) We know that one of the two factors n and $n + 1$ in the numerator is odd, and the other is even. If the triangular number

$$T_n = \frac{n(n+1)}{2}$$

is to be a **power of 2**, then any odd factor of T_n must be equal to 1, so $n < 3$: $n = 2$ does not give a power of 2. Hence $n = 1$, and $T_n = 1$ is the only triangular number which is also a power of 2.

(ii) If the triangular number T_n is to be **prime**, then either

* n is odd and one of n, $\frac{n+1}{2}$ is equal to 1 (so $n = 1$ and $T_1 = 1$ is not prime),

or

* n is even and one of $\frac{n}{2}$, $n+1$ is equal to 1, so $n = 2$, and $T_2 = 3$ is the only triangular number which is also prime.

(iii) The only immediately obvious "**square** triangular numbers T_n" are the first and the eighth – namely $T_1 = 1$ and $T_8 = 36$. But what seems obvious is rarely the whole truth. There are in fact *infinitely many* such "square triangular numbers" (e.g. $T_{49} = 1225$, $T_{288} = 41\,616$, $T_{1681} = 1\,413\,721$, ...). This is a consequence of the formula in part (b). For example:

When n is even, we notice that $a = \frac{n}{2}$ and $n + 1 = 2a + 1$ are integers with no common factors. We want their product to be a square. Because $HCF(a, 2a + 1) = 1$, this occurs precisely when both a ($= b^2$) and $2a + 1$ ($= c^2$) are both squares. So we see that solutions correspond to pairs of integers b, c which satisfy the *Pell* equation $c^2 = 2b^2 + 1$. Notice that $b = 2$, $c = 3$ is one solution, and that they satisfy the equation $c^2 - 2b^2 = 1$.

We have already met

$$a^2 + b^2 = (a + bi)(a - bi)$$

as the *norm* (or square of the length) of the complex number $a + bi$ (Problem 25). In a similar way, we can "factorise"

$$c^2 - 2b^2 = (c + b\sqrt{2})(c - b\sqrt{2}).$$

So once we have one solution of the equation $c^2 - 2b^2 = 1$, we can take powers to get more solutions:

$$[(c + b\sqrt{2})^2][(c - b\sqrt{2})^2] = 1^2 = 1, \quad \text{etc..}$$

Hence, for example,

$$(3 + 2\sqrt{2})^2 = 17 + 12\sqrt{2}$$

gives rise to the solution $b = 12$, $c = 17$ – corresponding to $a = 144$, $n = 288$. Similarly

$$(3 + 2\sqrt{2})^3 = \ldots + \ldots \sqrt{2}$$

gives rise to the solution $b = \ldots$, $c = \ldots$, corresponding to $(a = \ldots)$, $n = \ldots$

Note: If you are not yet familiar with complex numbers, or with the idea of a *norm*, don't worry. Make a note of it as something that seems to be powerful and is worth learning. It will reappear later.

(iv) The only obvious **cube** triangular number is the first one – namely $T_1 = 1$. Basic algebra leads quickly to an equation as in part (i):

$$\frac{n(n+1)}{2} = m^3,$$

which is equivalent to

$$(2n+1)^2 - 1 = (2m)^3.$$

So $(2m)^3$ and $(2m)^3 + 1$ are consecutive integers that are both proper **powers**. Catalan (1814–1894) conjectured in 1844 that $8 = 2^3$ and $9 = 3^2$ are the only consecutive powers (other than 0 and 1). This simple-sounding conjecture was finally proved only in 2004. It follows that $T_1 = 1$ is the only triangular number that is also a cube.

55.

(a)(i) 0, 1, 1, 2, 3, 5, 8, 13, 21, 34, 55, 89

(ii) 1, 0, 1, 1, 2, 3, 5, 8, 13, 21, 34; 1, 1, 0, 1, 1, 2, 3, 5, 8, 13

(iii) m^{th} term of kth sequence of differences $= F_{m-k}$

(b)(i) 1, 2, 4, 8, 16, 32, 64, 128, 256, 512, 1024, 2048

(ii) 1, 2, 4, 8, 16, 32, 64, 128, 256, 512, 1024; 1, 2, 4, 8, 16, 32, 64, 128, 256, 512

(iii) m^{th} term of kth sequence of differences $= 2^m$

56.

(a)(i) 1, 3, 7, 15, 31, 63, 127, 255, 511, 1023, ...

(ii) $x^{n+1} - 1 = (x-1)(x^n + x^{n-1} + \cdots + x + 1)$.
When $x = 2$, the first factor on the RHS $(x-1) = 1$, so

$$2^0 + 2^1 + 2^2 + \cdots + 2^n = 2^{n+1} - 1.$$

[Alternatively:

$$\begin{aligned}
2^0 + (2^0 + 2^1 + 2^2 + \cdots + 2^n) &= (2^0 + 2^0 \; [= 2^1]) + (2^1 + 2^2 + \cdots + 2^n) \\
&= (2^1 + 2^1 \; [= 2^2]) + (2^2 + 2^3 + \cdots + 2^n) \\
&= (2^2 + 2^2 \; [= 2^3]) + (2^3 + 2^4 + \cdots + 2^n) \\
&= \cdots \\
&= (2^n + 2^n) = 2^{n+1}.
\end{aligned}$$

(b)(i) 0, 1, 2, 4, 7, 12, 20, 33, 54, 88, ...

(ii) $F_0 + F_1 = F_2 = F_3 - F_1$
$F_0 + F_1 + F_2 = (F_3 - F_1) + F_2 = (F_3 + F_2) - F_1 = F_4 - F_1$
$F_0 + F_1 + F_2 + F_3 = (F_4 - F_1) + F_3 = (F_4 + F_3) - F_1 = F_5 - F_1.$

Claim:
$$F_0 + F_1 + F_2 + \cdots + F_{n-1} = F_{n+1} - F_1$$
holds for all $n \geqslant 1$.

Proof: When $n = 1$, the LHS $= F_0 = 0 = 1 - 1 = F_2 - F_1 =$ RHS.
We proved the next few case $n = 2$, $n = 3$, $n = 4$ above.
Suppose we have already proved the required relation holds all the way up to the $(k-1)^{\text{th}}$ equation:
$$F_0 + F_1 + F_2 + \cdots + F_{k-1} = F_{k+1} - F_1.$$
Then the kth equation follows like this:
$$\begin{aligned}(F_0 + F_1 + F_2 + \cdots + F_{k-1}) + F_k &= (F_{k+1} - F_1) + F_k \\ &= (F_{k+1} + F_k) - F_1 \\ &= F_{k+2} - F_1.\end{aligned}$$

So we have shown
* that the identity holds for the first few values, and
* that whenever we know it is true up to the $(k-1)^{\text{th}}$ identity, it also holds for the kth identity.

Hence the identity holds for all $n \geqslant 1$. QED

Alternatively:
$$\begin{aligned}F_1 + (F_0 + F_1 + \cdots + F_k) &= (F_1 + F_0\ [= F_2]) + (F_1 + F_2 + \cdots + F_k) \\ &= (F_2 + F_1\ [= F_3]) + (F_2 + F_3 + \cdots + F_k) \\ &= (F_3 + F_2\ [= F_4]) + (F_3 + F_4 + \cdots + F_k) \\ &= \cdots \\ &= F_{k+1} + F_k = F_{k+2}.\end{aligned}$$

Note: In **56**(a)(ii) we appealed directly to the factorisation of $x^{n+1} - 1$ as though this were a "known fact" which is easy to prove. And in the "alternative" proof, we repeatedly combined "$2^k + 2^k$" to make 2^{k+1}, inserting dots "$\ldots$" to indicate that this replacement operation is repeated $n + 1$ times. Both of these involved thinly veiled applications of the principle of *Mathematical Induction*, which is addressed in detail in Chapter 6. In **56**(b)(ii) we had no way of concealing the use of "proof by *Mathematical Induction*", which is likely to be lurking whenever we have

a proposition, or statement, $\mathbf{P}(n)$ involving the parameter n

and

we wish to prove the **infinite** collection of assertions:

"**P**(n) is true **for every** $n = 1, 2, 3, \ldots$ ".

The standard way of achieving this apparent miracle of proving infinitely many things at once is:

to check that **P**(1) holds (that is, to check that **P**(n) is true when $n = 1$);

then

to suppose that we have checked all of the instances **P**(1), **P**(2), ..., up to **P**(k) for some $k \geqslant 1$,

and

to show that the next instance **P**($k + 1$) must then also be true.

We then conclude that **P**(n) **is true for all** $n \geqslant 1$.

57.

(a)(i) 0, 2, 3, 10, 24, 65, 168, ...

(ii) **Guess:**
$$F_{n-1}F_{n+1} = F_n^2 + (-1)^n F_1, \text{ for all } n \geqslant 1.$$

Proof: By part (i), this identity holds for $n = 1, 2, 3, 4, 5, 6, 7$. Suppose we have checked it as far as the kth instance:
$$F_{k-1}F_{k+1} = F_k^2 + (-1)^k F_1.$$

Then the next instance follows, since
$$\begin{aligned}
F_{(k+1)-1}F_{(k+1)+1} &= F_k F_{k+2} \\
&= (F_{k+1} - F_{k-1})(F_k + F_{k+1}) \\
&= F_{k+1}^2 + (F_{k+1}F_k - F_{k-1}F_k) - F_{k-1}F_{k+1} \\
&= F_{k+1}^2 + (F_k^2 - F_{k-1}F_{k+1}) \\
&= F_{k+1}^2 + (-1)^{k+1} F_1.
\end{aligned}$$

So we have shown that the identity holds for the first few values of n, and whenever we know it is true up to the kth identity, it also holds for the $(k+1)^{\text{th}}$ identity. Hence the identity holds for all $n \geqslant 1$. QED

(b)(i) We suppose that
$$\frac{b}{a} < \frac{d}{c}$$
(if the inequality is reversed, the expression for the area is multiplied by "-1").

The lines $x = 0$, $y = 0$, $x = a + c$, $y = b + d$ form a rectangle of area $(a + c)(b + d)$, which surrounds the parallelogram.

To get from this to the area of the parallelogram, we must subtract

* the two external corner rectangles (top left, and bottom right) – each of area bc; and
* the four external right angled triangles–which fit together in pairs to make rectangles of areas ab and cd. Hence

$$\text{area}(OABC) = (a + c)(b + d) - 2bc - ab - cd = ad - bc.$$

(ii) 1

(iii) Half of the 2nd parallelogram is equal to half of the 1st – so both have the same area, namely 1.

Half of the 3rd parallelogram is equal to half of the 2nd – so they both have the same area, namely 1.

And so on. Hence the area of the nth parallelogram is equal to

$$|ad - bc| = |F_{n-1}F_{n+1} - F_n^2| = 1.$$

Part (a)(ii) is more precise in that it says that $F_{n-1}F_{n+1} - F_n^2 = (-1)^n$: this says that the relative positions of the generators (a, b), (c, d) for successive Fibonacci parallelograms alternate, with first $\frac{b}{a} > \frac{d}{c}$, and then $\frac{b}{a} < \frac{d}{c}$. (In fact the gradient of successive versions of the line OA, or the ratio of successive Fibonacci numbers, converges to the *Golden Ratio* τ, with successive Fibonacci points $A = (F_{n-1}, F_n)$ alternately above and below the line with equation $y = \tau x$.)

58.

(a) 1, 2, 5, 13, 34, ...

(b) **Guess**:
$$F_{n-1}^2 + F_n^2 = F_{2n-1}.$$

Note: When part (a) gives rise unexpectedly to "the odd-numbered terms of the Fibonacci sequence", it is almost impossible to believe that this is an accident. Yet the attempt to prove that this "Guess" is correct may well prove elusive – for it is hard to see how to relate the $(n-1)^{th}$ and n^{th} terms to the $(2n-1)^{th}$ term.

One approach is to

"try to prove something stronger than what seems to be required".

Claim: For each $n \geq 1$, **both** of the following are true:

$$F_{n-1}^2 + F_n^2 = F_{2n-1} \text{ and } F_{n+1}^2 - F_{n-1}^2 = F_{2n}.$$

Proof: We have already checked that the first relation holds for $n = 1, 2, 3, 4, 5$. And it is easy to check that

$$\begin{aligned} F_{1+1}^2 - F_{1-1}^2 &= 1 - 0 = 1 = F_2, \\ F_{2+1}^2 - F_{2-1}^2 &= 4 - 1 = 3 = F_4, \\ F_{3+1}^2 - F_{3-1}^2 &= 9 - 1 = 8 = F_6. \end{aligned}$$

So both identities hold for the first few values of n.

Now suppose we have checked that **both** relations hold all the way up to the kth pair of relations.

Then simply adding the two relations in the kth pair gives the first relation of the next pair:

$$\begin{aligned} F_k^2 + F_{k+1}^2 &= (F_{k-1}^2 + F_k^2) + (F_{k+1}^2 - F_{k-1}^2) \\ &= F_{2k-1} + F_{2k} \\ &= F_{2k+1} \end{aligned}$$

To see that the second relation of the next pair also follows, consider

$$\begin{aligned} F_{k+2}^2 - F_k^2 &= (F_k + F_{k+1})^2 - F_k^2 \\ &= F_{k+1}^2 + 2F_k F_{k+1} \\ &= (F_{k+1}^2 - F_{k-1}^2) + F_{k-1}^2 + 2F_k F_{k+1} \\ &= F_{2k} + (F_{k+1} - F_k)^2 + 2F_k F_{k+1} \\ &= F_{2k} + (F_{k+1}^2 + F_k^2) \\ &= F_{2k} + F_{2k+1} \\ &= F_{2k+2}. \end{aligned}$$

So we have shown

– that the identities hold for the first few values of n, and
– that whenever we know the kth pair of identities hold, the $(k+1)^{\text{th}}$ pair also hold.

Hence the two identities hold for all $n \geqslant 1$. QED

59.

(a)(i) $0, 5, 8, 26, 63, \ldots$

(ii) **Guess**: $F_{n-2} F_{n+2} = F_n^2 + (-1)^{n+1}$.
Proof: By part (i), this identity holds for $n = 2, 3, 4, 5, 6$.
Suppose we have checked it as far as the kth instance:

$$F_{k-2} F_{k+2} = F_k^2 + (-1)^{k+1}.$$

Then the next instance follows using **57**, since

$$\begin{aligned} F_{(k+1)-2} F_{(k+1)+2} &= F_{k-1} F_{k+3} \\ &= F_{k-1}(F_{k+1} + F_{k+2}) \\ &= F_{k-1} F_{k+1} + F_{k-1} F_{k+2} \\ &= F_k^2 + (-1)^k + (F_{k+1} - F_k)(F_k + F_{k+1}) \\ &= (-1)^k + F_{k+1}^2. \end{aligned}$$

(b)(i) $0, 13, 21, 68, \ldots$

(ii) **Guess**:
$$F_{n-3}F_{n+3} = F_n^2 + (-1)^{n+3-1}F_3^2.$$

This suggests that we should reinterpret our previous guesses, and that the "correction terms" on the RHS:

* in Problem **57**(a) should have been written as "$(-1)^{n+0-1}F_0^2$",
* in Problem **57**(a)(ii) should have been written as "$(-1)^{n+1-1}F_1^2$", and
* in Problem **59**(a)(ii) should have been written as "$(-1)^{n+2-1}F_2^2$".

We leave the proof (or otherwise) of this conjecture as an exercise for the reader.

60.

(i) 10%

(ii) 21% – notice that
$$(1.1a)(1.1b) = (1+0.1)^2 ab = (1+0.2+0.01)ab = 1.21ab.$$

(iii) 0% – notice that
$$\frac{1.1a}{1.1b} = \frac{a}{b}.$$

61. If x is doubled in the expression "x", then the value of the expression doubles.
If y is doubled in the expression $x \div y$, then the value of the expression is halved.
If z is doubled in the expression $x \div (y \div z)$, then the bracket is halved, and the expression is doubled.

Replacing "x, y, z" by "d, e, f" we see that, if the value of f is doubled, the value of the bracket $(d \div (e \div f))$ is also doubled.

If we now take $x = b$, $y = c$, $z = (d \div (e \div f))$, then, when f is doubled, z is doubled, and the value of $(b \div (c \div (d \div (e \div f))))$ is doubled.

Hence the value of the whole expression
$$a \div (b \div (c \div (d \div (e \div f))))$$
is halved.

62.

(a) The fact that one can add the entries in any order depends on the commutative and associative laws of addition. Expressing the subtotal in the second row as $2(1+2+3+4)$ uses the distributive law. And expressing the overall sum
$$(1+2+3+4) + 2(1+2+3+4) + 3(1+2+3+4) + 4(1+2+3+4)$$
as $(1+2+3+4)^2$ uses the distributive law again.

(b)(i) $1 = 1^3$, $8 = 2^3$, $27 = 3^3$, $64 = 4^3$.

(ii) $(4+8+12+16) + (12+8+4) = 4T_4 + 4T_3$. Similarly, the kth reverse L-shape has sum
$$k \cdot T_k + k \cdot T_{k-1} = \frac{1}{2}k^2(k+1) + \frac{1}{2}k^2(k-1) = k^3.$$

Hence
$$C_n = 1^3 + 2^3 + 3^3 + \cdots + n^3 = (1 + 2 + 3 + \cdots + n)^2 = \frac{1}{4} \cdot n^2(n+1)^2.$$

63.

(a) The terms are 1×2, 2×3, 3×4, etc,; so the rth term is $r(r+1)$, and the last term is $(n-1)((n-1)+1)$.

The rth term can be expressed as "$r^2 + r$", so the sum
$$1 \times 2 + 2 \times 3 + 3 \times 4 + \cdots + r(r+1) + \cdots + (n-1)n$$
can be expressed as
$$(1^2 + 2^2 + 3^2 + \cdots + (n-1)^2) + (1 + 2 + 3 + \cdots + (n-1)) = S_{n-1} + T_{n-1}.$$

(b)(i) * $n = 2$: $6 = 1 \times 2 \times 3$.
* $n = 3$: $6 + 18 = 24 = 2 \times 3 \times 4$.
* $n = 4$: $6 + 18 + 36 = 60 = 3 \times 4 \times 5$.

Guess: $3(S_{n-1} + T_{n-1}) = (n-1)n(n+1)$.

Proof: This is true for $n = 1, 2, 3, 4$.

Suppose we have checked the claim for all values up to
$$3(S_{k-1} + T_{k-1}) = (k-1)k(k+1).$$

Then
$$\begin{aligned} 3(S_k + T_k) &= 3([S_{k-1} + k^2] + [T_{k-1} + k]) \\ &= (k-1)k(k+1) + 3k(k+1) \\ &= k(k+1)(k+2). \end{aligned}$$

Hence our guess is true for all $n \geqslant 1$.

(ii)
$$S_n + T_n = \frac{n(n+1)(n+2)}{3},$$
so
$$S_n = \frac{n(n+1)(n+2)}{3} - T_n = \frac{n(n+1)(2n+1)}{6}.$$

64. If one tries to apply the usual algorithms for decimals, then one is likely to get in something of a mess. But if we re-interpret each decimal as a fraction, then things are much easier.

(a) $\frac{5}{9} + \frac{6}{9} = \frac{11}{9} = 1.22222\cdots$.
(b) $0.99999\cdots = \frac{9}{9} = 1$; $1 + \frac{1}{9} = 1.11111\cdots$.
(c) $\frac{10}{9} - \frac{2}{9} = \frac{8}{9} = 0.88888\cdots$
(d) $\frac{1}{3} \times \frac{2}{3} = \frac{2}{9} = 0.22222\cdots$.
(e) $\frac{11}{9} \times \frac{9}{11} = 1$.

65.

(a) Such a decimal is by definition equal to the fraction with numerator
$$b_n b_{n-1} \cdots b_1 b_0 b_{-1} b_{-2} \cdots b_{-k}$$
(an integer with $n + k + 1$ decimal digits) and with and denominator 10^k.

(b) If $\frac{p}{q}$ is equivalent to a fraction with numerator
$$m = b_n b_{n-1} \cdots b_1 b_0 \text{ base } 10$$
and denominator 10^k, then m has decimal representation
$$b_n b_{n-1} \cdots b_k . b_{k-1} \cdots b_1 b_0.$$

(c) Parts (a) and (b) show that fractions $\frac{p}{q}$ with $HCF(p,q) = 1$, whose decimals terminate are precisely those which are equivalent to fractions having denominator a power of 10: that is, those for which the denominator q is a factor of some integer of the form $10^k = 2^k \times 5^k$.

(d) If q does not divide some power of 10, then its decimal does not terminate. Hence, when carrying out the division of p by q we never get remainder 0. So the only possible remainders are $1, 2, \ldots, q-1$. The first remainder after the decimal point is equal to $p \pmod{q}$. In the ensuing q decimal places, there are just $q-1$ distinct possible remainders, so some remainder (say r) must occur for the second time by the qth step, and the outputs (and remainders) thereafter will then be the same as they were the first time that the remainder r occurred.

(e) Suppose d has a decimal with a repeating block of length b starting in the $(k+1)^{\text{th}}$ decimal place. (e.g. $d = 1234.567890909090909090\cdots$ has $b = 2$, $k = 4$). Then the infinite decimal tails cancel when we subtract $M = 10^b d - d$, and the difference M becomes an integer N if we multiply by 10^k: $N = M \times 10^k$. Hence $d(10^b - 1)10^k = N$, and d is equal to a fraction with denominator $(10^b - 1)10^k$.

66.

(a) (i) $\frac{1}{27}$; (ii) $\frac{10}{27}$; (iii) $\frac{19}{27}$

(b) (i) $\frac{a}{9}$; (ii) $\frac{ab}{99}$; (iii) $\frac{abc}{999}$

67.

(a) $0.166666\cdots$ (block length 1); $0.833333\cdots$ (block length 1)

(b) All have block length 6:

$$0.142857142857142857\cdots;$$
$$0.285714285714285714\cdots;$$
$$0.428571428571428571\cdots;$$
$$0.571428571428571428\cdots;$$
$$0.714285714285714285\cdots;$$
$$0.857142857142857142\cdots.$$

Note: The repeating blocks are all cyclically related: e.g. the block for $\frac{2}{7}$ is the same as for $\frac{1}{7}$, but starting at "2" instead of at "1".

(c) All have block length 2:

0.090909···; 0.181818···; 0.272727···; 0.363636···; 0.454545···;
0.545454···; 0.636363···; 0.727272···; 0.818181···; 0.909090····.

Note: The repeating blocks are not all cyclically the same, but fall into five pairs:

- $\frac{1}{11}$ and $\frac{10}{11}$ are cyclically related;
- as are those for $\frac{2}{11}$ and $\frac{9}{11}$;
- and those for $\frac{3}{11}$ and $\frac{8}{11}$;
- and those for $\frac{4}{11}$ and $\frac{7}{11}$;
- and those for $\frac{5}{11}$ and $\frac{6}{11}$.

(d) All have block length 6.

Note: They fall into two families of six, where each family is cyclically related:

$$\frac{1}{13} = 0.076923076923076923\cdots,$$
$$\frac{3}{13} = 0.230769230769230769\cdots,$$
$$\frac{4}{13} = 0.307692307692307692\cdots,$$
$$\frac{9}{13} = 0.692307692307692307\cdots,$$
$$\frac{10}{13} = 0.769230769230769230\cdots,$$
$$\frac{12}{13} = 0.923076923076923076\cdots;$$

and

$$\frac{2}{13} = 0.153846153846153846\cdots;$$
$$\frac{5}{13} = 0.384615384615384615\cdots,$$
$$\frac{6}{13} = 0.461538461538461538\cdots,$$
$$\frac{7}{13} = 0.538461538461538461\cdots,$$
$$\frac{8}{13} = 0.615384615384615384\cdots,$$
$$\frac{11}{13} = 0.846153846153846153\cdots.$$

68.

(a) Does not recur. (If it did, it would have a recurring block of length b say. But by the time the counting sequence $1, 2, 3, \ldots$ reaches 10^{2b} the decimal will contain a periodic block of $2b$ zeros, so the recurring block must consist of 0s, in which case the decimal terminates.)

(b) Does not recur. (Similar to part (a).)

(c) Does not recur. (If it did recur, then $\sqrt{2}$ would be a rational number: see Problems **267**, **268**, **270**.)

69. Claim Decimal fractions have two decimal representations. All other numbers have exactly one decimal representation.

Proof: Every "decimal fraction" (that is, any fraction which can be written with denominator a power of 10) has two representations – one that terminates and one that ends with an endless string of 9s: if the last non-zero digit of the terminating decimal is k, then the second representation of the same number is obtained by changing the "k" to "$k-1$" and following it with an endless string of 9s.

Consider an unknown number with two different decimal representations α and β. Since they are "different", α and β must differ in at least one position. Suppose the first, or left-most, position in which they differ is that in the kth decimal place (corresponding to 10^{-k}), and that the two digits in that position are a_k (for α) and b_k (for β).

We may suppose that $a_k < b_k$. Then $b_k = a_k + 1$ (otherwise $b_k - a_k > 1$, and $\beta - \alpha > 10^{-k}$, so $\alpha \neq \beta$).

Moreover, since β is not larger than α, the digits following b_k must all be equal to 0, and the digits following a_k must all be equal to 9. QED

70. In case (d), A only has to choose a recurring block (such as "$55555\cdots$", or "$090909\cdots$", or "$123123123\cdots$") for his/her positions – no matter where they are. B's control terminates at some stage, after which A's recurring block guarantees that the resulting number is rational.

The other parts all offer a guaranteed strategy for B. Let the positions chosen by B be numbered

$$n_1, n_2, n_3, n_4, \ldots, n_k, \ldots.$$

Now exploit the fact that the positive rationals are *countable* – that is, can be included in a **single list**. To see this we can use Cantor's (1845–1918) diagonal enumeration

$$\frac{0}{1}; \frac{1}{1}; \frac{1}{2}, \frac{2}{1}; \frac{1}{3}, \frac{3}{1}; \frac{1}{4}, \frac{2}{3}, \frac{3}{2}, \frac{4}{1}; \frac{1}{5}, \frac{5}{1}; \frac{1}{6}, \frac{2}{5}, \frac{3}{4}, \frac{4}{3}, \frac{5}{2}, \frac{6}{1}; \frac{1}{7}, \ldots,$$

which lists all rationals $\frac{p}{q}$ with $HCF(p,q) = 1$

- first those with $p + q = 1$,
- then those with $p + q = 2$,
- then those with $p + q = 3$,

and so on.

All B needs to do is to make sure that the resulting decimal is not the decimal of any number in this list, and s/he can do this by choosing a digit in the n_k^{th} position which is different from the digit which the kth rational in the above list has in that position. The resulting real number is then different from every number in the list – and hence must be irrational.

71.

(a) 101 010 (in each column (i) $0 + 0 = 0$, (ii) $1 + 0 = 1$, (iii) $1 + 1 =$ "0 and carry 1").

(b) (i) 1 010 100   (ii) 101 010   (iii) 101 010

(c) 2

2.10. Chapter 2: Comments and solutions   87

Note: Trying to do this should make it clear how easily we confound "the fourteenth positive integer" with its familiar base 10 representation. It takes time and effort to learn to see "$14_{\text{base }10}$" as "2×7", and "$21_{\text{base }10}$" as 3×7, and hence to spot the common multiple "$42_{\text{base }10}$". In *base 2* the same numbers evoke no such familiar echoes.

72. Let $N = (a_k a_{k-1} \cdots a_1 a_0)_{\text{base 2}}$.

(i) N is divisible by 2 precisely when the units digit a_0 is equal to 0.

(ii) N is divisible by 3 precisely when the alternating sum

$$\text{"}a_0 - a_1 + a_2 - a_3 + \cdots \pm a_k\text{"}$$

is divisible by 3.

Proof

$$\begin{aligned} N &= (a_k a_{k-1} \cdots a_1 a_0)_{\text{base 2}} \\ &= 2^k a_k + 2^{k-1} a_{k-1} + \cdots + 2 a_1 + a_0. \end{aligned}$$

For each odd suffix m, *increase* the coefficient 2^m by 1: then

$$2^m + 1 = (2+1)(2^{m-1} - 2^{m-2} + \cdots - 2 + 1)$$

has 3 as a factor.

For each even suffix $m = 2n$, *decrease* the coefficient by 1: then

$$2^{2n} - 1 = (2^2 - 1)(2^{2n-2} + 2^{2n-4} + \cdots + 2^2 + 1)$$

has 3 as a factor.

Hence

$$\begin{aligned} N &= 2^k a_k + 2^{k-1} a_{k-1} + \cdots + 2 a_1 + a_0 \\ &= (\text{multiple of 3}) + (a_0 - a_1 + a_2 - \cdots \pm a_k). \end{aligned}$$

(iii) N is divisible by 4 precisely when the last two digits a_1 and a_0 are both equal to 0.

(iv) N is divisible by 5 precisely when the alternating sum

$$\text{"}a_1 a_0\text{"} - \text{"}a_3 a_2\text{"} + \text{"}a_5 a_4\text{"} - \cdots$$

is divisible by 5.

Proof:

$$\begin{aligned} N &= (a_k a_{k-1} \cdots a_1 a_0)_{\text{base 2}} \\ &= 2^k a_k + 2^{k-1} a_{k-1} + \cdots + 2 a_1 + a_0 \\ &= (2 a_1 + a_0) + 2^2 (2 a_3 + a_2) + 2^4 (2 a_5 + a_4) + \cdots \\ &= (2^2 + 1)(2 a_3 + a_2) + (2^4 - 1)(2 a_5 + a_4) + \cdots \\ &\quad + [(2 a_1 + a_0) - (2 a_3 + a_2) + (2 a_5 + a_4) - \cdots] \\ &= (\text{a multiple of 5}) + [\text{"}a_1 a_0\text{"} - \text{"}a_3 a_2\text{"} + \text{"}a_5 a_4\text{"} - \cdots]. \end{aligned}$$

73.

(a) To weigh an object with weight 1, we must have $w_0 = 1$.
To weigh an object with weight 2, we must have $w_1 = 2$. We can then weigh any object of weight 3, but not one of weight 4.

(i) Now assume each positive weight w can be balanced in exactly one way. Then we cannot have $w_2 = 3$, so $w_2 = 4$.
Suppose that, continuing in this way, we have deduced that $w_i = 2^i$ for each $i = 0, 1, 2, \ldots, k$.
Then the binary numeral system reveals precisely that every weight w from 0 up to
$$2^{k+1} - 1 = 1 + 2 + 2^2 + \cdots + 2^k$$
can be uniquely represented, but 2^{k+1} cannot. Hence
$$w_{k+1} = 2^{k+1}.$$
The result follows by induction.

(ii) If the representation of each integer is not unique, then the sequence
$$w_0, w_1, w_2, \ldots$$
need not include the powers of 2. For example, it could begin
$$1, 2, 3, 5, \ldots$$

(b) If each integer w is to be weighed in this way, then w has to be represented in the form
$$w = a_1 w_1 + a_2 w_2 + a_3 w_3 + \cdots$$
where each coefficient $a_i = 0$ (if the weight w_i is not used to weigh w), or $= 1$ (if the weight w_i is used to balance w), or $= -1$ (if the weight w_i is used to supplement w).
If each representation is to be unique, then one can prove as in (a)(i) that the sequence of weights must be the successive powers of 3.

74. Write m in "base 2":
$$m = (a_{n-1} \cdots a_1 a_0)_{\text{base 2}},$$
where each $a_k = 0$ or 1. Then
$$\frac{m}{2^n} = \frac{a_0}{2^n} + \frac{a_1}{2^{n-1}} + \cdots + \frac{a_{n-1}}{2}.$$
That is,
$$\frac{m}{2^n} = (0.a_{n-1} \cdots a_1 a_0)_{\text{base 2}}.$$

75. We give an example, starting with $N = 110\,111\,001_{\text{base 2}}$.

Write N, and pair off the digits, starting at the units digit.

2.10. Chapter 2: Comments and solutions   89

$$1 \parallel 10 \parallel 11 \parallel 10 \parallel 01$$

The left-most digit stands for 2^8, so the square root is at least 2^4 (and less than 2^5). Hence the required square root has five digits (one for each "pair" of digits of N), and starts with a 1.

Root $1 \parallel ? \parallel ? \parallel ? \parallel ?$

[We can also see that the final units digit will have to be a "1". But this is not the time to add such information.]

Let $x = 10\,000$, and subtract $x^2 = 100\,000\,000$ from N:

$$\begin{array}{c} 1 \quad 00 \quad 00 \quad 00 \quad 00 \\ \parallel 10 \parallel 11 \parallel 10 \parallel 01 \end{array}$$

This residue has to be equal to "$2xy + y^2$". However, as with long division, our immediate interest is in determining the **next digit** of our "partial square root". If the next digit is a 1 (contributing 2^3), then $2xy \geqslant 2^8$, which would spill over and change the digit we have already determined. Hence the next digit is a 0.

Root $1 \parallel 0 \parallel ? \parallel ? \parallel ?$

So we can again let $x = 10\,000$ giving the same remainder, which has to be equal to "$2xy + y^2$", but this time $y < 2^3$ has at most three digits.

The remainder

$$\parallel 10 \parallel 11 \parallel 10 \parallel 01$$

is greater than 2^7, so $y \geqslant 2^2$ and the next digit must be a "1".

Root $1 \parallel 0 \parallel 1 \parallel ? \parallel ?$

Now let $x = 10\,100$, and subtract $x^2 = 110\,010\,000$ from N, leaving

$$\parallel 10 \parallel 10 \parallel 01$$

This residue has to equal $2xy + y^2$, with $x = 10\,100$.
If the next digit in the square root is 1, then $2xy \geqslant 2^6 > 101\,001 = 2xy + y^2$.
Hence the next digit is 0, and the last digit is then 1.
Hence the required square root is equal to:

Root $1 \parallel 0 \parallel 1 \parallel 0 \parallel 1$

76.

(b)(i) The fact that
$$n_1 = p_1 + 1$$
says that

(c)(i) The fact that
$$n_2 = p_1 \times p_2 + 1$$
says that
"n_2 is equal to a multiple of p_1 with remainder = 1",

and that

"n_2 is equal to a multiple of p_2 with remainder = 1".

Hence neither p_1 nor p_2 are factors of n_2.

(d) The fact that
$$n_k = p_1 \times p_2 \times \cdots \times p_k + 1$$
says that

"n_k is equal to a multiple of p_i with remainder = 1"

for each suffix i, $1 \leqslant i \leqslant k$. Hence none of the primes $p_1, p_2, p_3, \ldots, p_k$ is a factor of n_k.

So the smallest prime factor of n_k always gives us a new prime p_{k+1}.

(e) If we start with $p_1 = 2$, then $n_1 = p_1 + 1 = 3$, so $p_2 = 3$.
Then $n_2 = p_1 \times p_2 + 1 = 7$, so $p_3 = 7$.
Then $n_3 = p_1 \times p_2 \times p_3 + 1 = 43$, so $p_4 = 43$.
Then $n_4 = p_1 \times p_2 \times p_3 \times p_4 + 1 = 1807 = 13 \times 139$, so $p_5 = 13$.

77.

(a) We write $\lceil x \rceil$ for the "first integer $\geqslant x$". Then

$$\begin{aligned}
\pi(\lceil e^1 \rceil) &= \pi(3) = 2; \\
\pi(\lceil e^2 \rceil) &= \pi(8) = 4; \\
\pi(\lceil e^3 \rceil) &= \pi(21) = 8; \\
\pi(\lceil e^4 \rceil) &= \pi(55) = 16; \\
\pi(\lceil e^5 \rceil) &= \pi(149) = 35; \\
\pi(\lceil e^6 \rceil) &= \pi(404) = 79; \\
\pi(\lceil e^7 \rceil) &= \pi(1097) = 184; \\
\pi(\lceil e^8 \rceil) &= \pi(2981) = 429; \\
\pi(\lceil e^9 \rceil) &= \pi(8104) = 1019.
\end{aligned}$$

(b) The initial "doubling" is an accident of small numbers, which soon turns into "slightly more than doubling".

The observation that should (eventually) jump out at you concerns the ratio $e^N : \pi(N)$, which seems to be surprisingly close to $N - 1$. This suggests the possible

Conjecture: $\pi(x) \sim \frac{x}{\ln(x) - 1}$ (where $\ln(x) = \log_e(x)$).

III. Word Problems

> *All the evidence suggests that*
> *the shapes of reality*
> *are mathematical.*
> George Steiner (1929–)

The previous chapter focused on aspects of the arithmetic of *pure numbers* – mostly without any surrounding context. However, our mathematical experience does not begin with pure numbers. At school level, mathematical concepts, and the reasoning we bring to understanding and using them, have their roots in *language*. And in real life, every application of mathematics starts out with a situation which is described **in words**, and which has to be reformulated mathematically before we can begin to *calculate*, and to draw meaningful mathematical conclusions. *Word problems* play an important, if limited, role in helping students to appreciate, and to handle the subtleties involved in

> the art of *using the mathematics we know*
> to solve problems *given in words.*

This art of using mathematics involves two distinct – but interacting – processes, which we refer to here as "simplifying" and "recognising structure".

- To identify the mathematical heart of a problem arising in the real world, one may first have to *simplify* – that is, to side-line details that seem unimportant or irrelevant, and then simplify as much as possible *without changing the underlying problem* (e.g. by replacing some awkward feature by a different quantity which is easier to measure, or by an approximation which is easier to work with).

This "simplifying" stage is well-illustrated by the tongue-in-cheek title of the classic textbook *Consider a spherical cow* ... by John Harte (1985):

> Milk production at a dairy farm was low, so [...] a multidisciplinary team of professors was assembled. [...After] two weeks of intensive on-site investigation [...] the farmer received the write up, and opened it to read [...] "Consider a spherical cow ...".

The point to emphasise here is that the judgements needed when "simplifying" are subtle, depend on an understanding of the particular situation being modelled, and may lead to a model which at first sight seems to be counterintuitive, but which may not be as silly as it seems – and which therefore needs to be explained sensitively to non-mathematicians.

In contrast *word problems* by-pass the "simplifying" stage, and focus instead on "recognising structure": they present the solver with a problem which is already essentially mathematical, but where the inner structure is contextualised, and is described in words. All the solver has to do is to interpret the verbal description in a way that extracts the structure just beneath the surface, and to translate it into a familiar mathematical form. That is, *word problems* are designed to develop facility with the process of "recognising structure", while avoiding the complication of expecting students to make modelling judgements of the kind required by the subtler "simplifying" process.

Because *word problems* focus on the second process of "recognising structure", they tend to incorporate the relevant mathematical structure *isomorphically*. The underlying structure still needs to be identified and interpreted, but the interpretations are likely to be standard, with no need for imaginative assumptions and simplifications before the structure can be discerned. For example, if a problem in primary school refers to an unknown number of "sweets" to be "shared" between six children, then the collection of "sweets" is isomorphic to a pure number (the number of sweets); and the act of "sharing" is a thinly veiled reference to numerical division.

The story in a word problem may be a purely mathematical problem in disguise. But the art of identifying the *correspondence* between

> the *data* given in the story line, and

> the *mathematical entities* to which they correspond

and between

> the *actions* in the story line, and

> the corresponding *mathematical operations* on those mathematical entities

is non-trivial, and has to be learned the hard way. The first problem below illustrates the remarkable variety of instances of even the simplest subtraction, or difference.

As in Chapters 1 and 2 the "essence of mathematics" is to be found in the problems themselves. Some discussion of this "essence" is presented in the text between the problems; but most of the relevant observations are either to be found in the solutions (or in the **Notes** which follow many of the solutions), or are left for readers to extract for themselves.

3.1. Twenty problems which embody "$3 - 1 = 2$"

The answer to every one of the questions in Problem **78** is the same – at least, as a 'pure number'. The goal is therefore not to "solve" each problem, but to distinguish between, and to reflect upon, the different ways in which the very simple mathematical structure "$3 - 1 = 2$" turns out to be the relevant "model" in each case.

Problem 78

(a) I was given three apples, and then ate two of them. How many were left?

(b) A barge-pole three metres long stands upright on the bottom of the canal, with one metre protruding above the surface. How deep is the water in the canal?

(c) Tanya said: "I have three more brothers than sisters". How many more boys are there in Tanya's family than girls?

(d) How many cuts do you have to make to saw a log into three pieces?

(e) A train was due to arrive one hour ago. We are told that it is three hours late. When can we expect it to arrive?

(f) A brick and a spade weigh the same as three bricks. What is the weight of the spade?

(g) The distance between each successive pair of milestones is 1 mile. I walk from the first milestone to the third one. How far do I walk?

(h) The *arithmetic mean* (or average) of two numbers is 3. If half their difference is 1, what is the smaller number?

(i) The distance from our house to the train station is 3 km. The distance from our house to Mihnukhin's house along the same road is 1 km. What is the distance from the station to Mihnukhin's house?

(j) In one hundred years' time we will celebrate the tercentenary of our university. How many centuries ago was it founded?

(k) In still water I can swim 3 km in three hours. In the same time a log drifts 1 km downstream in the river. How many kilometres would I be able to swim in the same time travelling upstream in the same river?

(l) December 2nd fell on a Sunday. How many working days preceded the first Tuesday of that month?[4]

---

[4] This question is historically correct. In 1946, in the Soviet Union, when these problems were formulated, Saturday was a working day.

(m) I walk with a speed of 3 km per hour. My friend is some distance ahead of me, and is walking in the same direction pushing his broken down motorbike at 1 km per hour. At what rate is the distance between us diminishing?

(n) A trench 3 km long was dug in a week by three crews of diggers, all working at the same rate as each other. How many such crews would be needed to dig a trench 1 km shorter in the same time?

(o) Moscow and Gorky are cities in adjacent time zones. What is the time in Moscow when it is 3 pm in Gorky?[5]

(p) An old 'rule-of-thumb' for anti-aircraft gunners stated that: To hit a plane from a stationary anti-aircraft gun, one should aim at a point exactly three plane's lengths ahead of the moving plane. Now suppose that the gun was actually moving in the same direction as the plane with one third of the plane's speed. At what point should the gunner aim his fire?

(q) My brother is three times as old as I am. How many times my present age was his age when I was born?

(r) I add 1 to a number and the result is a multiple of 3. What would the remainder be if I were to divide the original number by 3?

(s) It takes 1 minute for a train 1 km long to completely pass a telegraph pole by the track side. At the same speed the train passes right through a tunnel in 3 minutes. What is the length of the tunnel?

(t) Three trams operate on a two-track route, with trams travelling in one direction on one track and returning on the other track. Each tram remains a fixed distance of 3 km behind the tram in front. At a particular moment one tram is exactly 1 km away from the tram on the opposite track. How far is the third tram from its nearest neighbour? △

3.2. Some classical examples

Problem 79 Katya and her friends stand in a circle in such a way that the two neighbours of each child are of the same gender. If there are five boys in the circle, how many girls are there? △

Problem 80 How much pure water must be added to a vat containing 10 litres of 60% solution of acid to dilute it into a 20% solution of acid? △

---

[5] Gorky (now the city of Nizhny Novgorod) lies to the east of Moscow.

Problem 81 A mother is $2\frac{1}{2}$ times as old as her daughter. Six years ago the mother was 4 times as old as her daughter. How old are mother and daughter? △

Problem 82

(a) Tom takes 2 hours to complete a job. Dick takes 3 hours to complete the same job. Harry takes 4 hours to complete the same job. How long would they take to complete the job, all working together (at their own rates)?

(b) Tom and Dick take 2 hours to complete a job working together. Dick and Harry take 3 hours to complete the same job. Harry and Tom take 4 hours to complete the same job. How long would they take to complete the same job, all working together? △

Problem 83 A team of mowers had to mow two fields, one twice as large as the other. The team spent half-a-day mowing the larger field. After that the team split: one half continued working on the big field and finished it by evening; the other half worked on the smaller field, and did not finish it that day – but the remaining part was mowed by one mower in one day. How many mowers were there? △

3.3. Speed and acceleration

Problem 84 Jack and Jill went up the hill, and averaged 2 mph on the way up. They then turned round and went straight back down by the same route, this time averaging 4 mph. What was their average speed for the round trip (up and down)? △

Problem 85

(a)(i) A cycling road race requires one to complete 3 laps of a long road circuit. On the first lap I average 40 km/h; on the second lap I average 30 km/h; and on the third lap I only average 20 km/h. What is my average speed for the whole race?

(ii) I cycle for 3 hours round the track of a velodrome, averaging 40 km/h for the first hour, 30 km/h for the second hour, and 20 km/h for the final hour. What is my average speed over the whole 3 hours?

(b) Two cyclists compete in an endurance event.

(i) The first cyclist pedals at 60 km/h for half the time and then at 40 km/h for the other half. The second cyclist pedals at 60 km/h for half of the total distance and then at 40 km/h for the remaining half. Who wins?

(ii) In a two hour event, the first cyclist pedals at u km/h for the first hour and then at v km/h for the second hour. The second cyclist pedals at u km/h for half of the total distance and then at v km/h for the remaining half. Who wins?

(c)(i) Apply your argument in (b)(ii) to prove an inequality between
* the **arithmetic** mean
$$\frac{u+v}{2}$$
of two positive quantities u, v, and
* the **harmonic** mean
$$\frac{2}{\frac{1}{u}+\frac{1}{v}}.$$

(ii) Give a purely algebraic proof of your inequality in (i). △

Problem 86 A train started from a station and, moving with a constant acceleration, covered a distance of 4 km, finally reaching a speed of 72 km/hour. Find the acceleration of the train, and the time taken for the 4 km. △

Problem 87 (Average speed of an accelerating car) A typical car (and maybe also a typical train!) does not move with constant acceleration. Starting from a standstill, a car moves through the gears and "accelerates more quickly" in lower gears, when travelling at lower speeds, than it does in higher gears, when travelling at higher speeds. Use this empirical fact to prove that the *average speed* of a car accelerating from rest is *more than half* of its final measured speed after the acceleration. △

3.4. Hidden connections

Problem 88 Two old women set out at sunrise and each walked with a constant speed. One went from A to B, and the other went from B to A. They met at noon, and continuing without a stop, they arrived respectively at B at 4 pm and at A at 9 pm. At what time was sunrise on that day? △

Problem 89 A paddle-steamer takes five days to travel from St Louis to New Orleans, and takes seven days for the return journey. Assuming that the rate of flow of the current is constant, calculate how long it takes for a raft to drift from St Louis to New Orleans. △

Problem 90 [From Paolo dell'Abbaco's *Trattato d'aritmetica*] "From here to Florence is 60 miles, and there is one who walks it in 8 days [in one direction], another in five days [in the opposite direction]. It is asked: Departing at the same time, in how many days will they meet?" △

Problem 91 Notice that in Problem **88** sunrise occurs $t = \sqrt{4 \times 9}$ hours before noon, and that $\sqrt{4 \times 9}$ is the **geometric mean** of 4 and 9. Once this is pointed out, can you reformulate your solution to Problem **88** to solve a more general problem? △

3.5. Chapter 3: Comments and solutions

78.

(a) This is the simplest form of all: 3 are given; 2 are removed; so what remains is "$3 - 2$".

(b) Length is a *continuous* quantity (rather than *discrete* quantity – like apples, or sweets). So we have to perceive a line segment (partially hidden beneath the water) rather than a quantity. We know the total length of the pole, and the length of the protruding portion. We can then infer the hidden length by subtraction.

Note: This kind of "geometrical subtraction" is needed in many contexts (such as: proving the general formula

$$\frac{1}{2}(\text{base} \times \text{height})$$

for the area of a triangle, or showing that the area of the parallelogram spanned by the origin and vectors (a, b), (c, d) is $|ad - bc|$, or in Euclid's *Elements*, Book I, Proposition 2). The idea can be strangely elusive.

(c) The situation here is significantly different. We start with Tanya's brothers and sisters, and finish with the related, but different, notion of "boys and girls in Tanya's family". The "3" does not represent anything specific: it is a numerical *excess* (of Tanya's brothers over her sisters). In contrast, the "1" seems to represent Tanya herself, who needs to be taken into account when we switch from the initial scenario (Tanya's brothers and sisters) to the final question about "boys and girls in Tanya's family".

(d) No doubt this can be solved by drawing a picture in which the underlying structure is only appreciated superficially. But beneath the surface, it seems to be a much more abstract representation of "$3 - 1 = 2$". The "3" certainly stands for the "three pieces". But the operation "-1" is not obviously subtracting anything.

The relevant observation is simply that, starting from one end, pieces and cuts *alternate*. So if we ignore the starting end, there must be the same number of pieces and cuts – except that if we start with a log (rather than a long tape from which we are cutting off pieces), the "last cut" is "the other end of the log", which has already been cut – so does not need to be cut again, and this obliges us to subtract 1 from the number of pieces to get the number of additional cuts.

Note: This idea arises in many settings, and is sometimes referred to as "Posts and gaps". Sometimes one has to "subtract 1" as here; at other times one has to "add 1" (e.g. when counting the number of "posts", if we are given the number of "gaps", or "fence panels").

(e) Once again we are dealing with a continuous quantity – *time*. On this occasion the problem invites us to construct a (horizontal?) diagram very like the pole and the water in (b). But this time, the origin is likely to be perceived as "now", with a time-line stretching back 1 hour (to the left?) to mark the time when the train was due, and then moving on 3 hours (to the right), passing through the origin to a point 2 hours from now.

(f) It is unclear how young children might tackle this with "bare hands". However, at some stage one would like them to see the words as evoking the powerful (and rather different) underlying image of "scales", or an imagined "equation". Once one 'sees' the two pans of a balance, with a "brick and a spade" on one side being balanced by "three bricks" on the other, one can imagine removing "1 brick" from each pan to be left with the spade on its own balanced by "$3 - 1 = 2$" bricks.

(g) This is in some ways a simpler version of "Posts and gaps". However, there is an additional step – since we are no longer merely counting the gaps, but translating this counting number into a *distance*. In this instance, if one does not pay too much attention to the extra step, both give an answer "2".

Note: The impact of the extra step (switching from discrete counting number to continuous distance) can be seen more clearly in the number of errors made when students are faced with such variations as:

> "There are ten lamp posts in my street, and they are 70 metres apart. How far is it from the first to the last?"

(h) One suspects that this superficially simple problem would prove inaccessible unless pupils have learned to represent word problems diagrammatically, or have already mastered simple algebra. The "3" and the "1" do not represent real-world entities; so one has to be prepared to mark a "3" on a number line, and to interpret "average" as indicating that the two unknown quantities lie equally spaced either side of it. "Half their difference" is then staring one in the face, and the smaller number (to the left) is clearly "$3 - 1$".

Note: This may look rather like the overdue train in part (e). We suggest that it is significantly different.

(i) The story line clearly adds layers of difficulty which we tend to overlook. Learning to "recognise structure" and to translate words into a form that allows one to calculate is clearly a non-trivial (and neglected) art. Distances in kilometres may convey something more active than the given "length of a barge pole" in part (b), or the reported times in part (e), even if the diagram – once constructed – is very similar (provided of course that "along the same road" is interpreted as meaning "in the same direction").

Note: Consider the following item from an authoritative international study TIMSS 2011[6] for pupils aged around 14:

"Points A, B, and C lie in a line and B is between A and C. If $AB = 10$ cm and $BC = 5.2$ cm, what is the distance between the midpoints of AB and BC?

A 2.4 cm B 2.6 cm C 5.0 cm D 7.6 cm"

The question is a multiple choice question, and the options represent different ways of failing to translate the words into a suitable diagram, or to interpret them correctly. The sampling (in around 50 countries) was done very carefully. So the different success rates in different countries (of which 5 are given below) suggest that some systems give far too little attention to helping pupils to learn the relevant underlying art:

Russia 60%, Hungary 41%, Australia 40%, England 38%, USA 29%

(j) The story line here has a different flavour. The time-line is the reverse of the overdue train in part (e), yet the measuring in centuries may make the question less immediately accessible. It may be harder to "feel" a natural interpretation, and so success may be more dependent on a willingness to represent the given information abstractly.

(k) Up to now, all problems were either static, or involved motion in a directly accessible form. Here we meet for the first time the need to interpret the words in terms of "relative motion". I may get as far as picturing myself swimming upstream in the river (against the current); but neither the "3" nor the "1" have any direct relevance to me at that time: they have to be **imagined** (as "me swimming in still water", and "the effect of the river in slowing me down"), and then interpreted in a way that allows a simple calculation.

(l) The words need to be interpreted from a very different kind of story line: if the 2nd is on a Sunday, then the "first Tuesday" must be the 4th. There are therefore "3" days preceding the first Tuesday – of which just "1" (Sunday) is not a working day. All that is needed is "counting"; but the wording requires a different kind of interpretation.

---

[6] Trends in International Mathematics and Science Study, https://timssandpirls.bc.edu/timss2011/index.html

100   Word Problems

(m) This is another example of "relative velocities" – but the need for subtraction no longer arises because of travel in opposite directions. In some ways it is simpler than (k); yet the final question relates to something less tangible – namely the "rate at which the distance between us is diminishing". Before one understands relative velocities, one has to choose to focus on "what happens during each hour", where I cover 3 km and my friend covers only 1 km, with the difference "3 − 1" measuring nothing tangible, but being *the amount by which our separation decreases* during that hour.

(n) Here it is even more important to translate the given information about "rates" into concrete form. "In the same time" should trigger the questions: "How many crews would be needed for (3 − 1) km?", which may then trigger the question: "How long a ditch could 1 crew dig *in the same time*?". Whatever approach is taken, it is worth asking "If the answer is "3 − 1", what exactly is the "3"? And what is the "1"?"

(o) This does presume a degree of fluency in "modelling" the given information (e.g. knowing that "adjacent time zones" almost always differ by 1 hour, and that the Earth's rotation is from West to East, so that the Sun "rises" first in the East). On the surface, if the "3" is interpreted as the "3" in 3 pm, then the calculation "3 − 1" is an *adjustment*, rather than a strict subtraction (the 3 pm and the "1 hour time difference" are not really comparable quantities with which one can do arithmetic). At a deeper level one can turn both the "3" and the "1" into comparable quantities, and so justify the arithmetic.

(p) Here we face full-on what has been lurking just below the surface of certain earlier problems (such as (n)) – namely that we are dealing with (approximate) *proportion*. We ignore marginal differences in the distance to a distant object at slightly different angles, and compare on the one hand

> distances along the plane's path (measured in "plane's lengths"),

and on the other hand

> the time taken by the anti-aircraft fire to reach the plane.

This comparison has to be made because of the added complication of the change in the relative velocity of the gun and the plane.

The given rule of thumb specifies the direction in which a stationary gunner should aim; and the reported (unrealistically fast, yet presumed to be steady) motion of the gun introduces a 2-dimensional (vector) version of "swimming upstream" – which suggests the expected answer "two thirds of 3 plane lengths", so that "1" of the "3 plane's lengths" is compensated by the gun's motion.

(q) A solution is again dependent on representing the given information in some form. Whether or not one uses symbols, the wording invites the solver to use "my present age (in years)" as a preferred unit, and to represent "my brother's present age" as "3" of these basic units. The "3 − 1" then represents how much older he is than I am – and hence how old he was when I was born, or "how many times my present age he was when I was born".

Note: The choice of unit may conceal the fact that the question and solution are rooted in ratio and proportion.

(r) The subtraction "3 − 1" here only makes sense in arithmetic (mod 3), where "remainder 0 (on division by 3)" and "remainder 3" are in some sense equivalent. Although the "1" in "3 − 1" may be taken to be the "1" that is added to the original number in the question, the "3" is an invented remainder – which is interchangeable with "0" when working (mod 3).

(s) In the usual answer "3 − 1", one could argue that the "1" **does** appear in the question, but that the "3" **does not**. Again we are dealing with *proportion*, where the times taken (at constant speed) are proportional to lengths, or to distances travelled. First the given length (1 km) of the train and the given time (1 minute) in relation to the "pole by the track side" gives a simple *constant of proportion* (= 1), which allows us to translate the time taken into the distance travelled (and hence to calculate speed). If we re-interpret the "endpoint of the tunnel" as being just like another "pole by the track side", then it takes 1 minute for the train to emerge from the tunnel, and hence "3 − 1 minutes" for the front buffers of the train to cover the full length of the tunnel, which is therefore "(3 − 1) km" long (given that the constant of proportionality = 1).

(t) It is not clear how to interpret the "3" and the "1" in "3 − 1" without getting one's hands dirty with the configuration described. In particular, somewhere along the line one has to interpret the "3 km" separation between trams as revealing that the total length of the track is 9 km, and hence that each of the two parallel stretches of track is 4.5 km.

The "tram on the opposite track" is travelling in the opposite direction, is 1 km away, and is "3 km ahead" (or "3 km behind"); so one of these trams is 1 km from the end of the track, and the other is on the other track and 2 km from one end (travelling in the opposite direction). There are exactly two possible configurations – each arising from the other if we reverse the direction of travel. By choosing the direction of travel (or by allowing "negative speed") we may assume that tram A is 2 km from the same end of the track and that tram B in front of it is 1 km beyond the end of the track on the opposite side. Tram C is 3 km ahead of B, and hence 4 km down that 4.5 km stretch of track (so has not yet "turned the corner"). Hence it is 1 km closer to its nearest neighbour (A) than it is to B.

79. If we ignore the first sentence, then there could be zero girls (and five boys). But the first sentence guarantees that there is at least one girl ("Katya and **her** friends"). So boys and girls must alternate, giving rise to 5 girls.

80. The problem requires a degree of "modelling" in that "60% solution of acid" suggests that the initial ratio

"acid : water" = 60 : 40.

Hence the initial 10 litres is made up of 4 litres of water and 6 litres of acid. Adding water does not change the amount of acid; so we want 6 litres to be 20% of the final mix – which must therefore be 30 litres. Hence we should add 20 litres.

81. The difference in ages is $\frac{3}{2} \times d$, where d is the daughter's age in years. Six years ago the difference was three times the daughter's age, which was then $d - 6$

years. Hence
$$3(d-6) = \frac{3}{2} \times d,$$
so $d = 12$.

82.

Note: Underpinning all such problems is the "**unitary method**", which here comes into its own. It is an essential tool, which is scarcely taught, and not sufficiently practised. (As a result many students mindlessly translate "Tom takes 2 hours" as "$T = 2$", etc..)

(a) When they all work together we need to know **not** how long each takes to do the job, but **at what rate** each contributor works.
Tom does the job in 2 hours, so works at the rate of "$\frac{1}{2}$ of a job in **1 hour**".
Dick works at a rate of "$\frac{1}{3}$ of a job in **1 hour**", and Harry works at the rate of "$\frac{1}{4}$ of a job in **1 hour**".
So working together, they can manage
$$\frac{1}{2} + \frac{1}{3} + \frac{1}{4} = \frac{13}{12}$$
of a job in 1 hour.
Hence, to complete 1 job they require $\frac{12}{13}$ of an hour.

(b) As in part (a), we need to know the rate at which each man works.
Suppose that Tom completes the fraction t of a job in **1 hour**, that Dick completes the fraction d of a job in **1 hour**, and that Harry completes the fraction h of a job in **1 hour**.
Then in 1 hour, working together, they complete $(t+d+h)$ jobs; so to complete 1 job takes them
$$\frac{1}{t+d+h} \text{ hours.}$$
We therefore need to find "$t + d + h$".
In 1 hour, Tom and Dick together complete $t + d$ jobs. And we are told that in 2 hours they complete 1 job, so $t + d = \frac{1}{2}$. Similarly $d + h = \frac{1}{3}$, and $h + t = \frac{1}{4}$.
Adding yields
$$2(t + d + h) = \frac{1}{2} + \frac{1}{3} + \frac{1}{4},$$
so
$$t + d + h = \frac{13}{24}.$$
Hence the time required for Tom, Dick and Harry to finish 1 job working together is
$$\frac{1}{t+d+h} = \frac{24}{13}$$
hours.

Note: Alternatively, one might let Tom take T hours to complete **1 job**, Dick take D hours to complete **1 job**, and Harry take H hours to complete **1 job**. Then
$$t = \frac{1}{T}, \quad d = \frac{1}{D}, \quad h = \frac{1}{H}.$$

83. Imagine the two fields as strips of equal width – with the larger field *twice as long* as the smaller one.

The large strip was completely mowed in two parts:

(i) by the whole team working for the first half day, and

(ii) by half the team working for the second half of the day.

Hence the whole team mowed *two thirds* of the large field and the half team mowed the remaining *one third*.

So the half team, who worked on the smaller field, mowed *the equivalent of one third of the larger field* – that is, *two thirds of the (half-size) smaller field*. Therefore the remaining one third of the smaller field was mowed by a single man on the second day.

The previous two thirds of the smaller field (twice as much) was mowed in half a day (half the time), so must have required 4 (= *four times as many*) men. So the whole team contained 8 mowers.

Alternatively, we may suppose that there are $2n$ mowers (since the team is said to split into two halves), and that each mower mows at the rate of "r large fields per day".

The total work done in completing the larger field is then

(i) $(2n \times r) \times \frac{1}{2}$ in the morning and

(ii) $(n \times r) \times \frac{1}{2}$ in the afternoon

where each part is equal to

$$(\text{number of men} \times \text{rate of working}) \times (\text{length of time worked}).$$

That is $\frac{3}{2}nr$. So $\frac{3}{2}nr = 1$.

The total work done on the smaller field is

(i) $(n \times r) \times \frac{1}{2}$ in the afternoon of the first day, and

(ii) $(1 \times r)$ on the second day.

That is $\frac{n+2}{2} \times r$. So $\frac{n+2}{2} \times r = \frac{1}{2}$ (since the smaller field is half the larger field). Hence $\frac{3}{2}n = n + 2$.

84. The words "average speed" often provoke an unthinking assumption that one is simply being asked to find the average of the "speed **numbers**" given in the problem. A moment's thought should remind us that the "average speed" for a journey is **not** equal to the "average of the various speeds taken as pure numbers"; it is equal to

$$(\text{the total distance travelled}) \div (\text{the total time taken}).$$

If the distance up the hill is m miles, then the climb takes $\frac{m}{2}$ hours, and the descent takes $\frac{m}{4}$ hours. The total distance for the round trip is $2m$ miles, so Jack and Jill's average speed is

$$\frac{2m}{\frac{3m}{4}} = \frac{8}{3} \text{ mph.}$$

Word Problems

Note: We first meet averages for *discrete* quantities, or whole numbers, where the goal is to replace a collection of quantities, or numbers, by a single representative statistic. If n quantities contribute equally, then each contributes exactly $\left(\frac{1}{n}\right)^{\text{th}}$ to the average.

One way of looking at this is to represent each of the quantities being averaged in a bar chart – as rectangles of width 1, and with height corresponding to the quantity represented. "Adding all the quantities and dividing by n" is then the same as "calculating the total area under the graph and then dividing by the total length of the interval". In other words, we have replaced the complicated bar chart by a *constant function* (or a single rectangle), which has the same domain as the bar chart, and which has the same area under it (or integral) as the more complicated bar chart.

More generally, given a function $y = f(x)$ defined for values of x in the interval $[a, b]$, its *average* $f_{[a,b]}$ (over the interval $[a, b]$) is defined to be

$$f_{[a,b]} = \frac{\int_a^b f(x)dx}{|b - a|}.$$

When we talk about "average speed", we are thinking of speed changing *as a function of time*; and the total distance covered in any given time interval $[a, b]$ is equal to the area under the graph. We want a single "average speed" $v_{[a,b]}$ (a *constant function*) that would cover the same distance in the same time as the more complicated reality of varying speed. That is,

- we consider the speed $v(t)$ as a function of time t,
- then we integrate with respect to t over the specified time interval $[a, b]$, and
- finally we divide the result by the total length $|b - a|$ of the time interval:

$$v_{[a,b]} = \frac{\int_a^b v(t)dt}{|b - a|}.$$

In Problem **84** the walking speed is misleadingly given in terms of "up" and "down" – which represent the first *half **distance*** travelled, and the second *half **distance*** travelled. The careful solver knows that s/he has to find "total distance travelled" and divide by "total time taken"; but s/he may not notice that s/he has in fact reinterpreted the given information so that speed is seen as a function of *time* (rather than of distance).

85.

(a)(i) Let the distance covered on each lap be m km. Then the first lap takes me $\frac{m}{40}$ hours; the second lap takes me $\frac{m}{30}$ hours; the third lap takes me $\frac{m}{20}$ hours. So the total time taken for the three laps is

$$\frac{m}{40} + \frac{m}{30} + \frac{m}{20} = \frac{13m}{120} \text{ hours.}$$

Hence my average speed for the race covering $3m$ km is

$$\frac{3m}{\left(\frac{13m}{120}\right)} = \frac{360}{13} \text{ km/h.}$$

Note: Alternatively, because the two factors of m in the numerator and the denominator cancel each other, this answer may be formulated as the *harmonic mean* of the given speeds:
$$\frac{3}{\left[\frac{1}{40} + \frac{1}{30} + \frac{1}{20}\right]}.$$

(ii) In the first hour I cycle 40 km; in the second hour I cycle 30 km; in the third hour I cycle 20 km. So in the three hours I cycle $40 + 30 + 20 = 90$ km. So my average speed is 30 km/h.

Note: Alternatively, as long as the three time intervals t are equal, we land up with t as a factor in both the numerator and the denominator, so these common factors cancel out, and the answer is simply the *arithmetic mean* of the given speeds:
$$\frac{20 + 30 + 40}{3}.$$

(b)(i) The second cyclist spends more time cycling at 40 km/h than at 60 km/h, so the **first** cyclist spends **more time** cycling **at the higher speed**. Hence the first cyclist wins.

(ii) Again (unless $u = v$), the first cyclist spends more time cycling at the higher speed. Hence the first cyclist wins.

(c)(i) As in part (a)(ii), the first cyclist finishes with average speed $\frac{u+v}{2}$ km/h; and as in part (a)(i) the second cyclist finishes with average speed
$$\frac{2}{\left[\frac{1}{u} + \frac{1}{v}\right]} \text{ km/h}.$$

Hence, part (b)(ii) shows that
$$\frac{u+v}{2} \geq \frac{2}{\left[\frac{1}{u} + \frac{1}{v}\right]} = \frac{2uv}{u+v}.$$

(ii) If we rearrange the required inequality
$$\frac{u+v}{2} \geq \frac{2uv}{u+v},$$
then we see that it is equivalent to proving that $(u+v)^2 \geq 4uv$. This suggests that we should **start** with the universally true statement:
$$(u-v)^2 \geq 0 \text{ for all } u, v \geq 0.$$

Adding $4uv$ to both sides yields $(u+v)^2 \geq 4uv$.
Multiplying both sides by the non-negative quantity $\frac{1}{2(u+v)}$ then gives the required inequality.

86. The only "modelling" required here is to translate the problem using the standard equations of kinematics. For motion from rest we have

(i) $v = at$, where t is the time, a is the uniform acceleration, and v the final speed, and

(ii) $s = \frac{1}{2}at^2$, where s is the distance travelled.

There is a question as to what units we should use. For the moment we stick to measuring v in km/h as given, s in km, t in hours, and a in the (unfamiliar) units of km/h^2: so $72 = at$ and $4 = \frac{1}{2}at^2$.

Dividing the second equation by the first gives $\frac{1}{18} = \frac{1}{2}t$, so $t = \frac{1}{9}$ hours (= 400 seconds).

Substituting in the first equation gives $a = 72 \times 9$ km/h^2 ($= \frac{1}{20}$ m/sec^2).

Note: Equations (i) and (ii) can be summarised as saying that, under uniform acceleration a, the distance travelled is $s = (\frac{1}{2}at) \times t$. Hence the *average* speed for the complete journey is equal to exactly *half of the final speed* $v = at$.

In general, those tackling the problem may agree that the familiar units of speed and distance do not give us a very good gut feeling for the scale of acceleration. If we measure acceleration in km/h^2, then we get huge numbers for acceleration which one cannot easily relate to. And if we switch to m (metres), m/sec, and m/sec^2, then we get rather small numbers for the acceleration, which again convey relatively little.

[The original (Russian) version of this problem had the train travelling 2.1 km and reaching a speed of 54 km/h. This produces a nice answer for the time taken, but a relatively inscrutable answer for the acceleration. So we have changed the parameters.]

87.

"We explain why, when a vehicle accelerates from 0 to 20 mph, its average speed is **more than** *10 mph. In general, the average speed of an accelerating vehicle is more than half the final speed after the acceleration.*

Consider first the case when the acceleration is constant: this means that the graph which represents the speed of the vehicle as a function of time is a straight line:

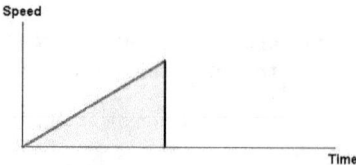

In that case, the distance travelled is equal to the area under the graph. But from the formula for the area of a triangle we know that this area equals the area of the rectangle with the same base and half the height of the triangle:

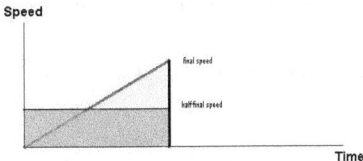

*This means that the **average speed** in that case is **exactly half** of the final (maximum) speed.*

*But **a car has higher acceleration in lower gears, that is, at smaller speeds.** Therefore the graph of speed as a function of time is concave, and the area under the graph is greater than in the case of constant acceleration. Hence, while reaching the same speed, the car travels further and its average speed is higher:*

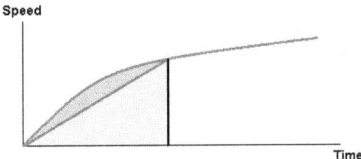

We come to the conclusion that the average speed of an accelerating car is greater than half its speed at the end of acceleration."

Note: The text of this solution is reproduced from the appendix to a document prepared for, and submitted to, the *Crown Prosecution Service* in England. This may partly explain why it contains not a single formula. It was written by a student studying economics, and the mixture of language and graphs used illustrates the typical economist's way of thinking. Economists rarely have complete data, so they tend to rely on a combination of common sense and the basic patterns of economic variables – such as the "convexity" or "concavity" of functions. Indeed some chapters of mathematical economics could be described as outlining "the kinematics of money", and have surprising similarities to mechanics.

88. Suppose sunrise was t hours before noon – so that the first woman covers the total distance in $t + 4$ hours, while the second covers the same distance in $t + 9$ hours.

We know nothing about the distance from A to B, so it makes sense to choose this distance as our unit.

Then the first woman's speed is $\frac{1}{t+4}$, while the second woman's speed is $\frac{1}{t+9}$ units per hour.

The relative speed of A and B (the speed with which the distance between them changes) is $\frac{1}{t+4} + \frac{1}{t+9}$.

They meet at noon, so in t hours, the distance between them reduces from 1 unit to 0.

Hence

$$1 = t \times \left(\frac{1}{t+4} + \frac{1}{t+9} \right);$$

that is, $t^2 = 36$, so $t = 6$, and sunrise was at $(12 - 6) = 6$ am.

89. Let us introduce a new measure of distance – which we call a *league*. (Readers may know from old documents or from poetry that this was an old measure of distance for journeys, without knowing exactly how far it was; so we feel free to use it as an *abstract* unit *of unknown size*.)

To mesh distance and time, the journey from St Louis to New Orleans needs to be some multiple of 7, and the journey from New Orleans to St Louis needs to be some multiple of 5. Hence we choose the distance to be equal to $5 \times 7 = 35$ "leagues".

Then the speed of the paddle-steamer upstream is:

$$\frac{35}{7} = 5 \text{ "leagues per day"}$$

and the speed downstream is:

$$\frac{35}{5} = 7 \text{ "leagues per day"}.$$

The speed of the current gets subtracted from the speed of the paddle-steamer going upstream, and gets added to the speed of the paddle-steamer going downstream; so the speed of the current is:

$$\frac{7-5}{2} = 1 \text{ "league per day"}.$$

Hence a raft will drift from St Louis to New Orleans in $\frac{35}{1} = 35$ days.

Note: This elegant solution involves the introduction of a *hidden intermediate parameter*, an unknown quantity which helps us reason about the problem. The parameter is apparently the *distance* (from St Louis to New Orleans); but it is in fact a measure of distance chosen so as to be *compatible with the time taken*.

The art of identifying, and choosing, relevant "hidden parameters", and the analysis of their relation to the data, and their mutual relations, constitute an important and challenging part of the mathematical modelling process.

Notice that if we reformulate the problem in more general terms, with the paddle-steamer taking "a days" downstream and "b days" upstream, then the answer "d days" (for the time to drift *downstream*) happens to be the *harmonic mean* of the quantities "a" and "$-b$":

$$d = \frac{2}{\frac{1}{a} + \frac{1}{-b}}.$$

90. [This is "Problem 108" in Paolo dell'Abbaco's *Trattato d'aritmetica* (c.1370), with a rough translation of the solution procedure given there courtesy of Roy Wagner.]

> "Do the following: multiply 5 by 8, which makes 40. Then say thus: in 40 days one will make the trip 8 times, and the other 5 times, so both together will make the trip 13 times.
>
> Now say: if 40 days equals 13 trips, how many days are needed [on average] for one trip? And so multiply 1 times 40, which makes 40; then divide this by 13, which makes 3 days and $\frac{1}{13}$ of a day.
>
> And so I say that in 3 days and $\frac{1}{13}$ of a day the two will come together.
>
> And as this is done, so all similar problems are done."

Note: The problem as stated conveys an air of reality by giving the distance "from here to Florence" in miles; but this fact is not mentioned in the solution! Instead, the solution starts by introducing a hidden parameter, measured by a *dimensionless* unit: a **trip**.

This move (to invent a natural unit of measurement) also featured in Problem **89** above and has deep mathematical reasons. Problem **89** was borrowed from an interview with Vladimir Arnold (*Notices of the AMS*, vol. 44, no. 4), where we read:

> **Interviewer**: *Please tell us a little bit about your early education. Were you already interested in mathematics as a child?*
>
> **Arnold**: *[...] The first real mathematical experience I had was when our schoolteacher I.V. Morotzkin gave us the following problem* [VA then formulated Problem **89**].
>
> *I spent a whole day thinking on this oldie, and the solution (based on what are now called scaling arguments, dimensional analysis, or toric variety theory, depending on your taste) came as a revelation.*
>
> *The feeling of discovery that I had then (1949) was exactly the same as in all the subsequent much more serious problems – be it the discovery of the relation between algebraic geometry of real plane curves and four-dimensional topology (1970), or between singularities of caustics and of wave fronts and simple Lie algebras and Coxeter groups (1972). It is the greed to experience such a wonderful feeling more and more times that was, and still is, my main motivation in mathematics.*

Arnold refers here to *scaling arguments* or *dimensional analysis*: that is, the mathematical art of choosing and analysing the use of units of measurement. This has its origins in, and includes as an integral part, Euclid's classical theory of proportion.

91. Suppose as before that the sun rises t hours before noon; but replace 4 pm (the time the woman starting at A arrived at B) by a pm, and replace 9 pm (the time the woman starting at B arrived at A) by b pm. Let C be the point where they meet (at noon).

Then, since each woman walks at a constant speed, we have

$$\frac{t}{a} = \frac{|AC|}{|CB|} \text{ (for the woman starting from } A\text{),}$$

and

$$\frac{t}{b} = \frac{|BC|}{|CA|} \text{ (for the woman starting from } B\text{).}$$

Hence

$$\frac{t}{a} = \frac{|AC|}{|CB|} = \frac{b}{t},$$

so $t^2 = ab$.

Note: This totally unexpected result validates the choice of the unknown t as the time in hours from sunrise to noon. Not knowing its significance in advance, this choice was motivated by the observation that "noon" occurs in the problem as the only common "origin", or reference point for time data.

IV. Algebra

> *The first rule of intelligent tinkering*
> *is to save all the parts.*
> Paul R. Ehrlich (1932–)

Many important aspects of serious mathematics have their roots in the world of arithmetic. However, when we implement an arithmetical procedure by combining numbers with very different meanings to produce *a single numerical output*, it becomes almost impossible to see how the separate ingredients contribute to the final answer. In other words, calculating exclusively with numbers contravenes Paul Ehrlich's "first rule of intelligent tinkering". This is why in Chapters 1 and 2 we stressed the need to move beyond blind calculation, and to begin to think *structurally* – even when calculating purely with numbers. *Algebra* can be seen as a remarkable way of "tinkering with numbers", so that we not only "keep all the parts", but manage to *keep them separate* (by giving them different names), and hence can see clearly what contribution each ingredient variable makes to the final output. To benefit from this feature of algebra, we need to learn to "read" algebraic expressions, and to interpret what they are telling us – in much the same way that we learn to read numbers (so that, where appropriate, 100 is seen as 10^2, and 10 is seen as $1 + 2 + 3 + 4$).

Before algebra proper was invented (around 1600), the ability to extract the general picture lying hidden inside each calculation was restricted to specialists. The ancient Babylonians (1700–1500 BC) described their general procedures as *recipes*, presented in the context of problems involving particular numbers. But they did this in such a way as to demonstrate convincingly that whoever formulated the procedure had managed to see "the general in the particular". The ancient Greeks used a geometrical setting to reveal generality, and encoded what we would see as "algebraic" methods in geometrical language. In the 9th century AD, Arabs such as Al-Khwarizmi (c.780–c.850), managed to encapsulate generality using a very limited kind of algebra, without the full symbolical language that would emerge later. The *abacists*, such as Paolo dell'Abbaco (1282–1374) who featured in Chapter 3, clearly saw that the power and spirit of mathematics was rooted in this generality. But modern algebraic symbolism – in particular, the idea that to express generality we need to use letters to represent not only variables, but also important *parameters* (such as the coefficients a, b, c in a general quadratic $ax^2 + bx + c$) – had to wait for the inscrutable writings of Viète (1540–1603), and especially for Fermat (1601–1665) and Descartes (1596–1650) who simplified and extended Viète's ideas in the 1630s.

Within a generation, the huge potential of this systematic use of symbols was revealed by the triumphs of Newton (1642–1727), Leibniz (1646–1716), and others in the years before 1700. Later, the refinements proposed by Euler (1707–1783) in his many writings throughout the 18th century, made this new language and its discoveries accessible to us all – much as Stevin's (1548–1620) version of place value for numbers made calculation accessible to Everyman.

Our coverage of algebra is necessarily selective. We focus on a few ideas that are needed in what follows, and which should ideally be familiar – but with an emphasis that may be less familiar. When working algebraically, the key mathematical messages are mostly implicit in the manipulations themselves. Hence many of the additional comments in this chapter are to be found as part of the solutions, rather than within the main text.

4.1. Simultaneous linear equations and symmetry

Problem 92 Dad took our new baby to the clinic to be weighed. But the baby would not stay still and caused the needle on the scales to wobble. So Dad held the baby still and stood on the scales, while nurse read off their combined weight: 78kg. Then nurse held the baby, while Dad read off their combined weight: 69kg. Finally Dad held the nurse, while the baby read off their combined weight: 137kg. How heavy was the baby? △

The situation described in Problem **92** is representative of a whole class of problems, where the given information incorporates a certain *symmetry*, which the solver would be wise to respect. Hence one should hesitate before applying systematic brute force (as when using the information from one weighing to substitute for one of the three unknown weights – a move which effectively reduces the number of unknowns, but which fails to respect the symmetry in the data).

A similar situation arises in certain puzzles like the following.

Problem 93 Numbers are assigned (secretly) to the vertices of a polygon. Each edge of the polygon is then labelled with the sum of the numbers at its two end vertices.

(a) If the polygon is a triangle ABC, and the labels on the three sides are c (on AB), b (on AC), and a (on BC), what were the numbers written at each of the three vertices?

(b) If the polygon is a quadrilateral $ABCD$, and the labels on the four sides are w (on AB), x (on BC), y (on CD), and z (on DA), what numbers were written at each of the four vertices?

(c) If the polygon is a pentagon $ABCDE$, and the labels on the five sides are d (on AB), e (on BC), a (on CD), b (on DE), and c (on EA), what numbers were written at each of the five vertices? △

In case any reader is inclined to dismiss such problems as "artificial puzzles", it may help to recall two familiar instances (Problems **94** and **96**) which give rise to precisely the above situation.

Problem 94 In the triangle ABC with sides of lengths a (opposite A), b (opposite B), and c (opposite C), we want to locate the three points where the *incircle* touches the three sides – at point P (on BC), Q (on CA), and R (on AB). To this end, let the two tangents to the incircle from A (namely AQ and AR) have length x, the two tangents from B (namely BP and BR) have length y, and the two tangents from C (namely CP and CQ) have length z. Find the values of x, y, z in terms of a, b, c. △

The second instance requires us first to review the basic properties of midpoints in terms of vectors.

Problem 95

(a) Write down the coordinates of the midpoint M of the line segment joining $Y = (a, b)$ and $Z = (c, d)$. Justify your answer.

(b) Position a general triangle XYZ so that the vertex X lies at the origin $(0, 0)$. Suppose that Y then has coordinates (a, b) and Z has coordinates (c, d). Let M be the midpoint of XY, and N be the midpoint of XZ. Prove the *Midpoint Theorem*, namely that

"MN is parallel to YZ and half its length".

(c) Given any quadrilateral $ABCD$, let P be the midpoint of AB, let Q be the midpoint of BC, let R be the midpoint of CD, and let S be the midpoint of DA. Prove that $PQRS$ is always a parallelogram. △

Problem 96

(a) Suppose you know the position vectors **p**, **q**, **r** corresponding to the midpoints of the three sides of a triangle. Can you reconstruct the vectors **x**, **y**, **z** corresponding to the three vertices?

(b) Suppose you know the vectors **p**, **q**, **r**, **s** corresponding to the midpoints of the four sides of a quadrilateral. Can you reconstruct the vectors **w**, **x**, **y**, **z** corresponding to the four vertices?

(c) Suppose you know the vectors **p**, **q**, **r**, **s**, **t** corresponding to the midpoints of the five sides of a pentagon. Can you reconstruct the vectors **v**, **w**, **x**, **y**, **z** corresponding to the five vertices? △

The previous five problems explore a common structural theme – namely the link between certain sums (or averages) and the original, possibly unknown, data. However this algebraic link was in every case embedded in some practical, or geometrical, context. The next few problems have been stripped of any context, leaving us free to focus on the underlying structure in a purely algebraic, or arithmetical, spirit.

Problem 97 Solve the following systems of simultaneous equations.

(a)(i) $x + y = 1$, $y + z = 2$, $x + z = 3$
(ii) $uv = 2$, $vw = 4$, $uw = 8$

(b)(i) $x + y = 2$, $y + z = 3$, $x + z = 4$
(ii) $uv = 6$, $vw = 10$, $uw = 15$
(iii) $uv = 6$, $vw = 10$, $uw = 30$
(iv) $uv = 4$, $vw = 8$, $uw = 16$ △

Problem 98 Use what you know about solving two simultaneous linear equations in two unknowns to construct the general positive solution to the system of equations:
$$u^a v^b = m, \quad u^c v^d = n.$$

Interpret your result in the language of *Cramer's Rule*. (Gabriel Cramer (1704–1752)). △

Problem 99

(a) For which values b, c does the following system of equations have a unique solution?

$$x + y + z = 3, \quad xy + yz + zx = b, \quad x^2 + y^2 + z^2 = c$$

(b) For which values a, b, c does the following system of equations have a unique solution?

$$x + y + z = a, \quad xy + yz + zx = b, \quad x^2 + y^2 + z^2 = c \qquad \triangle$$

4.2. Inequalities and modulus

The transition from school to university mathematics is in many ways marked by a shift from simple variables, equations and functions, to conditions and analysis involving inequalities and modulus.

Problem 100 What is $|-x|$ equal to: x or $-x$? (What if x is negative?)
$\triangle$

4.2.1 Geometrical interpretation of modulus, of inequalities, and of modulus inequalities

Problem 101

(a) Mark on the coordinate line all those points x in the interval $[0, 1)$ which have the digit "1" immediately after the decimal point in their decimal expansion. What fraction of the interval $[0, 1)$ have you marked?

  Note: "$[0, 1)$" denotes the set of all points *between* 0 and 1, together with 0, but not including 1. $[0, 1]$ denotes the interval including *both* endpoints; and $(0, 1)$ denotes the interval *excluding* both endpoints.

(b) Mark on the interval $[0, 1)$ all those points x which have the digit "1" in *at least one* decimal place. What fraction of the interval $[0, 1)$ have you marked?

(c) Mark on the interval $[0, 1)$ all those points x which have a digit "1" in at least one position of their *base 2* expansion. What fraction of the interval $[0, 1)$ have you marked?

(d) Mark on the interval $[0, 1)$ all those points x which have a digit "1" in at least one position of their *base 3* expansion. What fraction of the interval $[0, 1)$ have you marked? $\triangle$

Problem 102 Mark on the coordinate line all those points x for which *two* of the following inequalities are true, and *five* are false:

$$x > 1,\ x > 2,\ x > 3,\ x > 4,\ x > 5,\ x > 6,\ x > 7.$$
△

Problem 103 Mark on the coordinate line all those points x for which

$$|x - 5| = 3.$$
△

Problem 104

(a) Mark on the coordinate line all those points x for which *two* of the following inequalities are true, and *five* are false:

$$|x| > 1,\ |x| > 2,\ |x| > 3,\ |x| > 4,\ |x| > 5,\ |x| > 6,\ |x| > 7.$$

(b) Mark on the coordinate line all those points x for which *two* of the following inequalities are true, and *five* are false:

$$|x-1| > 1,\ |x-2| > 2,\ |x-3| > 3,\ |x-4| > 4,\ |x-5| > 5,\ |x-6| > 6,\ |x-7| > 7.$$
△

Problem 105 Mark on the coordinate line all those points x for which

$$|x + 1| + |x + 2| = 2.$$
△

Problem 106 Find numbers a and b with the property that the set of solutions of the inequality

$$|x - a| < b$$

consists of the interval $(-1, 2)$.
△

Problem 107

(a) Mark on the coordinate plane all points (x, y) satisfying the inequality

$$|x - y| < 3.$$

(b) Mark on the coordinate plane all points (x, y) satisfying the inequality

$$|x - y + 5| < 3.$$

(c) Mark on the coordinate plane all points (x, y) satisfying the inequality

$$|x - y| < |x + y|.$$
△

4.2.2 Inequalities

Problem 108 Suppose real numbers a, b, c, d satisfy $\frac{a}{b} < \frac{c}{d}$.

(i) Prove that
$$\frac{a}{b} < \frac{\left(\frac{a}{b} + \frac{c}{d}\right)}{2} < \frac{c}{d}.$$

(ii) If $b, d > 0$, prove that
$$\frac{a}{b} < \frac{a+c}{b+d} < \frac{c}{d}. \qquad \triangle$$

Problem 109 (Farey series) When the fully cancelled fractions in $[0, 1]$ with denominator $\leq n$ are arranged in increasing order, the result is called the *Farey series* (or *Farey sequence*) of order n.

Order 1: $\frac{0}{1} < \frac{1}{1}$

Order 2: $\frac{0}{1} < \frac{1}{2} < \frac{1}{1}$

Order 3: $\frac{0}{1} < \frac{1}{3} < \frac{1}{2} < \frac{2}{3} < \frac{1}{1}$

Order 4: $\frac{0}{1} < \frac{1}{4} < \frac{1}{3} < \frac{1}{2} < \frac{2}{3} < \frac{3}{4} < \frac{1}{1}$

(a) Write down the full Farey series (or sequence) of order 7.

(b)(i) Imagine the points $0.1, 0.2, 0.3, \ldots, 0.9$ dividing the interval $[0, 1]$ into ten subintervals of length $\frac{1}{10}$. Now insert the eight points corresponding to
$$\frac{1}{9}, \frac{2}{9}, \frac{3}{9}, \ldots, \frac{8}{9}.$$
Into which of the ten subintervals do they fall?

(ii) Imagine the n points
$$\frac{1}{n+1}, \frac{2}{n+1}, \frac{3}{n+1}, \ldots, \frac{n}{n+1}$$
dividing the interval $[0, 1]$ into $n+1$ subintervals of length $\frac{1}{n+1}$. Now insert the $n-1$ points
$$\frac{1}{n}, \frac{2}{n}, \frac{3}{n}, \ldots, \frac{n-1}{n}.$$
Into which of the $n + 1$ subintervals do they fall?

(iii) In passing from the Farey series of order n to the Farey series of order $n + 1$, we insert fractions of the form $\frac{k}{n+1}$ between certain pairs of adjacent fractions in the Farey series of order n. If $\frac{a}{b} < \frac{c}{d}$ are adjacent fractions in the Farey series of order n, prove that, when adding fractions for the Farey series of order $n + 1$, **at most one** fraction is inserted between $\frac{a}{b}$ and $\frac{c}{d}$.

(c) **Note:** It is worth struggling to prove the two results in part (c). But do not be surprised if they prove to be elusive – in which case, be prepared to simply use the result in part (c)(ii) to solve part (d).

(i) In the Farey series of order n the first two fractions are $\frac{0}{1} < \frac{1}{n}$, and the last two fractions are $\frac{n-1}{n} < \frac{1}{1}$. Prove that every other adjacent pair of fractions $\frac{a}{b} < \frac{c}{d}$ in the Farey series of order n satisfies $bd > n$.

(ii) Let $\frac{a}{b} < \frac{c}{d}$ be adjacent fractions in the Farey series of order n. Prove (by induction on n) that $bc - ad = 1$.

(d) Prove that if
$$\frac{a}{b} < \frac{c}{d} < \frac{e}{f}$$
are three successive terms in any Farey series, then
$$\frac{c}{d} = \frac{a+e}{b+f}.$$
△

Problem 110 Solve the following inequalities.

(a) $x + \dfrac{1}{x} < 2$

(b) $x \leqslant 1 + \dfrac{2}{x}$

(c) $\sqrt{x} < x + \dfrac{1}{4}$ △

Problem 111

(a) The sum of two positive numbers equals 5. Can their product be equal to 7?

(b) (**Arithmetic mean, Geometric mean, Harmonic mean, Quadratic mean**) Prove that, if $a, b > 0$, then
$$\frac{2}{\left[\frac{1}{a} + \frac{1}{b}\right]} = \frac{2ab}{a+b} \leqslant \sqrt{ab} \leqslant \frac{a+b}{2} \leqslant \sqrt{\frac{a^2+b^2}{2}}$$
(HM $\leqslant$ GM $\leqslant$ AM $\leqslant$ QM)   △

Problem 112 The two hundred numbers

$$1, 2, 3, 4, 5, \ldots, 200$$

are written on the board. Students take turns to replace two numbers a, b from the current list by their sum divided by $\sqrt{2}$. Eventually one number is left on the board. Prove that the final number must be less than 2000. △

4.3. Factors, roots, polynomials and surds

Problem 113

(a)(i) Find a prime number which is one less than a square.

(ii) Find another such prime.

(b)(i) Find a prime number which is one more than a square.

(ii) Find another such prime.

(c)(i) Find a prime number which is one less than a cube.

(ii) Find another such prime.

(d)(i) Find a prime number which is one more than a cube.

(ii) Find another such prime. △

Problem 114 Factorise $x^4 + 1$ as a product of two quadratic polynomials with real coefficients. △

4.3.1 Standard factorisations

The challenge to factorise unfamiliar expressions, may at first leave us floundering. But if we assume that each such problem is solvable with the tools at our disposal, we then have no choice but to fall back on the standard tools we have available (in particular, the standard factorisation of a difference of two squares, in which "cross terms" cancel out). The next problem extends this basic repertoire of standard factorisations.

Problem 115

(a)(i) Factorise $a^3 - b^3$.

(ii) Factorise $a^4 - b^4$ as a product of one linear factor and one factor of degree 3, and as a product of two linear factors and one quadratic factor.

(iii) Factorise $a^n - b^n$ as a product of one linear factor and one factor of degree $n - 1$.

(b)(i) Factorise $a^3 + b^3$.

(ii) Factorise $a^5 + b^5$ as a product of one linear factor and one factor of degree 4.

(iii) Factorise $a^{2n+1} + b^{2n+1}$ as a product of one linear factor and one factor of degree $2n$. △

Problem **115** develops the ideas that were implicit in Problem **113**. The clue lies in Problem **113**(a), and in the comment made in the main text in Chapter 1 (after Problem **4** in Chapter 1), which we repeat here:

"The last part [of Problem **113**(a)] is included to emphasise a frequently neglected message:

Words and images are part of the way we communicate. But most of us cannot *calculate* with words and images.

To make use of mathematics, we must routinely translate *words* into *symbols*. So "numbers" need to be represented by symbols, and points in a geometric diagram need to be properly labelled before we can begin to calculate, and to reason, effectively."

As soon as one reads the words "one less than a square", one should instinctively translate this into the form "$x^2 - 1$". Bells will then begin to ring; for it is impossible to forget the factorisation

$$x^2 - 1 = (x - 1)(x + 1).$$

And it follows that:

for a number that factorises in this way to be prime, the smaller factor $x - 1$ must be equal to 1;
∴ $x = 2$, so there is only one such prime.

The integer factorisations in Problem **113**(c) – namely

$$3^3 - 1 = 2 \times 13,\ 4^3 - 1 = 3 \times 21,\ 5^3 - 1 = 4 \times 31,\ 6^3 - 1 = 5 \times 43,\ \ldots$$

may help one to remember (or to discover) the related factorisation

$$x^3 - 1 = (x - 1)(x^2 + x + 1).$$

∴ For a number that factorises in this way to be prime, the smaller factor "$x - 1$" must be equal to 1;
∴ $x = 2$, so there is only one such prime.

Problem **113** parts (a) and (c) highlight the completely general factorisation (Problem **115**(a)(iii)):

$$x^n - 1 = (x - 1)(x^{n-1} + x^{n-2} + \cdots + x^2 + x + 1).$$

This family of factorisations also shows that we should think about the factorisation of $x^2 - 1$ as $(\boldsymbol{x} - \boldsymbol{1})(x + 1)$, with the uniform factor $(\boldsymbol{x} - \boldsymbol{1})$ **first** (rather than as $(x + 1)(x - 1)$). Similarly, the results of Problem **115** show that we should think of the familiar factorisation of $a^2 - b^2$ as $(\boldsymbol{a} - \boldsymbol{b})(\boldsymbol{a} + \boldsymbol{b})$, (not as $(a + b)(a - b)$), but always with the factor $(\boldsymbol{a} - \boldsymbol{b})$ **first**).

The integer factorisations in Problem **113**(d) – namely

$$3^3+1 = 4\times 7, 4^3+1 = 5\times 13, 5^3+1 = 6\times 21, 6^3+1 = 7\times 31, 7^3+1 = 8\times 43, \ldots$$

may help one to remember (or to discover) the related factorisation

$$x^3 + 1 = (x + 1)(x^2 - x + 1).$$

∴ For such a number to be prime, one of the factors must be equal to 1.

This time one has to be more careful, because the first bracket may not be the "smaller factor" – so there are two cases to consider:

(i) if $x + 1 = 1$, then $x = 0$, and $x^3 + 1 = 1$ is not prime;

(ii) if $x^2 - x + 1 = 1$, then $x = 0$ or $x = 1$, so $x = 1$ and we obtain the prime 2 as the only solution.

The factorisation for $x^3 + 1$ works because "3 is **odd**", which allows the alternating $+/-$ signs to end in a "$+$" as required. Hence Problem **113**(d)(iii) highlights the completely general factorisation **for odd powers**:

$$x^{2n+1} + 1 = (x + 1)(x^{2n} - x^{2n-1} + x^{2n-2} - \cdots + x^2 - x + 1).$$

You probably know that there is no standard factorisation of $x^2 + 1$, or of $x^4 + 1$ (but see Problem **114** above).

Problem 116

(a) Derive a *closed formula* for the sum of the geometric series

$$1 + r + r^2 + r^3 + \cdots + r^n.$$

(The meaning of *closed formula* was discussed in the **Note** to the solution to Problem **54**(b) in Chapter 2.)

(b) Derive a closed formula for the sum of the geometric series

$$a + ar + ar^2 + ar^3 + \cdots + ar^n.$$ △

We started this subsection by looking for prime numbers of the form $x^2 - 1$. A simple-minded approach to the distribution of prime numbers might look for formulae that generate primes – all the time, or infinitely often, or at least much of the time. In Chapter 1 (Problem **25**) you showed that no prime of the form $4k + 3$ can be "represented" as a sum of two squares (i.e. in the form "$x^2 + y^2$"), and we remarked that every other prime can be so represented in exactly one way. It is true (but not obvious) that roughly half the primes fall into the second category; so it follows that substituting integers for the two variables in the polynomial $x^2 + y^2$ produces a prime number infinitely often.

Problem 117 Experiment suggests, and Goldbach (1690–1764) showed in 1752 that no polynomial in one variable, and with integer coefficients, can give prime values for all integer values of the variable. But Euler (1707–1783) was delighted when he discovered the quadratic

$$f(x) = x^2 + x + 41.$$

Clearly $f(0) = 41$ is prime. And $f(1) = 43$ is also prime. What is the first positive integer n for which $f(n)$ is **not** prime? △

Problem 117 should be seen as a particular instance of the question as to whether prime numbers can be captured by a *polynomial* with integer coefficients, and in particular by a *quadratic*. The next two problems consider the simplest instances of representing prime numbers by expressions involving *exponentials* (that is, where the variable is in the *exponent*).

Problem 118

(a)(i) Suppose $a^n - 1 = p$ is a prime. Prove that $a = 2$ and that n must itself be prime.

(ii) How many primes are there among the first five such numbers

$$2^2 - 1,\ 2^3 - 1,\ 2^5 - 1,\ 2^7 - 1,\ 2^{11} - 1?$$

(b)(i) Suppose $a^n + 1 = p$ is a prime. Prove that either $a = 1$, or a must be even and that n must then be a power of 2.

(ii) In the simplest case, where $a = 2$, how many primes are there among the first five such numbers

$$2^1 + 1,\ 2^2 + 1,\ 2^4 + 1,\ 2^8 + 1,\ 2^{16} + 1?$$ △

Primes of the form $2^p - 1$ are called *Mersenne primes* (after Marin Mersenne (1588–1648)). We now know at least fifty such primes (with the exponent p ranging up to around 80 million). Finding new primes is not in itself important, but the search for Mersenne primes has been used as a focus for many new developments in programming, and in number theory.

Primes of the form $2^n + 1$ are called *Fermat primes* (after Pierre de Fermat (1601–1665)). The story here is very different. We now refer to the number $2^n + 1$ with $n = 2^k$ as the kth *Fermat number* f_k. You showed in Problem 118 (as Fermat did himself) that f_0, f_1, f_2, f_3, f_4 are all prime. Fermat then rather rashly claimed that f_n is always prime. However, Euler showed (100 years later) that the very next Fermat number f_5 fails to be prime. And despite all the power of modern computers, we have still not found another Fermat number that is prime!

4.3.2 Quadratic equations

The general solution of quadratic equations dates back to the ancient Babylonians ($\approx$ 1700 BC). Our modern understanding depends on two facts:

- an equation of the form $x^2 = a$ where $a > 0$, has exactly two solutions: $x = \pm\sqrt{a}$;

- any product $X \cdot Y$ is equal to 0 precisely when one of the two factors X, Y is equal to 0.

Problem 119 Solve the following quadratic equations:

(a) $x^2 - 3x + 2 = 0$

(b) $x^2 - 1 = 0$

(c) $x^2 - 2x + 1 = 0$

(d) $x^2 + \sqrt{2}x - 1 = 0$

(e) $x^2 + x - \sqrt{2} = 0$

(f) $x^2 + 1 = 0$

(g) $x^2 + \sqrt{2}x + 1 = 0$ △

Problem 120 Let
$$p(x) = x^2 + \sqrt{2}x + 1.$$
Find a polynomial $q(x)$ such that the product $p(x)q(x)$ has integer coefficients. △

Problem 121

(a) I am thinking of two numbers, and am willing to tell you their sum s and their product p. Express the following procedure algebraically and explain why it will always determine my two unknown numbers.

> Halve the sum s, and square the answer.
> Then subtract the product p and take the square root of the result, to get the answer.
> Add "the answer" to half the sum and you have one unknown number; subtract "the answer" from half the sum and you have the other unknown number.

(b) I am thinking of the length of one side of a square. All I am willing to tell you are two numbers b and c, where when I add b times the side length to the area I get the answer c. Express the following procedure algebraically and explain why it will always determine the side length of my square.

> Take one half of b, square it and add the result to c.
> Then take the square root.
> Finally subtract half of b from the result.

(c) A regular pentagon $ABCDE$ has sides of length 1.

 (i) Prove that the diagonal AC is parallel to the side ED.
 (ii) If AC and BD meet at X, explain why $AXDE$ is a rhombus.
 (iii) Prove that triangles ADX and CBX are similar.
 (iv) If AC has length x, set up an equation and find the exact value of x.

△

Problem **121**(a), (b) link to Problem **111**(a) (and to Problem **129** below), in relating the roots and the coefficients of a quadratic. If we forget for the moment that the coefficients are usually known, while the roots are unknown, then we see that if α and β are the roots of the quadratic

$$x^2 + bx + c,$$

then

$$(x - \alpha)(x - \beta) = x^2 + bx + c,$$

so

$$\alpha + \beta = -b \text{ and } \alpha\beta = c.$$

In other words, the two coefficients b, c are equal to the two simplest *symmetric* expressions in the two roots α and β. Part (a) of the next problem is meant to suggest that all other symmetric expressions in α and

β (that is, any expression that is unchanged if we swap α and β) can then be written in terms of b and c. The full result proving this fact is generally attributed to Isaac Newton (1642–1727). Part (b) suggests that, provided one is willing to allow case distinctions, something similar may be true of *anti-symmetric* expressions (where the effect of swapping α and β is to multiply the expression by "-1").

Problem 122 Let α and β be the roots of the quadratic equation

$$x^2 + bx + c = 0.$$

(a)(i) Write $\alpha^2 + \beta^2$ in terms of b and c only.

  (ii) Write $\alpha^2\beta + \beta^2\alpha$ in terms of b and c only.

  (iii) Write $\alpha^3 + \beta^3 - 3\alpha\beta$ in terms of b and c only.

(b)(i) Write $\alpha - \beta$ in terms of b and c only.

  (ii) Write $\alpha^2\beta - \beta^2\alpha$ in terms of b and c only.

  (iii) Write $\alpha^3 - \beta^3$ in terms of b and c only. △

Problem 123 (Nested surds, simplification of surds)

(a)(i) For any positive real numbers a, b, prove that

$$\sqrt{a} + \sqrt{b} = \sqrt{a + b + \sqrt{4ab}}$$

  (ii) Simplify $\sqrt{5 + \sqrt{24}}$.

(b)(i) Find a similar formula for $\sqrt{a} - \sqrt{b}$.

  (ii) Simplify $\sqrt{5 - \sqrt{16}}$ and $\sqrt{6 - \sqrt{20}}$. △

Problem 124 (Integer polynomials with a given root) We know that $\alpha = 1$ is a root of the polynomial equation $x^2 - 1 = 0$; that $\alpha = \sqrt{2}$ is a root of $x^2 - 2 = 0$; and that $\alpha = \sqrt{3}$ is a root of $x^2 - 3 = 0$.

(a) Find a quadratic polynomial with integer coefficients which has

$$\alpha = 1 + \sqrt{2}$$

as a root.

(b) Find a quadratic polynomial with integer coefficients which has
$$\alpha = 1 + \sqrt{3}$$
as a root.

(c) Find a polynomial with integer coefficients which has
$$\alpha = \sqrt{2} + \sqrt{3}$$
as a root. What are the other roots of this polynomial?

(d) Find a polynomial with integer coefficients which has
$$\alpha = \sqrt{2} + \frac{1}{\sqrt{3}}$$
as a root. What are the other roots of this polynomial? △

Problem 125

(a) Prove that the number $\sqrt{2} + \sqrt{3}$ is irrational.

(b) Prove that the number $\sqrt{2} + \sqrt{3} + \sqrt{5}$ is irrational. △

Problem 126 (Polynomial long division) Find

(i) the quotient and the remainder when we divide $x^{10} + 1$ by $x^3 - 1$

(ii) the remainder when we divide $x^{2013} + 1$ by $x^2 - 1$

(iii) the quotient and the remainder when we divide $x^m + 1$ by $x^n - 1$, for $m > n \geqslant 1$. △

Problem 127 Find the remainder when we divide $x^{2013} + 1$ by $x^2 + x + 1$. △

4.4. Complex numbers

Up to this point, the chapter and solutions have largely avoided mentioning *complex numbers*. However, the present chapter would be incomplete were we not to interpret some of the earlier material in terms of complex numbers. Readers who have already met complex numbers will probably still find much

4.4. Complex numbers

in the next two sections that is new. Those for whom complex numbers are as yet unfamiliar should muddle through as best they can, and may then be motivated to learn more in due course.

We already know that the square x^2 of any real number x is ≥ 0.

- If $a = 0$, then the equation $x^2 = a$ has exactly one root, namely $x = 0$;
- if $a > 0$, then the equation $x^2 = a$ has exactly two roots – namely $\pm\sqrt{a}$, where $\sqrt{a}$ denotes the root that is *positive*;
- if $a < 0$, then the equation $x^2 = a$ has no real roots.

And that is where the matter would have rested.

From a modern perspective, we can see that *complex numbers* are implicit in the formula for the roots of a quadratic equation: complex numbers become explicit as soon as the coefficients of a quadratic $ax^2 + bx + c$ give rise to a negative *discriminant* $b^2 - 4ac < 0$.

But this may not have been quite how complex numbers were discovered. Contrary to oft-repeated myths, complex numbers may not have forced themselves on our attention by someone asking about "solutions to the *quadratic* equation $x^2 = -1$". As long as we inhabit the domain of *real* numbers, we can be sure that no known number x could possibly have such a square, so we are unlikely to go in search of it.

New ideas in the history of mathematics tend to emerge when a fresh analysis of *familiar* entities forces us to consider the possible existence of some previously unsuspected universe. In the time from the ancient world up to the fifteenth century, the idea of "number", and of calculation, was restricted to the world of *real* (usually positive) numbers. In such a world, quadratic equations with non-real solutions simply could not arise.

However, in the Brave New World of the Renaissance, where novelty, exploration, and discovery were part of the *Zeitgeist*, a general method for solving *cubic* equations was part of the as-yet-undiscovered "wild west" of mathematics, part of the mathematical New World which invited exploration. Notice that this was not a wildly speculative venture (like trying to solve the meaningless equation "$x^2 = -1$"), since a cubic polynomial **always** has at least one real root. After three thousand years in which little progress had been made, the first half of the sixteenth century witnessed an astonishing burst of progress, resulting in the solution not only of cubic equations, but also of quartic equations. We postpone the details until Section 4.5. All we note here is that,

> the general method for solving cubic equations published in 1545, was given as a procedure, illustrated by examples, that showed how to find *genuinely* **real** *solutions to equations of the third degree having genuinely real (and positive) coefficients.*

The procedure clearly worked. And it proceeded as follows:

> Construct the real solution x as the sum $x = u + v$ of two intermediate answers u and v – where the two summands u and v sometimes turned out to be what we would call "conjugate complex numbers", *whose imaginary parts cancelled out*, leaving a real result for the required root x.

Those who devised the procedure had no desire to leave the *real* domain: they were focused on a problem in the *real* domain (a cubic equation with *real* coefficients, having a *real* root), and devised a general procedure to find that genuinely *real* root. But the procedure they discovered led the solver on a journey that sometimes "passed into the complex domain", before returning to the real domain! (See Problem **129**.)

Working with complex numbers depends on two skills – one very familiar, and one less so.

- The familiar skill is a willingness to work with a "number" *in terms of its properties only*, without wishing to evaluate it.

We are thoroughly familiar with this when we work with $\frac{2}{3}$ and other fractions: we know that $\frac{2}{3} = 2 \times \frac{1}{3}$; and all we know about $\frac{1}{3}$ is that "whenever we have 3 copies of $\frac{1}{3}$, we can simplify this to 1". Much the same happens when we first learn to work with $\sqrt{2}$, where we carry out such calculations as $(1 + \sqrt{2})^2 = 3 + 2\sqrt{2}$, based only on collecting up like terms and the fact that $\sqrt{2} \times \sqrt{2}$ can always be replaced by 2.

- The less familiar skill is easily overlooked. When, for whatever reason, we decide to allow solutions to the equation $x^2 = -1$, three things need to be understood.

    - First, these new solutions come **in pairs**: if i *is one solution of* $x^2 = -1$ then $-i$ is another (because $(-1) \times (-1) = 1$ means that $(-x)^2 = x^2$ for all "numbers" x.

    - Second, the equation $x^2 = -1$ has **exactly two solutions** – one the negative of the other (if x and y are both solutions, then $x^2 = y^2$, so $x^2 - y^2 = (x-y)(x+y) = 0$, so either $x = y$, or $x = -y$).

    - Third, *we have no way of telling these two solutions apart*: we know that each is the negative of the other, but there is no way of singling out one of them as "the main one" (as we could when defining the square root of a positive real such as 2). We can call them $\pm i$, but they are *each as good as the other*. This important fact is often undermined by referring to one of these roots as $\sqrt{-1}$ (as if it were the dominant partner), and to the other as $-\sqrt{-1}$ (as if it were somehow just the "negative" of the main root).

The truth is that "$\sqrt{-1}$" is a serious abuse of notation, because there is no way to extend the definition of the *function* "$\sqrt{}$" in the way that this implies: when we try to "take square roots" of negative (or complex) numbers, the output is inescapably "two-valued", so "$\sqrt{}$" is no longer a function. The two roots of $x^2 = -1$ are like Tweedledum and Tweedledee: we know there are two of them, and we know how they are related; but we have no way of distinguishing them, or of singling one of them out.

Once we accept this, we can write complex numbers in the form $a + bi$, where a and b are real numbers (just as we used to write numbers in the form $a + b\sqrt{2}$, where a and b are rational numbers). And we can proceed to add, subtract, multiply, and divide such expressions, and then collect up the "real" and "imaginary" parts to tidy up the answer.

Problem 128

(a) Write the inverse $(a + bi)^{-1}$ in the form $c + di$.

(b) Write down a quadratic equation with real coefficients, which has $a + bi$ as one root (where a and b are real numbers). △

Problem 129 Divide 10 into two parts, whose product is 40. △

Problem **129** appears in Chapter XXXVII of Girolamo Cardano's (1501–1576) book *Ars Magna* (1545). Having previously presented the general methods for solving quadratic, cubic, and quartic equations, he honestly confronts the phenomenon that his method for solving cubic equations (see Problem **135**) produces the required real (and positive) solution x as a sum of *complex* conjugates u and v – involving not only *negative* numbers, but **square roots** *of negative numbers*. After presenting the formal solution of Problem **129**, and having shown that the calculation works exactly as it should, he adds the bemused remark:

> "*So progresses arithmetic subtlety,
> the end of which ... is as refined as it is useless.*"

Arithmetic with complex numbers in the form $a + bi$ is done by carrying out the required operations, and then collecting up the "real" and "imaginary" parts as separate components – just as with adding vectors (a, b). We treat the two parts as Cartesian coordinates, and so identify the complex number $a + bi$ with the point (a, b) in the complex plane.

The "Cartesian" representation $a + bi$ is very convenient for *addition*. But the essential definition (and significance) of complex numbers is rooted in *multiplication*. And for multiplication it is often much better to work with complex numbers written in *polar form*. Suppose we mark the complex number $w = a + bi$ in the complex plane.

The *modulus* $|w|$ of w (often denoted by r) is the distance $r = \sqrt{a^2 + b^2}$ of the complex number $a + bi$ from the origin in the complex plane.

The angle θ, measured anticlockwise from the positive real axis to the line joining the complex number w to the origin, is called the *argument*, $\text{Arg}(w) = \theta$, of w.

It is then easy to check that $a = r\cos\theta$, $b = r\sin\theta$, and that

$$w = r(\cos\theta + i\sin\theta).$$

This is the *polar form* for w. Instead of focusing on the Cartesian coordinates a, b, the polar form pinpoints w in terms of

- its *length*, or *modulus*, r (which specifies the circle, with centre at the origin, on which the complex number w lies), and

- the *argument* θ (which tells us where on this circle w is to be found).

Problem 130

(a) Given two complex numbers in polar form:

$$w = r(\cos\theta + i\sin\theta), \quad z = s(\cos\phi + i\sin\phi),$$

show that their product is precisely

$$wz = rs(\cos(\theta + \phi) + i\sin(\theta + \phi)).$$

(b) (**de Moivre's Theorem**: Abraham de Moivre (1667–1754)) Prove that

$$(\cos\theta + i\sin\theta)^n = \cos(n\theta) + i\sin(n\theta).$$

(c) Prove that, if

$$z = r(\cos\theta + i\sin\theta)$$

satisfies $z^n = 1$ for some integer n, then $r = 1$. △

The last three problems in this subsection look more closely at "roots of unity" – that is, roots of the polynomial equation $x^n = 1$. In the *real* domain, we know that:

(i) when n is odd, the equation $x^n = 1$ has exactly one root, namely $x = 1$; and

(ii) when n is even, the equation $x^n = 1$ has just two solutions, namely $x = \pm 1$.

In contrast, in the *complex* domain, there are n "n^{th} roots of unity". Problem **130**(c) shows that these "roots of unity" all lie on the unit circle, centered at the origin. And if we put $n\theta = 2k\pi$ in Problem **130**(b) we see that the n n^{th} roots of unity include the point "$1 = \cos 0 + i \sin 0$", and are then equally spaced around that circle with $\theta = \frac{2k\pi}{n}$ ($1 \leqslant k \leqslant n-1$), and form the vertices of a regular n-gon.

Problem 131

(a) Find all the complex roots of unity of degree 3 (that is, the roots of $x^3 = 1$) in surds form.

(b) Find all the complex roots of unity of degree 4 in surd form.

(c) Find all the complex roots of unity of degree 6 in surd form.

(d) Find all the complex roots of unity of degree 8 in surd form. △

Problem 132 Use Problem **131**(d) to factorise $x^4 + 1$ as a product of four linear factors, and hence as a product of two quadratic polynomials with real coefficients. △

Problem 133

(a) Find all the complex roots of unity of degree 5 in surd form.

(b) Factorise $x^5 - 1$ as a product of one linear and two quadratic polynomials with real coefficients. △

4.5. Cubic equations

The first recorded procedure for finding the positive roots of any given quadratic equation dates from around 1700 BC (ancient Babylonian). A corresponding procedure for cubic equations had to wait until the early sixteenth century AD. The story is a slightly complicated one – involving public contests, betrayal, and much else besides.

In Section 4.4 we saw that the cubic equation $x^3 = 1$ has three solutions – two of which are complex numbers. But in the sixteenth century, even negative numbers were viewed with suspicion, and complex numbers were still unknown. Moreover, symbolical algebra had not yet been invented, so everything was carried out in words: constants were "numbers"; a given multiple of the unknown was referred to as so many "things"; a given multiple of the square of the unknown was simply referred to as "squares"; and so on.

In short, we know that an improved method for sometimes finding a (positive) unknown which satisfied a cubic equation was devised by Scipione del Ferro (1465–1526) around 1515. He kept his method secret until just before his death, when he told his student Antonio del Fiore (1506–??). Niccolò Tartaglia (1499–1557) then made some independent progress in solving cubic equations. At some stage (around 1535) Fiore challenged Tartaglia to a public "cubic solving contest". In preparing for this event, Tartaglia managed to improve on his method, and he seems to have triumphed in the contest. Tartaglia naturally hesitated to divulge his method in order to preserve his superiority, but was later persuaded to communicate what he knew to Girolamo Cardano (1501–1576) after Cardano promised not to publish it (either never, or not before Tartaglia himself had done so). Cardano improved the method, and his student Ferrari (1522–1565) extended the idea to give a method for solving quartic equations – all of which Cardano then published, contrary to his promise, but with full attribution to the rightful discoverers, in his groundbreaking book *Ars Magna* (1545 – just two years after Copernicus (1473–1543) published his *De revolutionibus* ...). Problem **134** illustrates the necessary first move in solving any cubic equation. Problem **135** then illustrates the general method in a relatively simple case.

Problem 134

(a) Given the equation $x^3 + 3x^2 - 4 = 0$, choose a constant a, and then change variable by substituting $y = x + a$ to produce an equation of the form $y^3 + ky = $ constant.

(b) In general, given any cubic equation $ax^3 + bx^2 + c^x + d = 0$ with $a \neq 0$, show how to change variable so as to reduce this to a cubic equation with no quadratic term. △

Problem 135 The equation $x^3 + 3x^2 - 4 = 0$ clearly has "$x = 1$" as a positive solution. (The other two solutions are $x = -2$, and $x = -2$ - a repeated root; however negatives were viewed with suspicion in the sixteenth century, so this root might well have been ignored.) Try to understand how the following sequence of moves "finds the root $x = 1$":

(i) substitute $y = x + 1$ to get a cubic equation in y with no term in y^2;

(ii) imagine $y = u + v$ and interpret the identity for

$$(u+v)^3 = u^3 + 3uv(u+v) + v^3$$

as your cubic equation in y;

(iii) solve the simultaneous equations "$3uv = 3$", "$u^3 + v^3 = 2$" (not by guessing, but by substituting $v = \frac{1}{u}$ from the first equation into the second to get a quadratic equation in "u^3", which you can then solve for u^3 before taking cube roots);

(iv) then find the corresponding value of v, hence the value of $y = u + v$, and hence the value of x. △

The simple method underlying Problem **135** is in fact completely general. Given any cubic equation

$$ax^3 + bx^2 + cx + d = 0 \quad \text{(with } a \neq 0\text{)}$$

we can divide through by a to reduce this to

$$x^3 + px^2 + qx + r = 0$$

with leading coefficient $= 1$. Then we can substitute $y = x + \frac{p}{3}$ and reduce this to a cubic equation in y

$$y^3 - 3\left(\frac{p}{3}\right)^2 y + qy + \left[r + 2\left(\frac{p}{3}\right)^3 - q\left(\frac{p}{3}\right)\right] = 0$$

which we can treat as having the form

$$y^3 - my - n = 0.$$

So we can set $y = u + v$ (for some unknown u and v yet to be chosen), and treat the last equation as an instance of the identity

$$(u + v)^3 - 3uv(u + v) - (u^3 + v^3) = 0$$

which it will become if we simply choose u and v to solve the simultaneous equations

$$3uv = m, \quad u^3 + v^3 = n.$$

We can then solve these equations to find u, then v – and hence find $y = u + v$ and $x = y - \frac{p}{3}$.

4.6. An extra

Back in Chapter 1, Problem **6** we introduced the Euclidean algorithm for integers. The same idea was extended to polynomials with integer coefficients in Problem **126**. In both these settings one starts with a domain (whether the set of integers, or the set of all polynomials with integer coefficients)

134   Algebra

where there is a notion of divisibility: given two elements m, n in the relevant domain, we say

"n **divides** m" if there exists an element q in the domain such that $m = qn$.

The next problem invites you to think how one might extend the Euclidean algorithm to a new domain, namely the *Gaussian integers* $\mathbb{Z}[i]$ – the set of all complex numbers $a + bi$ in which the real and imaginary "coordinates" a and b are integers.

Problem 136 Complex numbers $a + bi$, where both a and b are integers, are called Gaussian integers. Try to formulate a version of the "division algorithm" for "division with remainder" (where the remainder is always "less than" the divisor in some sense) for pairs of Gaussian integers. Extend this to construct a version of the Euclidean algorithm to find the HCF of two given Gaussian integers. △

> It is a profoundly erroneous truism ...
> that we should cultivate the habit of thinking what we are doing.
> The precise opposite is the case.
> Civilisation advances by extending the number of important
> operations which we can perform without thinking about them.
>
> Alfred North Whitehead (1861–1947)

4.7.   Chapter 4: Comments and solutions

92. Answer: Humour aside, this is a common situation.

We know $d + b$, $n + b$, $d + n$ rather than the values of d, b, n.

The key is to exploit the symmetry in the given data, rather than solving blindly. Adding all three two-way totals gives $2(d + b + n) = 284$, whence $d + b + n = 142$. We can then subtract the given value of $d + n = 137$ to get the value of $b = 5$.

93.

(a) Let the numbers at the three vertices be A, B, C. Adding shows that

$$a + b + c = 2(A + B + C)$$

so

$$A = \frac{a+b+c}{2} - (B+C) = \frac{b+c-a}{2}$$

etc.

(c) **Note:** We postpone the "solution" of part (b), and address part (c) first. Let the numbers at the five vertices be A, B, C, D, E. Adding shows that

$$d + e + a + b + c = 2(A + B + C + D + E)$$

so

$$\begin{aligned} A &= \frac{d+e+a+b+c}{2} - (B+C) - (D+E) \\ &= \frac{d-e+a-b+c}{2} \end{aligned}$$

etc.

(b) The second part is different. The four given edge-values do not determine the four unknown vertex-values. It may look as though four pieces of information should suffice to find four unknowns; but there is a catch: the sum of the numbers on the two opposite edges AB and CD is just the sum of the numbers at the four vertices, and so is equal to the sum of the numbers on the edges BC and DA. Hence one of the given edge-values is determined by the other three.

Note: This distinction between polygons with an odd and an even number of vertices would arise in exactly the same way if each edge was labelled with the average ("half the sum") of the numbers at its two end vertices.

94. $a = BC = BP + PC = y + z$; $b = x + z$; $c = x + y$. Hence

$$a + b + c = 2(x + y + z)$$

so

$$x + y + z = \frac{a+b+c}{2}.$$

So

$$\begin{aligned} x &= \frac{a+b+c}{2} - (y+z) \\ &= \frac{b+c-a}{2} \end{aligned}$$

etc.

95.

(a) Let

$$M = \left(\frac{a+c}{2}, \frac{b+d}{2}\right).$$

The shift, or vector, from (a, b) to (c, d) goes

"along $c - a$ in the x-direction" and "up $d - b$ in the y-direction".

Draw the ordinate through Y and the abscissa through Z, to meet at P, so creating a right angled triangle with legs YP of length $|c - a|$ and PZ of length

$|d-b|$. The midpoint of YP clearly lies halfway along YP at
$$S = \left(a + \frac{c-a}{2}, b\right)$$
and the midpoint of PZ clearly lies halfway up PZ at
$$T = \left(c, d - \frac{d-b}{2}\right).$$
Then $\triangle YSM$ and $\triangle MTZ$ are both right-angled triangles and are congruent (by RHS congruence). Hence $YM = MZ$, so M is the midpoint of YZ.

(b)
$$M = \left(\frac{a}{2}, \frac{b}{2}\right), \quad N = \left(\frac{c}{2}, \frac{d}{2}\right)$$
so vector
$$\mathbf{MN} = \left(\frac{c-a}{2}, \frac{d-b}{2}\right) = \frac{1}{2}\mathbf{BC}.$$

(c) **Note:** We use the result from part (b), but not the method from part (b).
By part (b) applied to $\triangle BAC$, PQ is half the length of AC and parallel to AC.
By part (b) applied to $\triangle DAC$, SR is half the length of AC and parallel to AC.
Hence PQ is parallel to SR.
Similarly one can prove (applying part (b) twice – first to $\triangle ABD$, and then to $\triangle CBD$) that PS is parallel to QR.
Hence $PQRS$ is a parallelogram.

96.

(a) $\mathbf{p} = \frac{1}{2}(\mathbf{x}+\mathbf{y})$, $\mathbf{q} = \frac{1}{2}(\mathbf{y}+\mathbf{z})$, $\mathbf{r} = \frac{1}{2}(\mathbf{z}+\mathbf{x})$, so
$$\mathbf{p} + \mathbf{q} + \mathbf{r} = \mathbf{x} + \mathbf{y} + \mathbf{z}.$$
Hence
$$\begin{aligned} \mathbf{x} &= (\mathbf{p}+\mathbf{q}+\mathbf{r}) - (\mathbf{y}+\mathbf{z}) = \mathbf{p} - \mathbf{q} + \mathbf{r} \\ \mathbf{y} &= (\mathbf{p}+\mathbf{q}+\mathbf{r}) - (\mathbf{x}+\mathbf{z}) = \mathbf{p} + \mathbf{q} - \mathbf{r} \\ \mathbf{z} &= (\mathbf{p}+\mathbf{q}+\mathbf{r}) - (\mathbf{x}+\mathbf{y}) = \mathbf{q} + \mathbf{r} - \mathbf{p}. \end{aligned}$$

(b)
$$\mathbf{p} = \frac{1}{2}(\mathbf{w}+\mathbf{x}), \ \mathbf{q} = \frac{1}{2}(\mathbf{x}+\mathbf{y}), \ \mathbf{r} = \frac{1}{2}(\mathbf{y}+\mathbf{z}), \ \mathbf{s} = \frac{1}{2}(\mathbf{z}+\mathbf{w})$$
so
$$\mathbf{p} + \mathbf{q} + \mathbf{r} + \mathbf{s} = \mathbf{w} + \mathbf{x} + \mathbf{y} + \mathbf{z}.$$
Hence
$$\mathbf{w} = (\mathbf{p}+\mathbf{q}+\mathbf{r}+\mathbf{s}) - (\mathbf{x}+\mathbf{y}+\mathbf{z});$$
but there is no obvious way to pin down $(\mathbf{x}+\mathbf{y}+\mathbf{z})$.

In fact different quadrilaterals may give rise to the same four "midpoints". (It is an interesting exercise to identify the family of quadrilaterals corresponding to a given set of four midpoints.)

(c) As in parts (a) and (b),

$$\mathbf{p} = \frac{1}{2}(\mathbf{v}+\mathbf{w}), \mathbf{q} = \frac{1}{2}(\mathbf{w}+\mathbf{x}), \mathbf{r} = \frac{1}{2}(\mathbf{x}+\mathbf{y}), \mathbf{s} = \frac{1}{2}(\mathbf{y}+\mathbf{z}), \mathbf{t} = \frac{1}{2}(\mathbf{z}+\mathbf{v}).$$

Hence

$$\mathbf{p}+\mathbf{q}+\mathbf{r}+\mathbf{s}+\mathbf{t} = \mathbf{v}+\mathbf{w}+\mathbf{x}+\mathbf{y}+\mathbf{z}$$

so

$$\begin{aligned}
\mathbf{v} &= (\mathbf{p}+\mathbf{q}+\mathbf{r}+\mathbf{s}+\mathbf{t}) - (\mathbf{w}+\mathbf{x}) - (\mathbf{y}+\mathbf{z}) = \mathbf{p}-\mathbf{q}+\mathbf{r}-\mathbf{s}+\mathbf{t} \\
\mathbf{w} &= (\mathbf{p}+\mathbf{q}+\mathbf{r}+\mathbf{s}+\mathbf{t}) - (\mathbf{x}+\mathbf{y}) - (\mathbf{z}+\mathbf{v}) = \mathbf{p}+\mathbf{q}-\mathbf{r}+\mathbf{s}-\mathbf{t} \\
\mathbf{x} &= (\mathbf{p}+\mathbf{q}+\mathbf{r}+\mathbf{s}+\mathbf{t}) - (\mathbf{v}+\mathbf{w}) - (\mathbf{y}+\mathbf{z}) = -\mathbf{p}+\mathbf{q}+\mathbf{r}-\mathbf{s}+\mathbf{t} \\
\mathbf{y} &= (\mathbf{p}+\mathbf{q}+\mathbf{r}+\mathbf{s}+\mathbf{t}) - (\mathbf{w}+\mathbf{x}) - (\mathbf{z}+\mathbf{v}) = \mathbf{p}-\mathbf{q}+\mathbf{r}+\mathbf{s}-\mathbf{t} \\
\mathbf{z} &= (\mathbf{p}+\mathbf{q}+\mathbf{r}+\mathbf{s}+\mathbf{t}) - (\mathbf{v}+\mathbf{w}) - (\mathbf{x}+\mathbf{y}) = -\mathbf{p}+\mathbf{q}-\mathbf{r}+\mathbf{s}+\mathbf{t}.
\end{aligned}$$

97.

(a)(i) As in Problems **93-95** we instinctively add to get

$$2(x+y+z) = 6$$

so

$$x+y+z = 3.$$

Hence

$$\begin{aligned}
x &= 3-(y+z) = 1 \\
y &= 3-(x+z) = 0 \\
z &= 3-(x+y) = 2.
\end{aligned}$$

(ii) The same idea (replacing addition by multiplication) leads to

$$2 \times 4 \times 8 = 64 = uv \cdot vw \cdot wu = (uvw)^2$$

so $uvw = \pm 8$. Hence

$$\begin{aligned}
u &= \frac{uvw}{vw} = \frac{\pm 8}{4} = \pm 2 \\
v &= \frac{uvw}{uw} = \frac{\pm 8}{8} = \pm 1 \\
w &= \frac{uvw}{uv} = \frac{\pm 8}{2} = \pm 4.
\end{aligned}$$

$\therefore (u,v,w) = (2,1,4)$ or $(-2,-1,-4)$.

138   *Algebra*

Note: Alternatively, we may notice that u, v, w are either all positive, or all negative. If we restrict in the first instance to purely positive solutions, then we may set $u = 2^x$, $v = 2^y$, $w = 2^z$, translate (ii) into (i), and conclude that $(x, y, z) = (1, 0, 2)$, so that $(u, v, w) = (2, 1, 4)$. We must then remember the negative solution $(u, v, w) = (-2, -1, -4)$.

(b)(i) As in (a)(i) we add to get $2(x + y + z) = 9$, so $x + y + z = \frac{9}{2}$. Hence
$$x = \frac{9}{2} - (y + z) = \frac{3}{2}$$
$$y = \frac{9}{2} - (x + z) = \frac{1}{2}$$
$$z = \frac{9}{2} - (x + y) = \frac{5}{2}.$$

(ii) The same idea leads to
$$6 \times 10 \times 15 = 900 = uv \cdot vw \cdot wu = (uvw)^2,$$
so $uvw = \pm 30$.
Hence
$$u = \frac{uvw}{vw} = \frac{\pm 30}{10} = \pm 3$$
$$v = \frac{uvw}{uw} = \frac{\pm 30}{15} = \pm 2$$
$$w = \frac{uvw}{uv} = \frac{\pm 30}{6} = \pm 5.$$

Either u, v, w are all positive, or all negative.
$\therefore (u, v, w) = (3, 2, 5)$ or $(-3, -2, -5)$.

(iii) The same idea leads to
$$6 \times 10 \times 30 = 2 \times 900 = vw \cdot wu = (uvw)^2,$$
so $uvw = \pm 30\sqrt{2}$. Hence
$$u = \frac{uvw}{vw} = \frac{\pm 30\sqrt{2}}{10} = \pm 3\sqrt{2}$$
$$v = \frac{uvw}{uw} = \frac{\pm 30\sqrt{2}}{15} = \pm 2\sqrt{2}$$
$$w = \frac{uvw}{uv} = \frac{\pm 30\sqrt{2}}{6} = \pm 5\sqrt{2}.$$

Either u, v, w are all positive, or all negative.
$\therefore (u, v, w) = (3\sqrt{2}, 2\sqrt{2}, 5\sqrt{2})$ or $(-3\sqrt{2}, -2\sqrt{2}, -5\sqrt{2})$.

(iv) We could of course repeat the same method.

Or we could again look in the first instance for positive solutions, notice that $4 = 2^2$, $8 = 2^3$, $16 = 2^4$, and take logs (to base 2). Then

$$\log_2 u + \log_2 v = 2$$
$$\log_2 v + \log_2 w = 3$$
$$\log_2 u + \log_2 w = 4,$$

so (from part (i)) any positive solution satisfies

$$\log_2 u = \frac{3}{2}, \ \log_2 v = \frac{1}{2}, \ \log_2 w = \frac{5}{2},$$

so

$$(u, v, w) = (2\sqrt{2}, \sqrt{2}, 4\sqrt{2}).$$

We must then remember to include

$$(u, v, w) = (-2\sqrt{2}, -\sqrt{2}, -4\sqrt{2}).$$

98. The simplest idea is to take logs, and reduce the system to a familiar linear system:

$$a \cdot \log u + b \cdot \log v = \log m$$
$$c \cdot \log u + d \cdot \log v = \log n.$$

Multiplying the first equation by c and subtracting it from the second equation multiplied by a gives:

$$\log v = \frac{a \cdot \log n - c \cdot \log m}{ad - bc}.$$

Multiplying the first equation by d and subtracting b times the second equation gives:

$$\log u = \frac{d \cdot \log m - b \cdot \log n}{ad - bc}.$$

If the numerators and denominators are expressed in determinant form, we get the 2×2 version of *Cramer's Rule*. The original unknowns u, v can then be obtained by taking suitable powers.

What emerges looks interesting:

$$u = m^{\frac{d}{ad-bc}} \cdot n^{-\frac{b}{ad-bc}}$$
$$v = m^{-\frac{c}{ad-bc}} \cdot n^{\frac{a}{ad-bc}}$$

but it is not clear how it generalises.

99.

(a) $x + y + z = 3$ is the equation of a plane through the three points $(3, 0, 0)$, $(0, 3, 0)$, $(0, 0, 3)$.

$x^2 + y^2 + z^2 = c$ is the equation of a sphere, centered at the origin, with radius $\sqrt{c}$. The sphere misses the plane completely when $c < 3$, meets the plane in a single point when $c = 3$, and cuts the plane in a circle C when $c > 3$ (the circle lying in the positive octant provided $c < 9$).

If $xy + yz + zx = b$ meets this intersection at all, then any permutation of the three coordinates x, y, z produces another point which also satisfies the other two equations (since they are both symmetrical). Hence for the system to have a unique solution, the circle C must contain a point with $x = y = z$. Hence $c = 3$, and $b = 3$, and the unique solution is

$$(x, y, z) = (1, 1, 1).$$

(b) We must have $c \geqslant 0$ for any solution. If $c = 0$, then for a unique solution, we must have $x = y = z = 0$, so $a = b = 0$. If we exclude this case, then we may assume that $c > 0$.

$$x + y + z = a$$

is the equation of a plane through the three points $(a, 0, 0)$, $(0, a, 0)$, $(0, 0, a)$.

$$x^2 + y^2 + z^2 = c$$

is the equation of a sphere, centre the origin, with radius $\sqrt{c}$, which misses the plane completely when $c < \frac{a^2}{3}$, meets the plane in a single point when $c = \frac{a^2}{3}$, and cuts the plane in a circle C when $c > \frac{a^2}{3}$ (the circle lying in the positive octant provided $c < a^2$).
If

$$xy + yz + zx = b$$

meets this intersection at all, then any permutation of the three coordinates x, y, z produces another point which also satisfies the other two equations (since they are both symmetrical). Hence for the system to have a unique solution, the circle C must contain a point with $x = y = z$. Hence that point is $x = y = z = \frac{a}{3}$, so $c = \frac{a^2}{3} = b$, and the unique solution is

$$(x, y, z) = \left(\frac{a}{3}, \frac{a}{3}, \frac{a}{3}\right).$$

100. $|-x|$ is never negative. If $x \geqslant 0$, then $|-x| = x$; if x is negative, then $-x$ is positive, so $|-x| = -x$.

Note: We need to learn to see both x and $-x$ as *algebraic* entities, with x as a placeholder (which may well be negative, in which case $-x$ would be positive).

101.

(a) The interval $[0.1, 0.2]$. We have marked exactly $\frac{1}{10}$ of the interval $[0, 1)$.

(b) This needs a little thought. First we mark the interval $[0.1, 0.2)$, of length $\frac{1}{10}$. Then we mark 9 smaller intervals

$$[0.01, 0.02), [0.21, 0.22), \ldots, [0.91, 0.92)$$

of total length $9 \cdot \left(\frac{1}{10}\right)^2$. Then 9^2 smaller intervals

$$[0.001, 0.002), [0.021, 0.022), \ldots, [0.991, 0.992)$$

of total length $9^2 \cdot \left(\frac{1}{10}\right)^3$. And so on.

$[0.1, 0.2)$
$\cup \quad [0.01, 0.02) \cup [0.21, 0.22) \cup [0.31, 0.32) \cup [0.41, 0.42)$
$\cup [0.51, 0.52) \cup [0.61, 0.62) \cup [0.71, 0.72)$
$\cup [0.81, 0.82) \cup [0.91, 0.92)$
$\cup \quad [0.001, 0.002) \cup [0.021, 0.022) \cup [0.031, 0.032) \cup \cdots$
$\cup \quad \cdots$

It would seem that a vast number of points are left *unmarked* – namely, every point whose decimal representation uses only 0s, 2s, 3s, 4s, 5s, 6s, 7s, 8s, and 9s. However, **the total length** of the *marked* intervals is given by adding:

$$\frac{1}{10} + 9 \cdot \left(\frac{1}{10}\right)^2 + 9^2 \cdot \left(\frac{1}{10}\right)^3 + 9^3 \cdot \left(\frac{1}{10}\right)^4 + 9^4 \cdot \left(\frac{1}{10}\right)^5 + 9^5 \cdot \left(\frac{1}{10}\right)^6 + \cdots$$

That is an infinite geometric series with first term $a = \frac{1}{10}$ and common ratio $r = \frac{9}{10}$, and hence with sum $= 1$. In other words, the **total length of what remains unmarked is zero.**

(c) $(0, 1)$: every real number except 0 has an expansion in base 2 with a "1" in some position. So this time *nothing is left unmarked* (except 0). Hence the complement of the set of marked points consists simply of one point, namely $\{0\}$. So it is not surprising that the total of all the marked intervals has length 1.

(d) First we mark the interval $[0.1, 0.2)$, of length $\frac{1}{3}$. Then we mark 2 smaller intervals

$$[0.01, 0.02), [0.21, 0.22)$$

of total length $2 \cdot \left(\frac{1}{3}\right)^2$. Then 2^2 smaller intervals

$$[0.001, 0.002), [0.021, 0.022), [0.201, 0.202), [0.221, 0.222)$$

of total length $2^2 \cdot \left(\frac{1}{3}\right)^3$. And so on.

$[0.1, 0.2)$
$\cup \quad [0.01, 0.02) \cup [0.21, 0.22)$
$\cup \quad [0.001, 0.002) \cup [0.021, 0.022)) \cup [0.201, 0.202) \cup [0.2201, 0.02202)$
$\cup \quad \cdots$

The set of marked points is the complement of the famous *Cantor set* (Georg Cantor (1845–1918)) and has total length

$$\tfrac{1}{3} + 2 \cdot \left(\tfrac{1}{3}\right)^2 + 2^2 \cdot \left(\tfrac{1}{3}\right)^3 + 2^3 \cdot \left(\tfrac{1}{3}\right)^4 + 2^4 \cdot \left(\tfrac{1}{3}\right)^5 + 2^5 \cdot \left(\tfrac{1}{3}\right)^6 + \cdots$$

This is an infinite geometric series with first term $a = \tfrac{1}{3}$ and common ratio $r = \tfrac{2}{3}$, and so has sum $= 1$.
Hence, the **total length of what remains unmarked is zero**.

Note: The set described in (d) leaves as its complement a collection of points – the *Cantor set* – which consists of the "endpoints" of the intervals that have been removed; these are points whose base 3 expansion involves only 0s and 2s. This complement:

(i) is "the same size" as the whole interval $[0, 1]$ (since if we interpret the 2s in the base 3 expansion as 1s, we get a correspondence between the set of "endpoints" and the set of all possible *base* 2 expansions for real numbers in $[0, 1]$);

(ii) is "nowhere dense" (since every pair of points in the complement is separated by some interval)

(ii) has total length $= 0$.

102. $(2, 3]$. Each inequality implies all the ones before it. Hence the two which are true must be the first two. Hence $x \leqslant 3$, and $x > 2$.

103. If $x - 5 \geqslant 0$, then we must solve $x - 5 = 3$; so $\boldsymbol{x = 8}$; if $x - 5 < 0$, we must solve $x - 5 = -3$, so $\boldsymbol{x = 2}$.

Note: The fact that $|x|$ denotes the positive value of the pair $\{x, -x\}$ can be rephrased as: $|x|$ is equal to the **distance** from x to 0.
In the same way, $|x - 5|$ denotes the positive member of the pair

$$\{x - 5, -(x - 5)\}$$

so $|x - 5|$ is equal to the distance from $x - 5$ to 0 (i.e. the distance from x to 5). This is a very important way to think about expressions like $|x - 5|$.

104.

(a) $[-3, -2) \cup (2, 3]$. (Each inequality implies all those that go before it. So we need solutions to $|x| > 1$ **and** $|x| > 2$, which satisfy $|x| \leqslant 3$.)

(b) $(4, 6]$. (Each inequality implies the one before it. To see this, think in terms of distances: we want points x whose distance from 1 is > 1, whose distance from 2 is > 2, etc.. So we need to find points x which solve the first two inequalities, but not the third. Points in the half line $(-\infty, 0)$ satisfy all seven inequalities, so we are left with $(4, 6]$.)

4.7. Chapter 4: Comments and solutions   143

105. $\{-\frac{5}{2}, -\frac{1}{2}\}$. (We need all points x for which

"the distance from x to -1" plus "the distance from x to -2"

equals 2. This excludes all points between -2 and -1, for which the sum is equal to 1; for points between $-\frac{5}{2}$ and $-\frac{1}{2}$ the sum is < 2; for points in $(-\infty, -\frac{5}{2})$ or $(-\frac{1}{2}, \infty)$ the sum is > 2.)

106. $a = \frac{1}{2}, b = \frac{3}{2}$. (For solutions to exist, we must have $b > 0$. The solutions of the given inequality then consist of all x such that

"the distance from x to a is less than b"

that is, all x in the interval $(a - b, a + b)$. Hence $a - b = -1, a + b = 2$.)

107.

(a) "The difference between the x- and y-coordinates is < 3", means that the point (x, y) lies in the infinite strip between the lines $x - y = -3$ and $x - y = 3$.

(b) Shifting the origin of coordinates to $(-5, 0)$ changes the x-coordinate to "$X = x + 5$" and leaves the y-coordinate unaffected (so $Y = y$). In this new frame we want "$|X - Y| < 3$", so the required points lie in the infinite strip between the lines $X - Y = -3$ and $X - Y = 3$; that is, between the lines $x - y + 5 = -3$ and $x - y + 5 = 3$.

(c) $x > 0$ and $y > 0$, or $x < 0$ and $y < 0$. (For any solution at all, we must have $|x + y| > 0$, which excludes points on the line $x + y = 0$. Divide both sides by $|x + y|$ and simplify to get

$$\left|1 - \frac{2y}{x + y}\right| < 1.$$

In other words:

$$0 < \frac{2y}{x + y} < 2.$$

If $y > 0$, then $x + y > 0$, so $2x + 2y > 2y$, whence $x > 0$ (so "$x > 0$ and $y > 0$").
If $y < 0$, then $x + y < 0$, so $2x + 2y < 2y$, whence $x < 0$ (so "$x < 0$ and $y < 0$").
If $x > 0$ and $y > 0$, or $x < 0$ and $y < 0$, then clearly $|x - y| < |x + y|$.)

108. Let

$$x = \frac{a}{b} < \frac{c}{d} = y.$$

(i) Since $x < y$, it follows that

$$x - \frac{x}{2} = \frac{x}{2} < \frac{y}{2},$$

so $x < \frac{x+y}{2}$; moreover $\frac{x}{2} < y - \frac{y}{2}$, so $\frac{x+y}{2} < y$.

(ii) Since $\frac{a}{b} < \frac{c}{d}$ and $b, d > 0$, we can multiply both sides by bd to get $ad < bc$. Therefore
$$a(b+d) = ab + ad < ba + bc = b(a+c),$$
and
$$(a+c)d = ad + cd < bc + dc = (b+d)c.$$
$\therefore \frac{a}{b} < \frac{a+c}{b+d}$, and $\frac{a+c}{b+d} < \frac{c}{d}$ (since b, d, and $b+d$ are all > 0, so we can divide the first inequality by $b(b+d)$ and the second by $d(b+d)$).

109.

(a)
$$\frac{0}{1} < \frac{1}{7} < \frac{1}{6} < \frac{1}{5} < \frac{1}{4} < \frac{2}{7} < \frac{1}{3} < \frac{2}{5} < \frac{3}{7} < \frac{1}{2} < \frac{4}{7} < \frac{3}{5} < \frac{2}{3} < \frac{5}{7} < \frac{3}{4} < \frac{4}{5} < \frac{5}{6} < \frac{6}{7} < \frac{1}{1}$$

(b)(i) It is tempting simply to consider the decimals
$$\frac{1}{9} = 0.111\ldots, \quad \frac{2}{9} = 0.222\ldots, \quad \frac{3}{9} = 0.333\ldots, \quad \ldots, \quad \frac{8}{9} = 0.888\ldots$$

in order to conclude that these fractions miss the first and last subinterval, and then fall one in each of the remaining subintervals. In preparation for part (ii), it is better to observe that

* $\frac{1}{10} < \frac{1}{9}$ and $\frac{8}{9} < \frac{9}{10}$, so none of the 9^{th}s land up in the first or last subintervals;
* then rewrite

$$\frac{1}{9} = \frac{1}{10} + \frac{1}{90}, \quad \frac{2}{9} = \frac{2}{10} + \frac{2}{90}, \quad \frac{3}{9} = \frac{3}{10} + \frac{3}{90}, \quad \ldots, \quad \frac{8}{9} = \frac{8}{10} + \frac{8}{90}$$

and notice that
$$\frac{0}{10} < \frac{1}{90} < \cdots < \frac{8}{90} < \frac{1}{10},$$
so that, for $1 \leqslant m \leqslant 9$,
$$\frac{m}{10} < \frac{m}{9} < \frac{m+1}{10};$$
hence **exactly one** 9th goes in each of the other eight subintervals.

(ii) Notice that

* $\frac{1}{n+1} < \frac{1}{n}$ and $\frac{n-1}{n} < \frac{n}{n+1}$, so none of the n^{th}s land up in the first or last subintervals;
* then rewrite

$$\frac{1}{n} = \frac{1}{n+1} + \frac{1}{n(n+1)}, \quad \frac{2}{n} = \frac{2}{n+1} + \frac{2}{n(n+1)}, \quad \ldots, \quad \frac{n-1}{n} = \frac{n-1}{n+1} + \frac{n-1}{n(n+1)}$$

and notice that
$$\frac{0}{n+1} < \frac{1}{n(n+1)} < \cdots < \frac{n-1}{n(n+1)} < \frac{1}{n+1},$$

so, for $1 \leqslant m \leqslant n$,
$$\frac{m}{n+1} < \frac{m}{n} < \frac{m+1}{n+1};$$
hence **exactly one** n^{th} goes in each of the other $n-1$ subintervals.

(iii) Suppose two (or more) fractions are inserted between $\frac{a}{b}$ and $\frac{c}{d}$. Then these two fractions would have to be successive multiples of $\frac{1}{n+1}$; but then they would have a multiple of $\frac{1}{n}$ between them (by part (ii)), and this would be a term of the Farey series of order n sitting between $\frac{a}{b}$ and $\frac{c}{d}$. Since there is no such term, at most one fraction can be inserted between $\frac{a}{b}$ and $\frac{c}{d}$.

(c) **Note 1:** This problem was included because the idea of Farey series seems so simple, and their curious properties are so intriguing. While this remains true, it turns out that Farey series also have something different, and slightly unexpected, to teach us about "the essence of mathematics". Part of us expects that simple-looking results should have short and accessible proofs – even though we know that *Fermat's Last Theorem* shows otherwise. In the case of Farey series, the relevant properties can be proved in ways that should be accessible (in principle); but the proofs are not easy. So do not be upset if, after all your efforts, you land up trying to absorb the solution given here – and the underlying idea that

> simple objects and "elementary" proofs can sometimes be more intricate than one anticipates.

Note 2: If
$$\frac{a}{b} < \frac{c}{d}$$
are consecutive terms in a Farey series, then "$bc - ad$" must be an integer > 0. The fact that this difference is always equal to 1 is easily checked in any *particular* case, but it is unclear exactly why this is *necessarily* true (rather than an accident) – or even how one would go about proving it. Every treatment of Farey series has to find its own way round this difficulty. We give the simplest proof we can (in the sense that it assumes no more than we have already used: a little about numbers and some algebra). But it is not at all "easy". We indicate a different approach in the "**Notes**" at the end of part (d).

(i) [The fact that, except for the two end intervals, we have $bd > n$ will be needed in the proof of part (c)(ii).]

We proceed by mathematical induction on n (the "order" of the Farey series) – a technique which we have already met in Chapter 2 (Problems **54–59, 76**) and which will be addressed more fully in Chapter 6.

∗ When $n = 1$, the Farey series of order n is just:
$$\frac{0}{1} < \frac{1}{1}$$
and this subinterval is both the first and the last, so the claim is "vacuously true" (because there is "nothing to check"). When $n = 2$, the Farey series of order n is:
$$\frac{0}{1} < \frac{1}{2} < \frac{1}{1},$$

and again the only subintervals are the first and the last, so again there is nothing to check.

* We now *suppose that we know the claim is true* for the Farey series of order k, for some $k > 1$, and show that it must then also be true for the Farey series of order $k + 1$. (Since we know it is true for $n = 2$, it will then be true for $n = 3$; and once we know it is true for $n = 3$, it must then be true for $n = 4$; and so on.)

To show that the claim is true for the Farey series of order $k+1$, we consider any adjacent pair of fractions

$$\frac{a}{b} < \frac{c}{d}$$

(*other than the first pair and the last pair*) in the Farey series of order $k + 1$.

Claim $bd > k + 1$.

Proof Note first that, since we are avoiding the two end subintervals, both b and d are > 1.

Suppose first that the pair $\frac{a}{b} < \frac{c}{d}$ are not adjacent in the previous Farey series of order k. Then at least one of the two fractions has been inserted in creating the Farey series of order $k+1$, and so has denominator $= k+1$. (The fractions inserted are precisely those with denominator "$k + 1$" which cannot be reduced by cancelling.) Hence the product

$$bd \geqslant 2(k + 1) > k + 1.$$

Thus we may assume that the pair $\frac{a}{b} < \frac{c}{d}$ were already adjacent in the Farey series of order k. But then by our "induction hypothesis" (namely that the desired result is already known to be true for the Farey series of order k), we know that $bd > k$. If $bd > k + 1$, then the pair $\frac{a}{b} < \frac{c}{d}$ satisfies the required condition. Hence we only have to worry about the possibility that $bd = k+1$. Suppose that $bd = k + 1$. Then the interval $\frac{a}{b} < \frac{c}{d}$ has length

$$\frac{bc - ad}{bd} = \frac{r}{k + 1}$$

for some positive integer $r = bc - ad$.

If $r > 1$, then the interval would have length $> \frac{1}{k+1}$, so $\frac{a}{b} < \frac{c}{d}$ would **not** be successive terms in the series (for we would have inserted some additional term when moving from the Farey series of order k to the Farey series of order $k + 1$).

Hence we can be sure that $r = 1$, that the subinterval $\frac{a}{b} < \frac{c}{d}$ has length exactly $\frac{1}{k+1}$. Now successive fractions with denominator $k + 1$ differ by exactly $\frac{1}{k+1}$, so some fraction with denominator $k+1$ must lie in this subinterval. Since no additional fraction is inserted between them in passing from the series of order k to the series of order $k+1$, $\frac{a}{b}$ and $\frac{c}{d}$ must both be "cancelled versions" of two successive fractions with denominator $k + 1$. But then, by part (b)(ii), there would have to be a fraction with denominator k in the interval $\frac{a}{b} < \frac{c}{d}$ – which is not the case.

Therefore the possibility $bd = k+1$ does not in fact occur.

So we can be sure that in every case, $bd > k+1$.

Hence whenever the result is true for the Farey series of order k, it must then also be true for the Farey series of order $k+1$.

It follows that the result is true for the Farey series of order n, for all $n \geqslant 1$.

QED

(ii) We proceed by induction on n.

* If $\frac{a}{b} < \frac{c}{d}$ are adjacent fractions in the Farey series of order 1, then $\frac{a}{b} = \frac{0}{1}$ and $\frac{c}{d} = \frac{1}{1}$, so $bc - ad = 1$.
* Now suppose that, for some $k \geqslant 1$, we already know that the result holds for any adjacent pair in the Farey series of order k.

Let $\frac{a}{b} < \frac{c}{d}$ be adjacent fractions in the Farey series of order $k+1$.

If $\frac{a}{b} < \frac{c}{d}$ were already adjacent fractions in the Farey series of order k (i.e. if no fraction has been inserted between $\frac{a}{b}$ and $\frac{c}{d}$ in passing from the series of order k to the series of order $k+1$), then we already know (by the induction hypothesis) that $bc - ad = 1$.

Thus we may concentrate on the case where $\frac{a}{b} < \frac{c}{d}$ are not adjacent fractions in the Farey series of order k. By (b)(iii), at most one fraction with denominator $k+1$ is inserted between any two adjacent fractions in the Farey series of order k, so we have either

$$\frac{a}{b} < \frac{c}{d} < \frac{e}{f},$$

with $\frac{a}{b} < \frac{e}{f}$ being adjacent fractions in the Farey series of order k (so $be - af = 1$), or

$$\frac{e}{f} < \frac{a}{b} < \frac{c}{d},$$

with $\frac{e}{f} < \frac{c}{d}$ being adjacent fractions in the Farey series of order k (so $fc - ed = 1$). We consider the first of these possibilities (the second is entirely similar).

Suppose

$$\frac{a}{b} < \frac{c}{d} < \frac{e}{f},$$

with $\frac{a}{b} < \frac{e}{f}$ being adjacent fractions in the Farey series of order k. By part (i) we know that $bf \geqslant k+1$; and by induction we know that $be - af = 1$. Hence the interval $\frac{a}{b} < \frac{e}{f}$ has length at most $\frac{1}{k+1}$. We have to prove that $bc - ad = 1$.

Let $bc - ad = r > 0$, and $ed - fc = s > 0$.

Then $sa + re = c$, and $sb + rf = d$.

In particular, $HCF(r, s) = 1$ (since $HCF(c, d) = 1$).

Hence $\frac{c}{d}$ belongs to the family

$$S = \left\{ \frac{xa + ye}{xb + yf} : \text{ where } x, y \text{ are any positive integers with } HCF(x, y) = 1 \right\}.$$

Since everything is positive, easy algebra shows that
$$\frac{a}{b} < \frac{xa+ye}{xb+yf} < \frac{e}{f},$$
so every element of S lies between $\frac{a}{b}$ and $\frac{e}{f}$.

As long as we choose x, y such that $HCF(x,y) = 1$, any common factor of $xa + ye$ and $xb + yf$ would also divide both
$$b(xa+ye) - a(xb+yf) = (be-af)y = y,$$
and
$$e(xb+yf) - f(xa+ye) = (be-af)x = x.$$
Hence
$$HCF(xa+ye, xb+yf) = 1$$
so each element of S is in lowest terms (i.e. no further cancelling is possible). We have shown that "$\frac{c}{d}$ belongs to the family S", and that *all* elements of S fit *between* $\frac{a}{b}$ and $\frac{e}{f}$; which are *adjacent* fractions in the Farey series of order k. So none of the elements of S can have arisen before the series of order $k+1$. But each fraction in S arises at some stage in a Farey series. And the first to *arise* (because it has the smallest denominator) is "$\frac{a+e}{b+f}$". Hence
$$\frac{c}{d} = \frac{a+e}{b+f},$$
so $r = s = 1$, and $bc - ad = 1$ as required. QED

(d) Let
$$\frac{a}{b} < \frac{c}{d} < \frac{e}{f}$$
be three successive terms in any Farey sequence. By (c) we know that $bc - ad = 1$, and that $de - cf = 1$. In particular, $bc - ad = de - cf$, so
$$\frac{c}{d} = \frac{a+e}{b+f}.$$

Note 1: It may not be clear why we are proving this result "again" – since it appeared in the final line of the solution to part (c). However, in part (c) the statement that
$$\frac{c}{d} = \frac{a+e}{b+f}$$
was arrived at *within the induction step*, and so was *subject to other assumptions*. In contrast, now that the result in part (c) has been clearly established, we can use it to prove part (d) without any hidden assumptions.

Note 2: If we represent each fraction $\frac{a}{b}$ in the Farey series of order n by the point (b, a), then each point lies in the right angled triangle joining $(0,0)$, $(n,0)$, and (n,n), and each fraction in the Farey series is equal to the gradient of the line, or vector, joining the origin to the integer lattice point (b, a). The ordering of the fractions in the Farey series corresponds to the sequence of increasing gradients, from $\frac{0}{1}$ up to $\frac{1}{1}$. If $\frac{a}{b}$ and $\frac{e}{f}$ are adjacent fractions in some Farey series, then the result in (d) says that the next fraction to be inserted between them is $\frac{a+e}{b+f}$ corresponding to the vector sum of (b,a) and (f,e). And the result in (c) says that the area of the parallelogram with vertices $(0,0)$, (b,a), (f,e), $(b+f, a+e)$ is equal to 1 (see Problem **57**(b)). Hence the result in (c) reduces to the fact that

> **Theorem** Any parallelogram, whose vertices are integer lattice points (i.e. points (b,a) where both coordinates are integers), and with no additional lattice points inside the parallelogram or on the four sides, has area 1.

110.

(a) Suppose that x satisfies $x + \frac{1}{x} < 2$. Then $x \neq 0$ (or $\frac{1}{x}$ is not defined).
$\therefore \frac{x^2+1}{x} < 2$.
If $x > 0$, then $x^2 - 2x + 1 = (x-1)^2 < 0$, which has no solutions.
$\therefore x < 0$, in which case x satisfies $x + \frac{1}{x} < 0 < 2$, so every $x < 0$ is a solution of the original inequality.

(b) Suppose x satisfies $x \leqslant 1 + \frac{2}{x}$. Again $x \neq 0$ (or $\frac{1}{x}$ is not defined).

(i) If $x > 0$, then $x^2 - x - 2 = (x-2)(x+1) \leqslant 0$.
$\therefore -1 \leqslant x \leqslant 2$ (and $x > 0$); hence $0 < x \leqslant 2$, and all such x satisfy the original inequality.

(ii) If $x < 0$, then $x^2 - x - 2 \geqslant 0$, so $(x-2)(x+1) \geqslant 0$.
$\therefore$ either $x \leqslant -1$, or $x \geqslant 2$ (and $x < 0$); hence $x \leqslant -1$, and all such x satisfy the original inequality.

(c) Suppose x satisfies $\sqrt{x} < x + \frac{1}{4}$.
$\therefore 4\left(\sqrt{x}\right)^2 - 4\sqrt{x} + 1 > 0$
$\therefore (2\sqrt{x} - 1)^2 > 0$, so $x \neq \frac{1}{4}$, and all such x satisfy the original inequality.

111.

(a) If $a + b = 5$ and $ab = 7$, then a, b are solutions of
$$(x-a)(x-b) = x^2 - 5x + 7 = 0.$$
But the roots of this quadratic equation are
$$\frac{5 \pm \sqrt{25-28}}{2} = \frac{5 \pm \sqrt{-3}}{2},$$
so a and b cannot be "positive reals".

(b) We abbreviate the "arithmetic mean" by AM, the "geometric mean" by GM, the "harmonic mean" by HM, and the "quadratic mean" by QM.
$$\left(\sqrt{a} - \sqrt{b}\right)^2 \geqslant 0$$

150   *Algebra*

so
$$a+b \geqslant 2\sqrt{ab}$$
therefore
$$\sqrt{ab} \geqslant \frac{2ab}{a+b} \qquad (\text{GM} \geqslant \text{HM})$$
and
$$\sqrt{ab} \leqslant \frac{a+b}{2} \qquad (\text{GM} \leqslant \text{AM}).$$
Also
$$\left(\frac{a-b}{2}\right)^2 \geqslant 0,$$
so
$$\frac{a^2+b^2}{4} \geqslant \frac{2ab}{4}$$
whence
$$\frac{a^2+b^2}{2} \geqslant \frac{a^2+b^2+2ab}{4} = \left(\frac{a+b}{2}\right)^2.$$
Therefore
$$\sqrt{\frac{a^2+b^2}{2}} \geqslant \frac{a+b}{2} \qquad (\text{QM} \geqslant \text{AM}). \qquad \text{QED}$$

112. [This delightful problem was devised by Oleksiy Yevdokimov.] We need to find something which remains constant, or which does not increase, when we replace two terms a, b by $\frac{a+b}{\sqrt{2}}$.

Idea: If the two terms a, b were replaced each time by their sum $a+b$, then the sum of all the numbers in the list would be unchanged, so we could be sure that the final number after 199 such moves would have to be
$$1+2+3+\cdots+200 = \frac{200 \times 201}{2}.$$

This doesn't work here. However, in the spirit of this section on inequalities, one may ask:

> What happens to the sum of the squares of the terms in the list after each move?

When we move from one list to the next, only two terms are affected, and for these two terms, the previous sum of squares is replaced by $\left(\frac{a+b}{\sqrt{2}}\right)^2$. How does this affect the sum of all squares on the list?

We know that $a^2+b^2 \geqslant 2ab$ for all a, b. And it is easy to see that this is equivalent to:
$$a^2+b^2 \geqslant \left(\frac{a+b}{\sqrt{2}}\right)^2.$$

So when we replace two terms a, b by $\frac{a+b}{\sqrt{2}}$, the sum of the squares of all the terms in the list *never increases*. Hence the final term is less than or equal to the square root of the **initial** sum of squares

$$1^2 + 2^2 + 3^2 + \cdots + 200^2 = \frac{200 \times 201 \times 401}{6} \quad \text{(by Problem \textbf{62})}$$
$$< \frac{200 \times 300 \times 400}{6}$$
$$= 4 \times 10^6.$$

∴ the final term is $< \sqrt{4 \times 10^6} = 2000$.

113.

(a) (i) $3 = 2^2 - 1$.

(ii) It seems hard to find another.

(b) (i) $2 = 1^2 + 1$.

(ii) $5 = 2^2 + 1$ (or $17 = 4^2 + 1$; or $37 = 6^2 + 1$; or $101 = 10^2 + 1$; or ...). In other words, there seem to be lots.

Note: At first sight primes of this form "keep on coming". Given that we now know (see Problem **76**) that the list of all prime numbers "goes on for ever", it is natural to ask: Are there infinitely many prime numbers "one more than a square"? Or does the list run out?

This is one of the simplest questions one can ask to which the answer is **not yet known!**

(c) (i) $7 = 2^3 - 1$.

(ii) It seems hard to find another.

(d) (i) $2 = 1^3 + 1$.

(ii) It seems hard to find another.

Note: Parts (a), (c) and (d) should make one suspicious – provided one notices that:

(a) $63 = 7 \times 9$, $143 = 11 \times 13$, $323 = 17 \times 19$;

(c) $511 = 7 \times 73$, $1727 = 11 \times 157$;

(d) $217 = 7 \times 31$, $513 = 9 \times 57$, $1001 = 7 \times 143$.

This problem is so instructive that its solution is discussed in the main text following Problem **115**.

114.
$$x^4 + 1 = \left(x^2 + \sqrt{2} \cdot x + 1\right)\left(x^2 - \sqrt{2} \cdot x + 1\right).$$

(Suppose
$$x^4 + 1 = \left(x^2 + ax + b\right)\left(x^2 + cx + d\right).$$

It is natural to try $b = d = 1$ in order to make the constant term $bd = 1$, and then to try $c = -a$ (so that the coefficients of x^3 and of x are both 0). It then remains to choose the value of a so that the total coefficient "$2 - a^2$" of all terms in x^2 is equal to 0: that is, $a = \sqrt{2}$.)

115.

(a) (i)
$$a^3 - b^3 = (a-b)(a^2 + ab + b^2).$$

(ii)
$$\begin{aligned} a^4 - b^4 &= (a-b)\left(a^3 + a^2 b + ab^2 + b^3\right) \\ &= \left(a^2 - b^2\right)\left(a^2 + b^2\right) \\ &= (a-b)(a+b)\left(a^2 + b^2\right). \end{aligned}$$

(iii)
$$a^n - b^n = (a-b)\left(a^{n-1} + a^{n-2}b + a^{n-3}b^2 + \cdots + ab^{n-2} + b^{n-1}\right).$$

Note: The general factorisation
$$x^n - 1 = (x-1)\left(x^{n-1} + x^{n-2} + \cdots + x^2 + x + 1\right)$$
provides a fresh slant on the test for divisibility by 9 in base 10, or in general for divisibility by $b - 1$ in base b (see Problem **51**):

"an integer written in base b is divisible by $b - 1$ precisely when its digit sum is divisible by $b - 1$".

(b) (i)
$$a^3 + b^3 = (a+b)\left(a^2 - ab + b^2\right).$$

(ii)
$$a^5 + b^5 = (a+b)(a^4 - a^3 b + a^2 b^2 - ab^3 + b^4).$$

(iii)
$$a^{2n+1} + b^{2n+1} = (a+b)\left(a^{2n} - a^{2n-1}b + a^{2n-2}b^2 - a^{2n-3}b^3 + \cdots - ab^{2n-1} + b^{2n}\right).$$

116.

(a) Replace a by 1, b by r, and n by $n+1$ in the answer to **115**(a)(iii), to see that:
$$1 + r + r^2 + r^3 + \cdots + r^n = \frac{1 - r^{n+1}}{1 - r}.$$

(b) Multiply the closed formula in (a) by "a" to see that:
$$a + ar + ar^2 + ar^3 + \cdots + ar^n = a \cdot \frac{1 - r^{n+1}}{1 - r}.$$

117. When $x = 40$,

$$f(x) = x^2 + (x + 40) + 1 = 40^2 + 2 \times 40 + 1 = 41^2$$

is not prime. So the sequence of prime outputs must stop some time before $f(40)$. But it in fact keeps going as long as it possibly could, so that

$$f(0), f(1), f(2), \ldots, f(39)$$

are all prime. (This may explain Euler's delight.)

Note: The links between polynomials with integer coefficients (even lowly quadratics) and prime numbers are still not fully understood. For example, you might like to look up *Ulam's spiral*. (Ulam (1909–1984) plotted the positive integers in a square spiral and found the primes arranging themselves in curious patterns that we still do not fully understand.)

Interest in the connections between polynomials and primes was revived in the second half of the 20th century. It was eventually proved that there exists a polynomial in 10 variables, with *integer* coefficients, which takes both positive and negative values when the variables run through all possible non-negative *integer* values, but which does so in such a way that it generates **all** the primes as the set of positive outputs.

118.

(a)(i) For

$$a^n - 1 = (a - 1)\left(a^{n-1} + a^{n-2} + \cdots + a + 1\right)$$

to be prime, the smaller factor must be $= 1$, so $a = 2$.

If n is not prime, we can factorise $n = rs$, with $r, s > 1$. Then

$$2^n - 1 = 2^{rs} - 1 = (2^r)^s - 1 = (2^r - 1)\left(2^{r(s1)} + 2^{r(s2)} + \cdots + 2 + 1\right);$$

Hence $2^n 1$ also factorises, so could not be prime. Hence n must be prime.

(ii) $2^2 - 1 = \mathbf{3}$, $2^3 - 1 = \mathbf{7}$, $2^5 - 1 = \mathbf{31}$, $2^7 - 1 = \mathbf{127}$ are all prime; $2^{11} - 1 = 2047 = 23 \times 89$ is not.

Note: This is a simple example of the need to distinguish carefully between the statement

"if $2^n - 1$ is prime, then n is prime" (which is true),

and its converse

"if n is prime, then $2^n - 1$ is prime" (which is false).

(b)(i) Suppose that $a > 1$. Then $a^n + 1 > 2$; so for $a^n + 1$ to be prime, it must be odd, so a must be even.

If n has an odd factor $m > 1$, we can write $n = km$. Then
$$\begin{aligned} a^n + 1 &= a^{km} + 1 \\ &= (a^k)^m + 1 \\ &= (a^k + 1)\left(a^{k(m-1)} - a^{k(m-2)} + \cdots - a^k + 1\right). \end{aligned}$$

Since m is odd and > 1, we have $m \geq 3$. It is then easy to show that
$$a^k + 1 \leq a^{k(m-1)} - a^{k(m-2)} + \cdots - a^k + 1.$$

And since $a > 1$, neither factor $= 1$, so $a^n + 1$ can never be prime. Hence n can have no odd factor > 1, which is the same as saying that $n = 2^r$ must be a power of 2.

(ii) $2^1 + 1 = \mathbf{3}$, $2^2 + 1 = \mathbf{5}$, $2^4 + 1 = \mathbf{17}$, $2^8 + 1 = \mathbf{257}$, $2^{16} + 1 = \mathbf{65\,537}$ are all prime. (The very next such expression
$$2^{32} + 1 = 4\,294\,967\,297 = 641 \times 6\,700\,417$$
is not prime.)

Note: The sad tale of Fermat's claim that "all Fermat numbers are prime" shows that mathematicians are not exempt from the obligation to distinguish carefully between a statement and its converse!

119.

(a) $x^2 - 3x + 2 = (x-2)(x-1) = 0$ precisely when one of the brackets $= 0$; that is, $x = 2$, or $x = 1$.

(b) $x^2 - 1 = (x-1)(x+1) = 0$ precisely when $x = 1$ or $x = -1$.

(c) $x^2 - 2x + 1 = (x-1)^2 = 0$ precisely when $x = 1$ (a repeated root).

(d) $x^2 + \sqrt{2}x - 1 = 0$ requires us

   – to complete the square
$$x^2 + \sqrt{2}x - 1 = \left(x + \frac{\sqrt{2}}{2}\right)^2 - 1 - \frac{1}{2},$$
so
$$x + \frac{\sqrt{2}}{2} = \pm\sqrt{\frac{3}{2}},$$

   – or to use the quadratic formula:
$$x = \frac{-\sqrt{2} \pm \sqrt{2+4}}{2}.$$

(e) $x^2 + x - \sqrt{2} = 0$ requires us

- to complete the square

$$x^2 + x - \sqrt{2} = \left(x + \frac{1}{2}\right)^2 - \sqrt{2} - \frac{1}{4},$$

so

$$x + \frac{1}{2} = \pm\sqrt{\sqrt{2} + \frac{1}{4}}$$

- or to use the quadratic formula:

$$x = \frac{-1 \pm \sqrt{1 + 4\sqrt{2}}}{2}.$$

(f) $x^2 + 1 = 0$ yields $x = \pm\sqrt{-1}$.

(g) $x^2 + \sqrt{2}x + 1 = 0$ yields

$$x = \frac{-\sqrt{2} \pm \sqrt{2 - 4}}{2} = \frac{-\sqrt{2} \pm \sqrt{-2}}{2}.$$

120. $q(x) = x^2 - \sqrt{2} \cdot x + 1$. (There is no obvious magic method here. However, it should be natural to try to insert a term $\sqrt{2} \cdot x$ in $q(x)$ to "resolve" the term $\sqrt{2} \cdot x$ in $p(x)$; and the familiar cancelling of cross terms in $(a + b)(a - b)$ should then suggest the possible benefit of trying $q(x) = x^2 - \sqrt{2} \cdot x + 1$.)

Note: $p(x)q(x) = x^4 + 1$ (see Problem **114**).

121.

(a) Let the two unknown numbers be α and β. Then $s = \alpha + \beta$, and $p = \alpha\beta$. "The square of half the sum" $\left(\frac{s}{2}\right)^2 = \left(\frac{\alpha+\beta}{2}\right)^2$.

Subtracting $p = \alpha\beta$ produces $\left(\frac{\alpha-\beta}{2}\right)^2$ whose "square root" will be either $\frac{\alpha-\beta}{2}$, or $-\left(\frac{\alpha-\beta}{2}\right)$ – whichever is positive.

Adding this to "half the sum" gives one root; subtracting gives the other root.

(b) Let the length of one side be x. We are told that $x^2 + bx = c$.

"Take half of b, square it, and add the result to c"

translates as:

"Rewrite the equation as: $\left(x + \frac{b}{2}\right)^2 = c + \left(\frac{b}{2}\right)^2$."

That is, we have "completed the square" $\left(x + \frac{b}{2}\right)^2$. If we now take the (positive) square root and subtract $\frac{b}{2}$, we get a single value for x, which determines the side length of my square as required.

If the same method is applied to the general quadratic equation

$$ax^2 + bx + c = 0,$$

with the extra initial step of "multiply through by $\frac{1}{a}$", we produce first

$$x^2 + \frac{b}{a}x + \frac{c}{a} = 0,$$

then

$$\left(x + \frac{b}{2a}\right)^2 + \left(\frac{c}{a} - \left(\frac{b}{2a}\right)^2\right) = 0,$$

then

$$x + \frac{b}{2a} = \pm\sqrt{\left(\frac{b}{2a}\right)^2 - \frac{c}{a}} = \frac{\pm\sqrt{b^2 - 4ac}}{2a},$$

which leads to the familiar quadratic formula.

(c) See Problem **3**(c)(iv). $AD : CB = DX : BX$, so $x : 1 = 1 : (x-1)$. Hence $x^2 - x - 1 = 0$. If we use the quadratic formula derived in the answer to part (b) above, and realise that $x > 1$, then we obtain $x = \frac{1+\sqrt{5}}{2}$.

Note: The procedure given in (a) dates back to the ancient Babylonians ($\sim$ 1700 BC) and later to the ancient Greeks ($\sim$ 300 BC). Both cultures worked *without* algebra. The Babylonians gave their verbal procedures as recipes in words, in the context of particular examples. The Greeks expressed everything geometrically. In modern language, if we denote the unknown numbers by α and β, then

$$(x - \alpha)(x - \beta) = x^2 - (\alpha + \beta)x + \alpha\beta.$$

Being told the sum and product is therefore the same as being given the coefficients of a quadratic equation, and being asked to find the two roots.

Our method for factorizing a quadratic involves a mental process of 'inverse arithmetic', where we juggle possibilities in search of α and β, when all we know are the coefficients (that is, the sum $\alpha + \beta$, and the product $\alpha\beta$).

The procedure in (b) also dates back to the ancient Babylonians, and is essentially our process of completing the square. It was given as a procedure, without our algebraic notation. The Babylonians seem not to have been hampered (as the Greeks were) by the fact that it makes no sense to add a *length* and an *area*! They worded things geometrically, but seem to have understood that they were really playing *numerical* games (an idea which European mathematicians found elusive right up to the time of Descartes (1590–1656)).

Similarly, the modern use of symbols – allowing one to represent either positive or negative quantities – was widely resisted right into the nineteenth century. What we would write as a single family of quadratic equations, $ax^2 + bx + c = 0$, had to be split into separate cases where two *positive* quantities were equated. For example, the groundbreaking book *Ars Magna* in which Cardano (1501–1576) explained how to solve cubic and quartic equations begins with quadratics – where his procedure distinguishes four different cases: "squares equal to numbers", "squares equal to things", "squares and things equal to numbers", "squares and numbers equal to things".

122.

(a) (i) $\alpha^2 + \beta^2 = (\alpha + \beta)^2 - 2\alpha\beta = b^2 - 2c.$

(ii) $\alpha^2\beta + \beta^2\alpha = \alpha\beta(\alpha + \beta) = c \cdot (-b) = -bc.$

(iii) We rearrange
$$\begin{aligned} \alpha^3 + \beta^3 - 3\alpha\beta &= (\alpha + \beta)\left(\alpha^2 - \alpha\beta + \beta^2\right) - 3\alpha\beta \\ &= (-b) \cdot (b^2 - 3c) - 3c \\ &= -b^3 + 3bc - 3c. \end{aligned}$$

[Alternatively: $\alpha^3 + \beta^3 = (\alpha + \beta)^3 - 3\alpha\beta(\alpha + \beta)$, etc.]

(b) (i) [Cf **121**(a).] $(\alpha - \beta)^2 = (\alpha + \beta)^2 - 4\alpha\beta.$
Therefore
$$\alpha - \beta = \sqrt{b^2 - 4c} \text{ if } \alpha \geqslant \beta,$$
and
$$\alpha - \beta = -\sqrt{b^2 - 4c} \text{ if } \alpha < \beta.$$

(ii)
$$\alpha^2\beta - \beta^2\alpha = -\alpha\beta(\alpha - \beta) = -c\sqrt{b^2 - 4c} \text{ if } \alpha \geqslant \beta,$$
and
$$\alpha^2\beta - \beta^2\alpha = -\alpha\beta(\alpha - \beta) = c\sqrt{b^2 - 4c} \text{ if } \alpha < \beta.$$

(iii) $\alpha^3 - \beta^3 = (\alpha - \beta)(\alpha^2 + \alpha\beta + \beta^2).$
Therefore
$$\alpha^3 - \beta^3 = \left[\sqrt{b^2 - 4c}\right](b^2 - c) \text{ if } \alpha \geqslant \beta,$$
and
$$\alpha^3 - \beta^3 = \left[-\sqrt{b^2 - 4c}\right](b^2 - c) \text{ if } \alpha < \beta.$$

123.

(a) (i) $\sqrt{a} + \sqrt{b}$ and $\sqrt{a + b + \sqrt{4ab}}$ are both positive. And it is easy to check that they have the same square:
$$\left(\sqrt{a} + \sqrt{b}\right)^2 = a + b + 2\sqrt{ab},$$
and
$$\left(\sqrt{a + b + \sqrt{4ab}}\right)^2 = a + b + \sqrt{4ab}.$$
Hence
$$\sqrt{a} + \sqrt{b} = \sqrt{a + b + \sqrt{4ab}}.$$

(ii) $5 = 2 + 3$, and $24 = 4 \times 2 \times 3$;
Therefore
$$\sqrt{2 + 3 + \sqrt{4 \times 2 \times 3}} = \sqrt{2} + \sqrt{3}$$
(which is easy to check).

(b)(i) **Claim** If $a \geq b \, (\neq 0)$, then
$$\sqrt{a} - \sqrt{b} = \sqrt{a+b-\sqrt{4ab}}.$$

Proof $\sqrt{a} - \sqrt{b}$ and $\sqrt{a+b-\sqrt{4ab}}$ are both ≥ 0 (Why?). And it is easy to check that
$$\left(\sqrt{a} - \sqrt{b}\right)^2 = a + b - 2\sqrt{ab},$$
and
$$\left(\sqrt{a+b-\sqrt{4ab}}\right)^2 = a + b - \sqrt{4ab}. \qquad \text{QED}$$

(ii) Simplify $\sqrt{5 - \sqrt{16}}$ and $\sqrt{6 - \sqrt{20}}$.
$5 = 4 + 1$ and $16 = 4 \times 4 \times 1$, so $\sqrt{5 - \sqrt{16}} = \sqrt{4} - \sqrt{1} = 1$.
Actually, there is a simpler solution:
$$\sqrt{5 - \sqrt{16}} = \sqrt{5 - 4} = \sqrt{1} = 1.$$
$6 = 5 + 1$ and $20 = 4 \times 5 \times 1$, so $\sqrt{6 - \sqrt{20}} = \sqrt{5} - \sqrt{1} = \sqrt{5} - 1$.

124.

(a) Let $\alpha = 1 + \sqrt{2}$. Then $\alpha^2 = 3 + 2\sqrt{2}$. Hence $\alpha^2 - 2\alpha = 1$, so α satisfies the quadratic polynomial equation $x^2 - 2x - 1 = 0$.

Note: Observe that the resulting polynomial is equal to
$$\left(x - \left(1 + \sqrt{2}\right)\right)\left(x - \left(1 - \sqrt{2}\right)\right).$$
In other words, to rationalize the coefficients, we need a polynomial which has both $\alpha = 1 + \sqrt{2}$ and its "conjugate" $1 - \sqrt{2}$ as roots.

(b) Let $\alpha = 1 + \sqrt{3}$. Then $\alpha^2 = 4 + 2\sqrt{3}$. Hence $\alpha^2 - 2\alpha = 2$, so α satisfies the quadratic polynomial equation $x^2 - 2x - 2 = 0$.

Note: Observe that the resulting polynomial is equal to
$$\left(x - \left(1 + \sqrt{3}\right)\right)\left(x - \left(1 - \sqrt{3}\right)\right).$$
In other words, to rationalize the coefficients, we need a polynomial which has both $\alpha = 1 + \sqrt{3}$ and its "conjugate" $1 - \sqrt{3}$ as roots.

(c) Let $\alpha = \sqrt{2} + \sqrt{3}$. Then $\alpha^2 = 5 + 2\sqrt{6}$, so $\alpha^2 - 5 = 2\sqrt{6}$, and $\left(\alpha^2 - 5\right)^2 = 24$. Hence α satisfies the quartic polynomial equation $x^4 - 10x^2 + 1 = 0$.

Note: Observe that the resulting polynomial is equal to
$$\left(x - \left(\sqrt{2} + \sqrt{3}\right)\right)\left(x - \left(\sqrt{2} - \sqrt{3}\right)\right)\left(x - \left(-\sqrt{2} + \sqrt{3}\right)\right)\left(x - \left(-\sqrt{2} - \sqrt{3}\right)\right).$$
In other words, the roots are: $\sqrt{2} + \sqrt{3}$ (as required), and also $\sqrt{2} - \sqrt{3}$, $-\sqrt{2} - \sqrt{3}$, and $-\sqrt{2} + \sqrt{3}$.

(d) Let $\alpha = \sqrt{2} + \frac{1}{\sqrt{3}}$. Then
$$\alpha^2 = \frac{7}{3} + 2\sqrt{\frac{2}{3}},$$
so
$$\left(\alpha^2 - \frac{7}{3}\right)^2 = \frac{8}{3},$$
and α satisfies the quartic polynomial equation
$$x^4 - \frac{14}{3} \cdot x^2 + \frac{25}{9} = 0.$$

Note:
$$x^4 - \frac{14}{3} \cdot x^2 + \frac{25}{9} = \left(x - \left[\sqrt{2} + \frac{1}{\sqrt{3}}\right]\right)\left(x - \left[\sqrt{2} - \frac{1}{\sqrt{3}}\right]\right)$$
$$\cdot \left(x + \left[\sqrt{2} + \frac{1}{\sqrt{3}}\right]\right)\left(x + \left[\sqrt{2} - \frac{1}{\sqrt{3}}\right]\right),$$
so the roots are:
$$x = \sqrt{2} + \frac{1}{\sqrt{3}},\ \sqrt{2} - \frac{1}{\sqrt{3}},\ -\sqrt{2} - \frac{1}{\sqrt{3}},\ -\sqrt{2} + \frac{1}{\sqrt{3}}.$$

125. A direct approach can be made to work in both cases (but see the **Notes**).

(a) Suppose to the contrary that $\sqrt{2} + \sqrt{3} = \frac{p}{q}$, for some integers p, q with $HCF(p,q) = 1$. Then $(5 + 2\sqrt{6})q^2 = p^2$, so $\sqrt{6}$ is rational, and we can write $\sqrt{6} = \frac{r}{s}$ with $HCF(r,s) = 1$. But then $6s^2 = r^2$; hence $r = 2t$ must be even; so $3s^2 = 2t^2$, but then s must be even – contradicting $HCF(r,s) = 1$. Hence $\sqrt{2} + \sqrt{3}$ cannot be rational.

Note: It is slightly easier to rewrite the initial equation in the form
$$\sqrt{3} = \frac{p}{q} - \sqrt{2}$$
before squaring to get
$$\left(\frac{p}{q}\right)^2 - 1 = \frac{2p}{q}\sqrt{2},$$
which would imply that $\sqrt{2}$ is rational.

(b) Suppose to the contrary that $\sqrt{2} + \sqrt{3} + \sqrt{5} = \frac{p}{q}$, for some integers p, q with $HCF(p,q) = 1$. Then
$$10 + 2\left(\sqrt{6} + \sqrt{10} + \sqrt{15}\right) = \left(\frac{p}{q}\right)^2,$$
so $\sqrt{6} + \sqrt{10} + \sqrt{15}$ is rational. Squaring $\sqrt{6} + \sqrt{10} + \sqrt{15}$ then gives that
$$\sqrt{60} + \sqrt{90} + \sqrt{150} = 5\sqrt{6} + 3\sqrt{10} + 2\sqrt{15}$$

160   Algebra

is rational. Subtracting $2(\sqrt{6} + \sqrt{10} + \sqrt{15})$ then shows that $3\sqrt{6} + \sqrt{10}$ is rational, and we can proceed as in part (a) to obtain a contradiction. Hence $\sqrt{2} + \sqrt{3} + \sqrt{5}$ cannot be rational.

Note: It is simpler to rewrite the original equation in the form

$$\sqrt{2} + \sqrt{3} = \frac{p}{q} - \sqrt{5}$$

before squaring to obtain

$$5 + 2\sqrt{6} = \left(5 + \left(\frac{p}{q}\right)^2\right) - \frac{2p}{q}\sqrt{5},$$

whence $2\sqrt{6} + \frac{2p}{q}\sqrt{5}$ is rational, and we may proceed as in part (a).

126.

(i) We just have to fill in the missing bits of the partial factorisation

$$x^{10} + 1 = \left(x^3 - 1\right)\left(x^7 + x^4 + \cdots\right) + \text{remainder}.$$

To produce the required term x^{10} we first insert x^7. This then creates an unwanted term "$-x^7$", so we add $+x^4$ to cancel this out. This in turn creates an unwanted term "$-x^4$", so we add $+x$ to cancel this out. Hence the quotient is $x^7 + x^4 + x$, and the remainder is "$x + 1$":

$$x^{10} + 1 = \left(x^3 - 1\right)\left(x^7 + x^4 + x\right) + (x + 1).$$

Note: It is worth noting a short cut. The factorised term of the form $\left(x^3 - 1\right)\left(x^7 + \cdots\right)$ is equal to zero when $x^3 = 1$.
So one way to get the remainder is to "treat x^3 as if it were equal to 1". Then

$$x^{10} = \left(x^3\right)^3 \cdot x$$

is just like $1 \cdot x$, and $x^{10} + 1$ behaves as if it were equal to $x + 1$, which is the remainder.

(ii)
$$x^{2013+1} = \left(x^2 - 1\right)\left(x^{2011} + x^{2009} + x^{2007} + \cdots + x\right) + (x + 1),$$

so the remainder $= x + 1$.

Note: If we treat x^2 "as if it were equal to 1", then

$$x^{2013} + 1 = \left(x^2\right)^{1006} \cdot x + 1$$

behaves as if it were equal to $1 \cdot x + 1$.

(iii) Apply the Euclidean algorithm to m and n in order to write $m = qn + r$, where $0 \leqslant r < n$:

$$x^m = x^{qn+r} = \left(x^n\right)^q \cdot x^r.$$

Then
$$x^m + 1 = x^{qn+r} + 1$$
$$= (x^n - 1)\left(x^{n(q-1)+r} + x^{n(q-2)+r} + x^{n(q-3)+r} + \cdots + x^r\right) + x^r + 1.$$

So the remainder is $x^r + 1$.

Note: If we treat $x^n - 1$ as if were 0 – that is, if we treat x^n as if it were equal to 1 – then
$$x^m + 1 = x^{qn+r} + 1 = (x^n)^q \cdot x^r + 1$$
which behaves like $1^q \cdot x^r + 1$.

127. Suppose $x^{2013} + 1 = (x^2 + x + 1) q(x) + r(x)$, where $\deg(r(x)) < 2$. Then
$$(x^{2013} + 1)(x - 1) = x^{2014} - x^{2013} + x - 1$$
$$= (x^3 - 1) q(x) + (x - 1)r(x).$$

Now
$$x^{2014} - x^{2013} + x - 1 = (x^3 - 1)\left(x^{2011} - x^{2010} + x^{2008} - x^{2007} + x^{2004} - x^{2003} + \cdots + x\right) + 2x - 2$$

so the remainder $r(x) = 2$.

Note: If x satisfies $x^2 + x + 1 = 0$, then $x^3 - 1 = 0$ and $x \neq 1$.
$\therefore x^{2013} + 1 = (x^3)^{671} + 1$ behaves just like $1^{671} + 1 = 2$, so $r(x) = 2$.

128.

(a)
$$(a + bi)^{-1} = \frac{a}{a^2 + b^2} - \left[\frac{b}{a^2 + b^2}\right]i.$$

(b)
$$p(x) = x^2 - 2ax + \left(a^2 + b^2\right).$$

(Suppose that the quadratic equation
$$p(x) = x^2 + cx + d = 0,$$
with real coefficients c, d, has $x = a + ib$ as a root. Then take the complex conjugate of the equation $p(x) = 0$ to see that $x = a - ib$ is also a rooti of
$$p(x) = x^2 + cx + d = 0.$$

Therefore
$$p(x) = x^2 + cx + d$$
$$= (x - (a + ib))(x - (a - ib)),$$

162   Algebra

so $c = -2a$, and $d = a^2 + b^2$.)

129. Let the two unknown numbers be α and β. Then $10 = \alpha + \beta$, and $40 = \alpha\beta$, so α and β are roots of the quadratic equation $x^2 - 10x + 40 = 0$. Hence

$$\alpha, \beta = \frac{10 \pm \sqrt{100 - 160}}{2} = 5 \pm \sqrt{-15}.$$

130.

(a) Applying a simple rearrangement:

$$\begin{aligned} wz &= r(\cos\theta + i\sin\theta) \cdot s(\cos\phi + i\sin\phi) \\ &= rs[(\cos\theta \cdot \cos\phi - \sin\theta\sin\phi) + i(\cos\theta \cdot \sin\phi + \sin\theta \cdot \cos\phi)] \\ &= rs[\cos(\theta + \phi) + i\sin(\theta + \phi)] \end{aligned}$$

(by the usual addition formula: Problem **35**)

(b) By part (a),
$$(\cos\theta + i\sin\theta)^2 = \cos(2\theta) + i\sin(2\theta).$$

Hence

$$\begin{aligned} (\cos\theta + i\sin\theta)^3 &= (\cos\theta + i\sin\theta)^2(\cos\theta + i\sin\theta) \\ &= [\cos(2\theta) + i\sin(2\theta)] \cdot (\cos\theta + i\sin\theta) \\ &= \cos(3\theta) + i\sin(3\theta). \end{aligned}$$

Etc.

Note: This should really be presented as a "proof by mathematical induction", where (having established the initial cases) we "suppose the result holds for powers $n = 1, 2, 3, \ldots, k$", and then conclude that

$$\begin{aligned} (\cos\theta + i\sin\theta)^{k+1} &= (\cos\theta + i\sin\theta)^k (\cos\theta + i\sin\theta) \\ &= [\cos(k\theta) + i\sin(k\theta)](\cos\theta + i\sin\theta) \\ &= \cos((k+1)\theta) + i\sin((k+1)\theta). \end{aligned}$$

(c) $z^n = r^n(\cos(n\theta) + i\sin(n\theta))$. Hence if $z^n = 1$, then $|z^n| = r^n = 1$, so $r = 1$ (since $r \geqslant 0$).

131.

(a) We factorise: $x^3 - 1 = (x-1)(x^2 + x + 1)$, so the roots are $x = 1$; and

$$x = \frac{-1 \pm \sqrt{1-4}}{2} = \frac{-1 \pm \sqrt{-3}}{2} = -\frac{1}{2} \pm \frac{\sqrt{3}}{2}i$$

that is, the other two roots are
$$x = \cos\left(\frac{2\pi}{3}\right) + i\sin\left(\frac{2\pi}{3}\right)$$
and
$$x = \cos\left(-\frac{2\pi}{3}\right) + i\sin\left(-\frac{2\pi}{3}\right).$$

(b) We factorise:
$$x^4 - 1 = (x^2 - 1)(x^2 + 1) = (x-1)(x+1)(x^2+1),$$
so the roots are $x = 1$, $x = -1$, $x = i$, $x = -i$.

(c) We factorise:
$$\begin{aligned} x^6 - 1 &= \left[(x^2)^3 - 1\right] \\ &= (x^2 - 1)(x^4 + x^2 + 1) \\ &= (x-1)(x+1)\left[(x^2)^2 + x^2 + 1\right], \end{aligned}$$
so the roots are
- $x = 1$, $x = -1$, and
- four further values of x satisfying $x^2 = -\frac{1}{2} \pm \frac{\sqrt{3}}{2}i$: that is,
$$x = \cos\left(\frac{\pi}{3}\right) + i\sin\left(\frac{\pi}{3}\right) = \frac{1}{2} + \frac{\sqrt{3}}{2}i$$
and
$$x = \cos\left(\frac{2\pi}{3}\right) + i\sin\left(\frac{2\pi}{3}\right) = -\frac{1}{2} + \frac{\sqrt{3}}{2}i$$
and
$$x = \cos\left(\frac{-\pi}{3}\right) + i\sin\left(\frac{-\pi}{3}\right) = \frac{1}{2} - \frac{\sqrt{3}}{2}i$$
and
$$x = \cos\left(\frac{-2\pi}{3}\right) + i\sin\left(\frac{-2\pi}{3}\right) = -\frac{1}{2} - \frac{\sqrt{3}}{2}i.$$

(d) We factorise:
$$\begin{aligned} x^8 - 1 &= (x^4 - 1)(x^4 + 1) \\ &= (x^2 - 1)(x^2 + 1)\left(x^2 + \sqrt{2}\cdot x + 1\right)\left(x^2 - \sqrt{2}\cdot x + 1\right) \end{aligned}$$
so the roots are
- $x = 1$, $x = -1$;
- $x = i$, $x = -i$, and
- the roots of $x^2 + \sqrt{2}\cdot x + 1 = 0$ and $x^2 - \sqrt{2}\cdot x + 1 = 0$, which happen to be
$$x = \cos\left(\frac{\pi}{4}\right) + i\sin\left(\frac{\pi}{4}\right) = \frac{\sqrt{2}}{2} + \frac{\sqrt{2}}{2}i$$

and
$$x = \cos\left(-\frac{\pi}{4}\right) + i\sin\left(-\frac{\pi}{4}\right) = \frac{\sqrt{2}}{2} - \frac{\sqrt{2}}{2}i$$
and
$$x = \cos\left(\frac{3\pi}{4}\right) + i\sin\left(\frac{3\pi}{4}\right) = -\frac{\sqrt{2}}{2} + \frac{\sqrt{2}}{2}i$$
and
$$x = \cos\left(-\frac{3\pi}{4}\right) + i\sin\left(-\frac{3\pi}{4}\right) = -\frac{\sqrt{2}}{2} - \frac{\sqrt{2}}{2}.$$

132. [In Problem **114** you were left to work out the required factorisation with your bare hands – and a bit of inspired guesswork. The suggested approach here is more systematic.]

The roots of $x^4 + 1 = 0$ are complex numbers whose fourth power is equal to -1: that is,
$$x = \cos\left(\frac{\pi}{4}\right) + i\sin\left(\frac{\pi}{4}\right) = \frac{\sqrt{2}}{2} + \frac{\sqrt{2}}{2}i$$
and
$$x = \cos\left(-\frac{\pi}{4}\right) + i\sin\left(-\frac{\pi}{4}\right) = \frac{\sqrt{2}}{2} - \frac{\sqrt{2}}{2}i$$
and
$$x = \cos\left(\frac{3\pi}{4}\right) + i\sin\left(\frac{3\pi}{4}\right) = -\frac{\sqrt{2}}{2} + \frac{\sqrt{2}}{2}i$$
and
$$x = \cos\left(-\frac{3\pi}{4}\right) + i\sin\left(-\frac{3\pi}{4}\right) = -\frac{\sqrt{2}}{2} - \frac{\sqrt{2}}{2}i.$$

The first two are complex conjugates and give rise to two linear factors whose product is $x^2 + \sqrt{2} \cdot x + 1$; the other two are complex conjugates and give rise to two linear factors whose product is $x^2 - \sqrt{2} \cdot x + 1$. Hence
$$x^4 + 1 = \left(x^2 + \sqrt{2} \cdot x + 1\right)\left(x^2 - \sqrt{2} \cdot x + 1\right).$$

133.

(a) The roots of $x^5 - 1 = 0$ are precisely the five complex numbers of the form
$$\cos\left(\frac{2k\pi}{5}\right) + i\sin\left(\frac{2k\pi}{5}\right), \text{ for } k = 0, 1, 2, 3, 4:$$

that is,
$$x = 1$$
$$x = \cos\left(\frac{2\pi}{5}\right) + i\sin\left(\frac{2\pi}{5}\right)$$
$$x = \cos\left(\frac{4\pi}{5}\right) + i\sin\left(\frac{4\pi}{5}\right)$$
$$x = \cos\left(\frac{6\pi}{5}\right) + i\sin\left(\frac{6\pi}{5}\right)$$
$$x = \cos\left(\frac{8\pi}{5}\right) + i\sin\left(\frac{8\pi}{5}\right).$$

From Problem **3**(c) we know that

$$\cos\left(\frac{2\pi}{5}\right) = \frac{\sqrt{5}-1}{4} = \cos\left(\frac{8\pi}{5}\right)$$
$$\sin\left(\frac{2\pi}{5}\right) = \frac{\sqrt{10+2\sqrt{5}}}{4} = -\sin\left(\frac{8\pi}{5}\right)$$
$$\cos\left(\frac{4\pi}{5}\right) = -\cos\left(\frac{\pi}{5}\right) = -\frac{\sqrt{5}+1}{4} = \cos\left(\frac{6\pi}{5}\right)$$
$$\sin\left(\frac{4\pi}{5}\right) = \frac{\sqrt{10-2\sqrt{5}}}{4} = -\sin\left(\frac{6\pi}{5}\right).$$

(b) The linear factor is clearly $(x-1)$. Each quadratic factor arises as the product of two conjugate linear factors. We saw in Problem **128**(b) that two linear factors corresponding to roots $a+bi$ and $a-bi$ produce the quadratic factor $x^2 - 2ax + (a^2+b^2)$. Hence the two quadratic factors are:

$$x^2 - \frac{\sqrt{5}-1}{2}\cdot x + 1, \text{ and } x^2 + \frac{\sqrt{5}+1}{2}\cdot x + 1$$

(whose product is equal to $x^4 + x^3 + x^2 + x + 1$).

134.

(a) Put $a = 1$, $y = x+1$: then $x^3 + 3x^2 - 4 = 0$ becomes $y^3 - 3y = 2$.

(b) Divide through by a (which we may assume is non zero, since otherwise it would not be a *cubic* equation), to obtain a cubic equation

$$x^3 + px^2 + qx + r = 0.$$

If we now put $y = x + \frac{p}{3}$, then y^3 incorporates both the x^3 and the x^2 terms, and the equation reduces to:

$$y^3 + \left[q - 3\left(\frac{p}{3}\right)^2\right]y + \left[r + 2\left(\frac{p}{3}\right)^3 - q\left(\frac{p}{3}\right)\right] = 0.$$

135. Given the equation $x^3 + 3x^2 - 4 = 0$. Let $y = x + 1$.

(i) Then $y^3 = x^3 + 3x^2 + 3x + 1$, so $0 = x^3 + 3x^2 - 4 = y^3 - 3y - 2$.

(ii) Set $y = u + v$ and use the fact that

$$(u + v)^3 = u^3 + 3uv(u + v) + v^3$$

is an identity, and so holds for all u and v.

(iii) Solve "$3uv = 3$", "$u^3 + v^3 = 2$". Substitute $v = \frac{1}{u}$ from the first equation into the second to get the quadratic equation in $(u^3)^2 - 2u^3 + 1 = 0$: that is, $(u^3 - 1)^2 = 0$, so $u^3 = 1$.

(iv) **Hence $u = 1$ is certainly a solution.** (We know there are also complex cube roots of 1; these lead to the other two solutions of the original cubic, but to "solve the equation" it is enough to find one solution.) **Hence $v = 1$, so $y = u + v = 2$, and $x = 1$.**

136. The Euclidean algorithm for ordinary integers arises by repeating the division algorithm:

> given integers m, n ($\neq 0$), there exists unique integers q, r such that $m = qn + r$ where $0 \leqslant r < n$.

Here q is the *quotient* (the integer part of the division $m \div n$), and r is the *remainder*. If we then replace the initial pair (m, n) by the new pair (n, r) and repeat until we obtain the remainder 0, then the last non-zero remainder is equal to $HCF(m, n)$ (see Problem **6**). The same idea also works for polynomials with integer coefficients (see Problem **126**).

We start by clarifying what we mean by *divisibility* for Gaussian integers. Given two Gaussian integers, $m = a + bi$ and $n = c + di$, we say that $n = c + di$ **divides** $m = a + bi$ (exactly) precisely

> when there exists some other Gaussian integer $q = e + fi$ such that $m = qn$: that is, $a + bi = (e + fi)(c + di)$.

For example, $2 + 3i$ divides $-4 + 7i$ because $(1 + 2i)(2 + 3i) = -4 + 7i$.

If $m = a + bi$ and $n = c + di$ are any old Gaussian integers, then it will not in general be true that "n divides m", but we can imitate the division algorithm. The important idea here when carrying out particular calculations is to realize that "divide by $c + di$" is the same as "multiply by $\frac{c-di}{c^2+d^2}$"

- first carry out the division

$$m \div n = \frac{(a + bi)(c - di)}{c^2 + d^2};$$

- then take the "nearest" Gaussian integer $q = e + fi$, and let the difference $m - qn = r$ be the *remainder*.

As for ordinary integers, any Gaussian integer that is a "common factor of m and n" is then automatically a common factor of n and of $r = m - qn$, and conversely. That is, the common factors of m and n are precisely the same as the common factors of n and r. So we can repeat the process replacing m, n by n, r. Provided the "remainder" r is in some sense "smaller" than n, we can continue until we reach a stage where the remainder $r = 0$ – at which point, the last non-zero remainder is equal to the $HCF(m, n)$ (that is, the Gaussian integer which is the HCF of the two initial Gaussian integers m, n).

The feature of the remainders that gets progressively smaller is their *norm* (see Problem **25**, and Problem **54**). As so often, this becomes clearer when we look at an example.

Let us try to find the HCF of the two Gaussian integers $m = 14 - 42i$ and $n = 4 - 7i$.

- First do the division

$$m \div n = \frac{(14 - 42i)(4 + 7i)}{4^2 + 7^2} = \frac{350}{65} - \frac{70}{65}i.$$

- What is meant by the *nearest* Gaussian integer may require an element of judgment; but it is clear that the answer is fairly close to $5 - i = q$, where $qn = 13 - 39i$, with remainder $r = m - qn = 1 - 3i$.
- Now repeat the process with n, r:

$$n \div r = \frac{(4 - 7i)(1 + 3i)}{1^2 + 3^2} = \frac{5}{2} + \frac{1}{2}i.$$

- The *nearest* Gaussian integer is not well-defined, but the answer is fairly close to $3 = q'$. So $q'r = 3 - 9i$, with remainder $r' = n - q'r = 1 + 2i$.
- Now repeat the step with the pair $r = 1 - 3i$ and $r' = 1 + 2i$, to discover that

$$1 - 3i = -(1 + i)(1 + 2i)$$

with remainder 0. Hence

$$1 + 2i = HCF(14 - 42i, 4 - 7i).$$

Note: One way to picture the process is to learn to "see" the Gaussian integers *geometrically*. Every Gaussian integer (such as $a + bi$) can be written as an integer combination of the two basic Gaussian integers "1" and "i" – namely

$$a + bi = a \times 1 + b \times i.$$

Since 1 and i are both of length 1 and perpendicular to each other, this represents the set of all Gaussian integers as the dots in a "square dot lattice" generated by translations in the x- and y- directions of the basic unit square spanned by 0, 1, i, and $1 + i$.

Any other given Gaussian integer, such as $n = c + di$, then generates a "stretched and rotated" square lattice, which consists of all "Gaussian multiples" of $c + di$ – generated by the basic square which is spanned by

$$0, \ (c + di) \times 1, \ (c + di) \times i, \ \text{and} \ (c + di) \times (1 + i).$$

168 *Algebra*

Every Gaussian integer (or rather the point, or dot, which corresponds to it) lies either on the boundary, or inside, one of these larger "stretched and rotated" squares: if the diagonal of one of these larger squares has length $2k$, then any other Gaussian integer $m = a + bi$ lies inside one of these larger squares, and so lies *within distance* k (that is, half a diagonal) of some (Gaussian) multiple qn of $n = c + di$. And the difference $m - qn$ is precisely the required *remainder* r.

Extra: We interpret $\sqrt[3]{8} = 8^{\frac{1}{3}} = 2$. Prove that

$$\sqrt[\sqrt{-1}]{-1} \approx 23\frac{1}{7}$$

(where $\approx$ denotes "approximately equal to").

V. Geometry

> *Those who fear to experiment with their hands
> will never know anything.*
> George Sarton (1884–1956)

> *Mathematical truth is not determined arbitrarily
> by the rules of some 'man-made' formal system,
> but has an absolute nature and lies
> beyond any such system of specifiable rules.*
> Roger Penrose (1930–)

Geometry is in many ways the most natural branch of elementary mathematics through which to convey "the essence" of the discipline.

- The underlying subject matter is rooted in seeing, moving, doing, drawing, making, etc., and so is accessible to everyone.

- At secondary level this practical experience leads fairly naturally to a semi-formal treatment of "geometry as a mental universe"

  – a universe that is bursting with surprising facts, whose statements can be easily understood; and

  – which has a clear logical structure, in terms of which the proofs of these facts are accessible, if sometimes tantalisingly elusive.

This combination of elusive problems to be solved and the steady accumulation of proven results has provided generations of students with their first glimpse of serious mathematics. All readers can imagine the kind of experiences which lie behind the first bullet point above: many of the problems we have already met (such as Problems **4, 19, 20, 26, 27, 28, 29, 30, 31, 37, 38, 39**) do not depend on the "semi-formal treatment" referred to in the second bullet point, so can be tackled by anyone who is interested – *provided they accept the importance of learning to construct their own diagrams* (in the spirit of the George Sarton quotation).

> *The hand is the cutting edge of the mind.*
> Jacob Bronowski (1908–1974)

But there is a catch – which explains why the present chapter appears so late in the collection. For many problems to successfully convey "the essence of mathematics" there has to be some shared understanding of what constitutes a solution. And in geometry, many solutions require the construction of a **proof**. Yet many readers will never have experienced a coherent "semi-formal treatment" of elementary geometry in the spirit of the second bullet point. Hence in Problems **3**(c), **18**, **21**, **32**, **34**, **36** we committed the cardinal sin of leading the reader by the nose – breaking each problem into steps in order to impose a logical structure. This may have been excusable in Chapter 1; but in a chapter explicitly devoted to geometry, the underlying challenge has to be faced head on: that is, the raw experience of the *hand* has to be refined to provide a deductive structure for the *mind*.

As in Chapter 1, some of the problems listed from Section 5.3 onwards can be tackled without worrying too much about the logical structure of elementary geometry. But in many instances, the "essence" that is captured by a problem requires that the problem be seen within an agreed logical hierarchy – a sequencing of properties, results, and methods, which establishes *what* is a consequence of *what* – and hence, what can be used as part of a solution. In particular, we need to construct proofs that avoid circular reasoning.

> If B is a consequence of A, or if B is equivalent to A, then a 'proof' of A which makes use of B is at best dubious, and may well be a delusion.

The need to avoid such circular reasoning arose already in Problem **21** (the converse of Pythagoras' Theorem), where we felt the need to state explicitly that it would be inappropriate to use the Cosine Rule: (see Problem **192** below).

Such concerns may explain why this chapter on geometry is the last of the chapters relating to elementary 'school mathematics', and why we begin the chapter with

- an apparent digression (Section 5.1), and

- an outline of elementary Euclidean geometry (Section 5.2).

Those with a strong background in geometry may choose to skip these sections on a first reading, and move straight on to the problems which start in Section 5.3. But they may then fail to see how the cumulative architecture of Section 5.2 conveys a rather different aspect of the "essence of mathematics", deriving not just from the individual problems, but from the way a carefully crafted, systematic arrangement of simple "bricks" can create a much more significant mathematical structure.

5.1. Comparing geometry and arithmetic

The opening quotations remind us that the mental universe of formal mathematics draws much of its initial inspiration from human perception and activity – activity which starts with infants observing, moving around, and operating with objects in time and in space. Many of our earliest pre-mathematical experiences are quintessentially proto-geometrical. We make sense of visual inputs; we learn to recognise faces and objects; we crawl around; we learn to look 'behind' and 'underneath' obstructions in search of hidden toys; we sort and we build; we draw and we make; etc.. However, for this experience to develop into *mathematics*, we then need to

- identify certain semi-formal "objects" (points, lines, angles, triangles),
- pinpoint the key relations between them (bisectors, congruence, parallels, similarity), and then
- develop the associated language that allows us to encapsulate insights from prior experience into a coherent framework for calculation and deduction.

Too little attention has been given to achieving a consensus as to how this transition (from *informal experience*, to *formal reasoning*) can best be established for beginners in elementary geometry. In contrast, number and arithmetic move much more naturally

- from our early experience of time and quantity
- to the notation, the operations, the calculational procedures, and the rules of formal arithmetic and algebra.

Counting is rooted in the idea of a *repeated unit* – a notion that may stem from the ever-present, regular heartbeat that envelops every embryo (where the beat is presumably *felt* long before it is *heard*). Later we encounter repeated units with longer time scales (such as the cycles of day and night, and the routines of feeding and sleeping). The first months and years of life are peppered with instances of numerosity, of continuous quantity, of systematic ordering, of sequences, of combinations and partitions, of grouping and replicating, and of relations between quantities and operations – experiences which provide the raw material for the mathematics of number, of place value, of arithmetic, and later of 'internal structure' (or algebra).

The need for political communities to construct a formal school curriculum linking early infant experience and elementary formal mathematics is a recent development. Nevertheless, in the domain of number, quantity, and arithmetic (and later algebra), there is a surprising level of agreement about the steps that need to be incorporated – even though the details may differ in different educational systems and in different classrooms. For example:

- One must somehow establish the idea of a *unit*, which can be replicated to produce larger numbers, or *multiples*.

- One must then group units relative to a chosen base (e.g. 10), iterate this grouping procedure (by taking "ten tens", and then "ten hundreds"), and use *position* to create *place value* notation.

- One must introduce "0" – both as a number in its own right, and as a placeholder for expressing numbers using place value.

- One can then use combinations and differences, multiples and sharing (and partitions), to develop *arithmetic*.

- At some stage one introduces subunits (i.e. *unit fractions*) and submultiples (i.e. multiples of these subunits) to produce *general fractions*; one can then use *equivalence* and common submultiples to extend arithmetic to fractions.

- If we restrict to *decimal fractions*, then our ideas of place value for integers can be extended to the right of the decimal point to produce *decimals*.

- At every stage we need to
  - relate these ideas to *quantities*,
  - require pupils to interpret and solve *word problems*, and
  - cultivate both mental arithmetic and standard written algorithms.

- Towards the end of primary school, attention begins to move beyond bare hands computation, to consciously exploit internal *structure* in preparation for algebra.

Our early *geometrical* experience is just as natural as that relating to number; but it is more subtle. And there is as yet no comparable consensus about the path that needs to be followed if our primitive geometrical experience is to be formalised in a useable way.

The 1960s saw a drive to modernise school mathematics, and at the same time to make it accessible to all. Elementary geometry certainly needed a re-think. But the reformers in most countries simply dismissed the traditional mix (e.g. in England, where one found a blend of technical drawing, Euclidean, and coordinate geometry in different proportions for different groups of students) in favour of more modern-sounding alternatives. Some countries favoured a more abstract, deductive framework; some tried to exploit motion and transformations; some used matrices and groups; some used vectors and linear algebra; some even toyed with topology. More recently we have heard similarly ambitious claims on behalf of dynamic geometry software. And although each approach has its attractions,

none *of the alternatives has succeeded in helping* **more** *students to visualise, to reason, and to calculate effectively in geometrical settings.*

At a much more advanced level, geometry combines

- with abstract algebra (where the approach proposed by Felix Klein (1849–1925) shows how to identify each geometry with a group of transformations), and

- with analysis and linear algebra (where, following Gauss (1777–1855), Riemann (1826–1866) and Grassmann (1809–1877), calculus, vector spaces, and later topology can be used to analyse the geometry of surfaces and other spaces).

However, these subtle formalisms are totally irrelevant for beginners, who need an approach

- based on concepts which are relatively familiar (*points, lines, triangles* etc.), and

- whose basic properties can be formulated relatively simply.

The subtlety and flexibility of dynamic geometry software may be hugely impressive; but if students are to harness this power, they need *prior* mastery of some simple, semi-formal framework, together with the associated language and modes of reasoning. Despite the lack of an accepted consensus, the experience of the last 50 years would seem to suggest that the most relevant framework for beginners at secondary level involves some combination of:

- *static*, relatively traditional Euclidean geometry, and

- Cartesian, or coordinate (analytic) geometry.

5.2.   Euclidean geometry: a brief summary

Philosophy is written in this grand book – I mean the universe – which stands continually open to our gaze, but it cannot be understood unless one first learns to comprehend the language and to interpret the characters in which it is written. It is written in the language of mathematics, and its characters are triangles, circles, and other geometrical figures, without which it is humanly impossible to understand a single word of it;

> *without these, one is wandering about in a dark labyrinth.*
> Galileo Galilei (1564–1642)

This section provides a detailed, but compressed, outline of an initial formalisation of school geometry – of a kind that one would like good students and all teachers to appreciate. It is unashamedly a *semi*-formal approach for beginners, **not** a strictly formal treatment (such as that provided by David Hilbert (1862–1943) in his 1899 book *Foundations of Geometry*, or in the more detailed exposition by Edwin Moise (1918–1998) *Elementary Geometry from an Advanced Standpoint*, published in 1963). In particular:

- we work with relatively informal notions of *points*, *lines*, and *angles* in the plane;

- we focus attention on certain simple issues which really matter at school level (such as how points, lines, line segments, and angles are referred to; the notion of a triangle as an *ordered* triple of vertices; the fact that the vertices of a quadrilateral must be labelled cyclically; etc.);

- we limit the formal deductive structure to just three central criteria, namely the criteria for *congruence*, for *parallels*, and for *similarity*, and show how they allow one to develop results and methods in a logical sequence.

We begin with the intuitive idea of *points* and *lines* in the plane. Two points A, B determine

- the *line segment* $\underline{AB}$ (with endpoints A and B), and

- the *line* AB (which extends the line segment $\underline{AB}$ in both directions – beyond A, and beyond B).

Three points A, B, C determine an angle $\angle ABC$ (between the two line segments $\underline{BA}$ and $\underline{BC}$).

We can then begin to build more complicated figures, such as

- a triangle ABC (with three vertices A, B, C; three sides $\underline{AB}$, $\underline{BC}$, $\underline{CA}$; and three angles $\angle ABC$ at the vertex B, $\angle BCA$ at C, and $\angle CAB$ at A),

- a quadrilateral $ABCD$ (with four vertices A, B, C, D; and four sides $\underline{AB}$, $\underline{BC}$, $\underline{CD}$, $\underline{DA}$ which meet only at their endpoints).

And so on. Two given points A, B also allow us to construct the *circle* with centre A, and passing through B (that is, with *radius* $\underline{AB}$).

This very limited beginning already opens up the world of *ruler and compasses constructions*. In particular, given a line segment $\underline{AB}$, one can draw:

- the circle with centre A, and passing through B, and
- the circle with centre B, and passing through A.

If the two circles meet at C,

- then $\underline{AB} = \underline{AC}$ (radii of the first circle), and $\underline{BA} = \underline{BC}$ (radii of the second circle).

Hence we have constructed the *equilateral triangle* $\triangle ABC$ on the given segment $\underline{AB}$. This construction is the very first proposition in Book 1 of the *Elements* of Euclid (flourished c. 300 BC). Euclid's second proposition is presented next as a problem.

Problem 137 Given three points A, B, C, show how to construct – without measuring – a point D such that the segments $\underline{AB}$ and $\underline{CD}$ are equal (in length). △

Problem **137** looks like a simple starter (where the only available construction is to produce the third vertex of an equilateral triangle on a given line segment). However, to produce a valid solution requires a clear head and a degree of ingenuity.

Given two points A, B, the process of constructing an equilateral triangle $\triangle ABC$ illustrates how we are allowed to construct new points from old.

- Whenever we construct two lines or circles that cross, the points where they cross (such as the point C in the above construction of the equilateral triangle $\triangle ABC$) become available for further constructions. So, if points A and B are given, then once C has been constructed, we may proceed to draw the lines AC and BC.

However, the fact that we can construct a line segment $\underline{AB}$ does not allow us to 'measure' the segment with a ruler, and then to use the resulting measurement to 'copy' the segment $\underline{AB}$ to the point C in order to construct the required point D such that $\underline{AB} = \underline{CD}$. The "ruler" in *ruler and compasses constructions* is used only to draw the line through two known points – not to measure. (Measuring is an *approximate* physical action, rather than an *exact* "mental construction", and so is not really part of mathematics.) Hence in Problem **137** we have to find another way to produce a copy $\underline{CD}$ of the segment $\underline{AB}$ starting at the point C. Similarly, we can construct the circle with centre A and passing through B, but this does not allow us to use the pair of compasses to transfer distances physically (e.g. by picking up the compasses from $\underline{AB}$ and placing the compass point at C, like using the old geometrical drawing instrument that was called *a pair of dividers*). In seeking the construction required in Problem **137**, we are restricted to "exact mental constructions" which may be described in terms of:

- drawing (or constructing) the line joining any two known points,

- constructing the circle with centre at a known point and passing through a known point, and

- obtaining a new point D as the intersection of two constructed lines or circles (or of a line and a circle).

If on the line AB, the point X lies between A and B, then we obtain a *straight angle* $\angle AXB$ at X (or rather *two* straight angles at X – one on each side of the line AB). If we assume that all straight angles are equal, then it follows easily that "vertically opposite angles are always equal".

Problem 138 Two lines AB and CD cross at X, where X lies between A and B and between C and D. Prove that $\angle AXC = \angle BXD$. △

Define a *right angle* to be 'half a straight angle'. Then we say that two lines which cross at a point X are *perpendicular* if an angle at X is a right angle (or equivalently, if all four angles at X are equal). The next step requires us to notice two things – partly motivated by experience when coordinating hand, eye and brain to construct, and to think about, physical structures.

- First we need to recognise that *triangles* hold the key to the analysis of more complicated shapes.

- Then we need to realise that triangles in different positions can still be "equal", or *congruent* – which then focuses attention on the **minimal** conditions under which two triangles can be guaranteed to be congruent.

The first of these two bullet points has an important consequence – namely that solving any problem in 2- or in 3-dimensions generally reduces to working with **triangles**. In particular, solving problems in 3-dimensions reduces to working in some 2-dimensional *cross-section* of the given figure (since three points not only determine a triangle, but also determine the plane in which that triangle lies). It follows that 2-dimensional geometry holds the key to solving problems in 3-dimensions, and that **working with triangles is central in all geometry**.

The second bullet point forces us to think carefully about:

- what we mean by a *triangle* (and in particular, to understand why $\triangle ABC$ and $\triangle BCA$ are in some sense *different* triangles, even though they use the same three vertices and sides), and

- what it means for two triangles to be "the same".

A triangle $\triangle ABC$ incorporates six pieces of data, or information: the three sides $\underline{AB}$, $\underline{BC}$, $\underline{CA}$ and the three angles $\angle ABC$, $\angle BCA$, $\angle CAB$. We say that two (ordered) triangles $\triangle ABC$ and $\triangle A'B'C'$ are *congruent* (which we write as
$$\triangle ABC \equiv \triangle A'B'C',$$
where the order in which the vertices are listed matters) if their sides and angles "match up" in pairs, so that

$$\underline{AB} = \underline{A'B'},\ \underline{BC} = \underline{B'C'},\ \underline{CA} = \underline{C'A'},$$
$$\angle ABC = \angle A'B'C',\ \angle BCA = \angle B'C'A',\ \angle CAB = \angle C'A'B'.$$

As a result of drawing and experimenting with our hands, our minds may realise that certain subsets of these six conditions suffice to imply the others. In particular:

SAS-congruence criterion: if

$$\underline{AB} = \underline{A'B'}, \quad \angle ABC = \angle A'B'C', \quad \underline{BC} = \underline{B'C'},$$

then
$$\triangle ABC \equiv \triangle A'B'C'$$

(where the name "SAS" indicates that the three listed match-ups occur in the specified order S (side), A (angle), S (side) as one goes round each triangle).

SSS-congruence criterion: if

$$\underline{AB} = \underline{A'B'}, \quad \underline{BC} = \underline{B'C'}, \quad \underline{CA} = \underline{C'A'},$$

then
$$\triangle ABC \equiv \triangle A'B'C'.$$

ASA-congruence criterion: if

$$\angle ABC = \angle A'B'C', \quad \underline{BC} = \underline{B'C'}, \quad \angle BCA = \angle B'C'A',$$

then
$$\triangle ABC \equiv \triangle A'B'C'.$$

If in a given triangle $\triangle ABC$ we have $\underline{AB} = \underline{AC}$, then we say that $\triangle ABC$ is *isosceles* with **apex** A, and **base** $\underline{BC}$ (*iso* = same, or equal; *sceles* = legs).

Problem 139 Let $\triangle ABC$ be an isosceles triangle with apex A. Let M be the midpoint of the base $\underline{BC}$. Prove that $\triangle AMB \equiv \triangle AMC$ and conclude that AM is perpendicular to the base $\underline{BC}$. △

Problem 140 Construct two non-congruent triangles, $\triangle ABC$ and $\triangle A'B'C'$, where $\angle BCA = \angle B'C'A' = 30°$, $|\underline{CA}| = |\underline{C'A'}| = \sqrt{3}$, $|\underline{AB}| = |\underline{A'B'}| = 1$.
Conclude that there is in general no "ASS-congruence criterion". △

The congruence criteria allow one to prove basic results such as:

> **Claim** In any isosceles triangle $\triangle ABC$ with apex A (i.e. with $\underline{AB} = \underline{AC}$), the two base angles $\angle B$ and $\angle C$ are equal.
>
> **Proof 1** Let M be the midpoint of $\underline{BC}$.
> Then $\triangle AMB \equiv \triangle AMC$ (by the SSS-congruence criterion, since
> $\underline{AM} = \underline{AM}$,
> $\underline{MB} = \underline{MC}$ (by construction of M as the midpoint)
> $\underline{BA} = \underline{CA}$ (given)).
> $\therefore \angle B = \angle ABM = \angle ACM = \angle C.$ QED
>
> **Proof 2** $\triangle BAC \equiv \triangle CAB$ (by the SAS-congruence criterion, since
> $\underline{BA} = \underline{CA}$ (given),
> $\angle BAC = \angle CAB$ (same angle),
> $\underline{AC} = \underline{AB}$ (given)).
> $\therefore \angle B = \angle ABC = \angle ACB = \angle C.$ QED

We also have the converse result:

> **Claim** In any triangle $\triangle ABC$, if the base angles $\angle B$ and $\angle C$ are equal, then the triangle is isosceles with apex A (i.e. $\underline{AB} = \underline{AC}$).
>
> **Proof** $\triangle ABC \equiv \triangle ACB$ (by the ASA-congruence criterion, since
> $\angle ABC = \angle ACB$ (given),
> $\underline{BC} = \underline{CB}$, and
> $\angle BCA = \angle CBA$ (given)).
> $\therefore \underline{AB} = \underline{AC}.$ QED

Problem 141

(i) A circle with centre O passes through the point A. The line AO meets the circle again at B. If C is a third point on the circle, prove that $\angle ACB$ is equal to $\angle CAB + \angle CBA$.

(ii) Conclude that, if the angles in $\triangle ABC$ add to a straight angle, then $\angle ACB$ is a right angle. △

Once we introduce the parallel criterion, and hence can prove that the three angles in any triangle add to a straight angle, Problem **141** will guarantee that "the angle subtended on the circumference by a diameter is always a right angle".

Problem 142 Show how to implement the basic ruler and compasses constructions:

(i) to construct the midpoint M of a given line segment $\underline{AB}$;

(ii) to bisect a given angle $\angle ABC$;

(iii) to drop a perpendicular from P to a line AB (that is, to locate X on the line AB, so that the two angles that PX makes with the line AB on either side of PX are equal).

Prove that your constructions do what you claim. △

Problem 143 Given two points A and B.

(a) Prove that each point X on the perpendicular bisector of $\underline{AB}$ is equidistant from A and from B (that is, that $\underline{XA} = \underline{XB}$).

(b) Prove that, if X is equidistant from A and from B, then X lies on the perpendicular bisector of $\underline{AB}$. △

Problem **143** shows that, given a line segment $\underline{AB}$, the perpendicular bisector of $\underline{AB}$ is the locus of all points X which are equidistant from A and from B. This observation is what lay behind the construction of the circumcentre of a triangle (back in Chapter 1, Problem **32**(a)):

> Given any $\triangle ABC$.
> Let O be the point where the perpendicular bisectors of $\underline{AB}$ and $\underline{BC}$ meet.
> Then $\underline{OA} = \underline{OB}$
> and $\underline{OB} = \underline{OC}$.
> $\therefore \underline{OA} = \underline{OB} = \underline{OC}$.
> Hence O is the centre of a circle passing through all three vertices A, B, C.
> Moreover O also lies on the perpendicular bisector of $\underline{CA}$.

This circle is called the *circumcircle* of △ABC, and O is called the *circumcentre* of △ABC. As indicated back in Problem **32**, the radius of the circumcircle of △ABC (called the *circumradius* of the triangle) is generally denoted by R. Later we will meet other circles and "centres" associated with a given triangle △ABC.

Before moving on it is worth extending Problem **143** to three dimensions.

Problem 144 Given any two points N, S in 3D space, prove that the locus of all points X which are equidistant from N and from S form the plane perpendicular to the line NS and passing through the midpoint M of $\underline{NS}$. △

The next two fundamental results are often neglected.

Problem 145 Given any △ABC, if we extend the side $\underline{BC}$ beyond C to a point X, then the "exterior angle" ∠ACX at C is greater than each of the "two interior opposite angles" ∠A and ∠B. △

Problem 146

(a) If in △ABC we have $\underline{AB} > \underline{AC}$, then ∠ACB > ∠ABC. ("In any triangle, the larger angle lies opposite the longer side.")

(b) If in △ABC we have ∠ACB > ∠ABC, then $\underline{AB} > \underline{AC}$. ("In any triangle, the longer side lies opposite the larger angle.")

(c) **(The triangle inequality)** Prove that in any triangle △ABC,

$$\underline{AB} + \underline{BC} > \underline{AC}.$$
△

The results in Problems **145** and **146** have surprisingly many consequences. For example, they allow one to prove the converse of the result in Problem **141**.

Problem 147 Suppose that in △ABC, ∠C = ∠A + ∠B. Prove that C lies on the circle with diameter $\underline{AB}$.

(In particular, if the angles of △ABC add to a straight angle, and ∠ACB is a right angle, then C lies on the circle with diameter $\underline{AB}$.) △

We come next to a result whose justification is often fudged. At first sight it is unclear how to begin: there seems to be so little information to work with – just two points and a line through one of the points.

Problem 148 A circle with centre O passes through the point P. Prove that the tangent to the circle at P is perpendicular to the radius $\underline{OP}$. △

Problem **148** is an example of a result which implies its own converse – though in a backhanded way. Suppose a circle with centre O passes through the point P. If OP is perpendicular to a line m passing through P, then m must be tangent to the circle (because we know that the tangent at P is perpendicular to OP, so the angle between m and the tangent is "zero", which forces m to be equal to the tangent). This converse will be needed later, when we meet the *incircle*.

Problem 149 Let P be a point and m a line not passing through P. Prove that, among all possible line segments $\underline{PX}$ with X on the line m, a perpendicular from P to the line m is the shortest. △

The result in Problem **149** allows us to define the "distance" from P to the line m to be the length of any perpendicular from P to m. (As far as we know at this stage of the development, there could be more than one perpendicular from P to m.)

Note that all the results mentioned so far have avoided using the Euclidean "parallel criterion" (or – equivalently – the fact that the three angles in any triangle add to a straight angle). So results proved up to this point should still be "true" in any geometry where we have points, lines, triangles, and circles satisfying the congruence criteria – whether or not the geometry satisfies the Euclidean "parallel criterion".

The idea that there is only one "shortest" distance from a point to a line may seem "obvious"; but it is patently *false* on the sphere, where *every* line (i.e. 'great circle') from the North pole P to the equator is perpendicular to the equator (and all these lines have the same "length"). The proof that there is just one such perpendicular from P to m depends on the *parallel criterion* (see below) – a criterion which fails to hold for geometry on the sphere.

Euclid's *Elements* started with a few basic axioms that formalised the idea of ruler and compasses constructions. He then added a simple axiom that allowed one to compare angles in different locations. He made the forgivable mistake of omitting an axiom for congruence of triangles – imagining that it can be *proved*. (It can't.) However he then stated, and carefully developed the consequences of, a much more subtle axiom about parallel lines (two lines m, n in the plane are said to be *parallel* if they never meet, no matter how far they are extended). For reasons that remain unclear, instead of appreciating that Euclid's "parallel postulate" constituted a profound insight into the foundations of geometry, mathematicians in later ages saw the complexity of Euclid's postulate as some kind of flaw, and so tried to show that it could be derived from the other, simpler postulates. The attempt to "correct" this perceived flaw became a kind of Holy Grail.

The story is instructive, but too complicated to summarise accurately here. The situation was eventually clarified by two nineteenth century mathematicians (more-or-less at the same time, but working independently). In the revolutionary, romantic spirit of the nineteenth century, János Bolyai (Hungarian: 1802–1860) and Nikolai Lobachevski (Russian: 1792–1856) each allowed himself to consider what would happen if one adopted a *different* assumption about how "parallel lines" behave. Both discovered that one can then derive an apparently coherent theory of a completely novel kind, with its own beautiful results: that is, a geometry which seemed to be internally "consistent" – but different from Euclidean geometry. Lobachevski published brief notes of his work in 1829–30 (in Kazan); Bolyai knew nothing of this and published incomplete notes of his researches in 1832. Lobachevski published a more detailed booklet in 1840.

Neither mathematician got the recognition he might have anticipated, and it was only much later (largely after their deaths) that others realised how to show that the fantasy world they had each dreamt up was just as "internally consistent" as traditional Euclidean geometry. The story is further complicated by the fact that the dominant mathematician of the time – namely Gauss (1777–1855) – claimed to have proved something similar (and he may well have done so, but exactly what he knew has to be inferred from cryptic remarks in occasional letters, since he published nothing on the subject). If there is a moral to the story, it could be that success in mathematics may not be recognised, or may only be recognised after one's death: so those who spend their lives exploring the mathematical universe had better appreciate the delights of the mathematical journey, rather than being primarily motivated by a desire for immediate recognition and acclaim!

Two lines m, n in the plane are said to be *parallel* if they never meet – no matter how far they are extended. We sometimes write this as "$m \parallel n$".

Given two lines m, n in the plane, a third line p which crosses both m and n is called a *transversal* of m and n.

> **Parallel criterion**: Given two lines m and n, if some transversal p is such that the two "internal" angles on one side of the line p (that is the two angles that p makes with m and with n, and which lie between the two lines m and n) add to *less than* a straight angle, then the lines m and n must meet on that side of the line p.

If the internal angles on one side of p add to *more than* a straight angle, then internal angles on the *other side* of p add to less than a straight angle, so the lines m and n must meet on the other side of p. It follows

- that two lines m and n are parallel precisely when the two internal angles on one side of a transversal add to **exactly** a straight angle.

Parallel lines can be thought of as "all having the same direction"; so it is convenient to insist that "every line is parallel to itself" (even though it has lots of points in common with itself). It then follows

- that, given three lines k, m, n, if k is parallel to m and m is parallel to n, then k is parallel to n; and

- that given a line m and a point P, there is a unique line n through P which is parallel to m.

All this then allows one

- to conclude that, if m and n are any two lines, and p is a transversal, then m and n are parallel if and only if *alternate angles* are equal (or equivalently, if and only if *corresponding angles* are equal); and

- to extend the basic ruler and compasses constructions to include the construction:

    "given a line AB and a point P,
    construct the line through P which is parallel to AB"

(namely, by first constructing the line PX through P, perpendicular to AB, and then the line through P, perpendicular to PX).

One can then prove the standard result about the angles in any triangle.

Claim The angles in any triangle $\triangle ABC$ add to a straight angle.

Proof Construct the line m through A that is parallel to BC. Then AB and AC are transversals, which cross both the line m and the line BC, and which make three angles at the point A on m:

- one being just the angle $\angle A$ in the triangle $\triangle ABC$,

- one being equal to $\angle B$ (alternate angles relative to the transversal AB) and

- one being equal to $\angle C$ (alternate angles relative to the transversal AC).

The three angles at A clearly add to a straight angle, so the three angles $\angle A$, $\angle B$, $\angle C$ also add to a straight angle. QED

Once we know that the angles in any triangle add to a straight angle, we can prove all sorts of other useful facts. One is a simple reformulation of the above **Claim**.

Problem 150 Given any triangle $\triangle ABC$, extend $\underline{BC}$ beyond C to a point X. Then the exterior angle

$$\angle XCA = \angle A + \angle B.$$

("In any triangle, each exterior angle is equal to the sum of the two interior opposite angles.") △

Another important consequence is the result which underpins the sequence of "circle theorems".

Problem 151 Let O be the circumcentre of $\triangle ABC$. Prove that

$$\angle AOB = 2 \cdot \angle ACB.$$
△

Problem **151** implies that

> "the angles subtended by any chord $\underline{AB}$ on a given arc of the circle are all equal",

and are equal to exactly one half of the angle subtended by $\underline{AB}$ at the centre O of the circle. This leads naturally to the familiar property of *cyclic quadrilaterals*.

Problem 152 Let $ABCD$ be a quadrilateral inscribed in a circle (such a quadrilateral is said to be *cyclic*, and the four vertices are said to be *concyclic* – that is, they lie together on the same circle). Prove that opposite angles (e.g. $\angle B$ and $\angle D$) must add to a straight angle. (Two angles which add to a straight angle are said to be *supplementary*.) △

These results have lots of lovely consequences: we shall see one especially striking example in Problem **164**. Meantime we round up our summary of the "circle theorems".

Problem 153 Suppose that the line XAY is tangent to the circumcircle of $\triangle ABC$ at the point A, and that X and C lie on opposite sides of the line AB. Prove that $\angle XAB = \angle ACB$. △

Problem 154

(a) Suppose C, D lie on the same side of the line AB.

   (i) If D lies inside the circumcircle of $\triangle ABC$, then $\angle ADB > \angle ACB$.

(ii) If D lies outside the circumcircle of $\triangle ABC$, then $\angle ADB < \angle ACB$.

(b) Suppose C, D lie on the same side of the line AB, and that $\angle ACB = \angle ADB$. Then D lies on the circumcircle of $\triangle ABC$.

(c) Suppose that $ABCD$ is a quadrilateral, in which angles $\angle B$ and $\angle D$ are supplementary. Then $ABCD$ is a cyclic quadrilateral. △

Another result which follows now that we know that the angles of a triangle add to a straight angle is a useful additional congruence criterion – namely the **RHS-congruence criterion**. This is a 'limiting case' of the failed ASS-congruence criterion (see the example in Problem **140**). In the failed ASS criterion the given data correspond to two *different* triangles – one in which the angle opposite the first specified side (the first "S" in "ASS") is acute, and one in which the angle opposite the first specified side is obtuse. In the RHS-congruence criterion, the angle opposite the first specified side is a *right angle*, and the two possible triangles are in fact congruent.

RHS-congruence criterion: If $\angle ABC$ and $\angle A'B'C'$ are both right angles, and $\underline{BC} = \underline{B'C'}$, $\underline{CA} = \underline{C'A'}$, then

$$\triangle ABC \equiv \triangle A'B'C'.$$

Proof Suppose that $\underline{AB} = \underline{A'B'}$. Then

$\underline{AB} = \underline{A'B'}$,
$\angle ABC = \angle A'B'C'$,
$\underline{BC} = \underline{B'C'}$.

Hence we may apply the SAS-congruence criterion to conclude that $\triangle ABC \equiv \triangle A'B'C'$.

If on the other hand $\underline{AB} \neq \underline{A'B'}$, we may suppose that $\underline{BA} > \underline{B'A'}$. Now construct A'' on BA such that $\underline{BA''} = \underline{B'A'}$. Then

$\underline{A''B} = \underline{A'B'}$,
$\angle A''BC = \angle A'B'C'$,
$\underline{BC} = \underline{B'C'}$,
$\therefore \triangle A''BC \equiv \triangle A'B'C'$ (by SAS-congruence).
Hence $\underline{A''C} = \underline{A'C'} = \underline{AC}$, so $\triangle CAA''$ is isosceles.
$\therefore \angle CA''A = \angle CAA''$.

However, $\angle CA''A > \angle CBA$ (since the exterior angle $\angle CA''A$ in $\triangle CBA''$ must be greater than the interior opposite angle $\angle CBA$, by Problem **145**).

But then the two base angles in the isosceles triangle $\triangle CAA''$ are each greater than a right angle – so the angle sum of $\triangle CAA''$ is greater than a straight angle, which is impossible. Hence this case cannot occur. QED

RHS-congruence seems to be needed to prove the basic result (Problem **161** below) about the area of parallelograms, and this is then needed in the proof of Pythagoras' Theorem (Problem **18**). In one sense RHS-congruence looks like a special case of SSS-congruence (as soon as two pairs of sides in two right angled triangles are equal, Pythagoras' Theorem guarantees that the third pair of sides are also equal). However this observation cannot be used to justify RHS-congruence if RHS-congruence is needed to justify Pythagoras' Theorem.

Problem 155 Given a circle with centre O, let Q be a point outside the circle, and let QP, QP' be the two tangents from Q, touching the circle at P and at P'. Prove that $\underline{QP} = \underline{QP'}$, and that the line OQ bisects the angle $\angle PQP'$. △

Problem 156 You are given two lines m and n crossing at the point B.

(a) If A lies on m and C lies on n, prove that each point X on the bisector of angle $\angle ABC$ is equidistant from m and from n.

(b) If X is equidistant from m and from n, prove that X must lie on one of the bisectors of the two angles at B. △

Problem **156** shows that, given two lines m and n that cross at B, the bisectors of the two pairs of vertically opposite angles formed at B form the *locus* of **all** points X which are equidistant from the two lines m and n. This allows us to mimic the comments following Problem **143** and so to construct the *incentre* of a triangle.

> Given any $\triangle ABC$, let I be the point where the angle bisectors of $\angle ABC$ and $\angle BCA$ meet.
>
> Let the perpendiculars from I to the three sides AB, BC, CA meet the sides at P, Q, R respectively. Then
>
> $\underline{IP} = \underline{IQ}$ (since I lies on the bisector of $\angle ABC$) and
> $\underline{IQ} = \underline{IR}$ (since I lies on the bisector or $\angle BCA$).
>
> Hence the circle which has centre I and which passes through P also passes through Q and R.
>
> Moreover, I also lies on the bisector of $\angle CAB$; and since the radii $\underline{IP}, \underline{IQ}, \underline{IR}$ are perpendicular to the sides of the triangle, the circle is tangent to the three sides of the triangle (by the comments following Problem **148**).

This circle is called the *incircle* of $\triangle ABC$, and I is called the *incentre* of $\triangle ABC$. The radius of the incircle of $\triangle ABC$ is called the *inradius*, and is generally denoted by r.

A quadrilateral $ABCD$ in which $AB \parallel DC$ and $BC \parallel AD$ is called a *parallelogram*. A parallelogram $ABCD$ with a right angle is a *rectangle*. A parallelogram $ABCD$ with $\underline{AB} = \underline{AD}$ is called a *rhombus*. A rectangle which is also a rhombus is called a *square*.

Problem 157 Let $ABCD$ be a parallelogram.

(i) Prove that $\triangle ABC \equiv \triangle CDA$, so that each triangle has area exactly half of area($ABCD$).

(ii) Conclude that opposite sides of $ABCD$ are equal in pairs and that opposite angles are equal in pairs.

(iii) Let AC and BD meet at X. Prove that X is the midpoint of both $\underline{AC}$ and $\underline{BD}$. △

Problem 158 Let $ABCD$ be a parallelogram with centre X (where the two main diagonals $\underline{AC}$ and $\underline{BD}$ meet), and let m be any straight line passing through the centre. Prove that m divides the parallelogram into two parts of equal area. △

We defined a parallelogram to be "a quadrilateral $ABCD$ in which $AB \parallel DC$ and $BC \parallel AD$"; however, in practice, we need to be able to recognise a parallelogram even if it is not presented in this form. The next result hints at the variety of other conditions which allow us to recognise a given quadrilateral as being a parallelogram "in mild disguise".

Problem 159

(a) Let $ABCD$ be a quadrilateral in which $AB \parallel DC$, and $\underline{AB} = \underline{DC}$. Prove that $BC \parallel AD$, and hence that $ABCD$ is a parallelogram.

(b) Let $ABCD$ be a quadrilateral in which $\underline{AB} = \underline{DC}$ and $\underline{BC} = \underline{AD}$. Prove that $AB \parallel DC$, and hence that $ABCD$ is a parallelogram.

(c) Let $ABCD$ be a quadrilateral in which $\angle A = \angle C$ and $\angle B = \angle D$. Prove that $AB \parallel DC$ and that $BC \parallel AD$, and hence that $ABCD$ is a parallelogram. △

The next problem presents a single illustrative example of the kinds of things which we know in our bones must be true, but where the reason, or proof, may need a little thought.

Problem 160 Let $ABCD$ be a parallelogram. Let M be the midpoint of AD and N be the midpoint of BC. Prove that $MN \parallel AB$, and that MN passes through the centre of the parallelogram (where the two diagonals meet). △

Problem 161 Prove that any parallelogram $ABCD$ has the same area as the rectangle on the same base DC and "with the same height" (i.e. lying between the same two parallel lines AB and DC). △

The ideas and results we have summarised up to this point provide exactly what is needed in the proof of Pythagoras' Theorem outlined back in Chapter 1, Problem **18**. They also allow us to identify two more "centres" of a given triangle $\triangle ABC$.

Problem 162 Given any triangle $\triangle ABC$, draw the line through A which is parallel to BC, the line through B which is parallel to AC, and the line through C which is parallel to AB. Let the first two constructed lines meet at C', the second and third lines meet at A', and the first and third lines meet at B'.

(a) Prove that A is the midpoint of $B'C'$, that B is the midpoint of $C'A'$, and that C is the midpoint of $A'B'$.

(b) Conclude that the perpendicular from A to BC, the perpendicular from B to CA, and the perpendicular from C to AB all meet in a single point H. (H is called the *orthocentre* of $\triangle ABC$.) △

Let the foot of the perpendicular from A to BC be P, the foot of the perpendicular from B to CA be Q, and the foot of the perpendicular from C to AB be R. Then $\triangle PQR$ is called the *orthic triangle* of $\triangle ABC$. The "circle theorems" (especially Problems **151** and **154**(c)) lead us to discover that this triangle has two quite unexpected properties. As a partial preparation for one of the properties we digress slightly to introduce a classic problem.

Problem 163 My horse is tethered at H some distance away from my village V. Both H and V are on the same side of a straight river. How should I choose the shortest route to lead the horse from H to V, if I want to water the horse at the river en route? △

Problem 164 Let $\triangle ABC$ be an acute angled triangle.

(a) Prove that, among all possible triangles $\triangle PQR$ inscribed in $\triangle ABC$, with P on BC, Q on CA, R on AB, the orthic triangle is the one with the shortest perimeter.

(b) Suppose that the sides of $\triangle ABC$ act like mirrors. A ray of light is shone along one side of the orthic triangle PQ, reflects off CA, and the reflected beam then reflects in turn off AB. Where does the ray of light next hit the side BC? (Alternatively, imagine the sides of the triangle as billiard table cushions, and explain the path followed by a ball which is projected, without spin, along PQ.) $\triangle$

We come next to the fourth among the standard "centres of a triangle".

Problem 165 Given $\triangle ABC$, let L be the midpoint of the side $\underline{BC}$. The line AL is called a *median* of $\triangle ABC$. (It is not at all obvious, but if we imagine the triangle as a lamina, having a uniform thickness, then $\triangle ABC$ would exactly balance if placed on a knife-edge running along the line AL.) Let M be the midpoint of the side $\underline{CA}$, so that BM is another median of $\triangle ABC$. Let G be the point where AL and BM meet.

(a)(i) Prove that $\triangle ABL$ and $\triangle ACL$ have equal area. Conclude that $\triangle ABG$ and $\triangle ACG$ have equal area.

(ii) Prove that $\triangle BCM$ and $\triangle BAM$ have equal area. Conclude that $\triangle BCG$ and $\triangle BAG$ have equal area.

(b) Let N be the midpoint of $\underline{AB}$. Prove that CG and GN are the same straight line (i.e. that $\angle CGN$ is a straight angle). Hence conclude that the three medians of any triangle always meet in a point G. $\triangle$

The point where all three medians meet is called the *centroid* of the triangle. For the geometry of the triangle, this is all you need to know. However, it is worth noting that the centroid is the point that would be the 'centre of gravity' of the triangle if the triangle is thought of as a thin lamina with a uniform distribution of mass.

Next we revisit, and reprove in the Euclidean spirit, a result that you proved in Problem **95** using coordinates – namely the *Midpoint Theorem*.

Problem 166 (**The Midpoint Theorem**) Given any triangle $\triangle ABC$, let M be the midpoint of the side $\underline{AC}$, and let N be the midpoint of the side $\underline{AB}$. Draw in $\underline{MN}$ and extend it beyond N to a point M' such that $\underline{MN} = \underline{NM'}$.

(a) Prove that $\triangle ANM \equiv \triangle BNM'$.

(b) Conclude that $BM' = CM$ and that $BM' \parallel CM$.

(c) Conclude that $MM'BC$ is a parallelogram, so that $CB = MM'$. Hence MN is parallel to CB and half its length. △

The *Midpoint Theorem* can be reworded as follows:

> Given $\triangle AMN$.
> Extend AM to C such that $AM = MC$ and extend AN to B such that $AN = NB$.
> Then $CB \parallel MN$ and $CB = 2 \cdot MN$.

This rewording generalizes SAS-congruence in a highly suggestive way, and points us in the direction of "SAS-similarity".

> **SAS-similarity** (×2): if $A'B' = 2 \cdot AB$, $\angle BAC = \angle B'A'C'$, and $A'C' = 2 \cdot AC$, then
> $B'C' = 2 \cdot BC$, $\angle ABC = \angle A'B'C'$, and $\angle BCA = \angle B'C'A'$.

> **Proof** Extend AB to the point B'' such that $AB'' = A'B'$, and extend AC to the point C''' such that $AC''' = A'C'$.
> Then $\triangle B''AC''' \equiv \triangle B'A'C'$ (by SAS-congruence), so $B''C''' = B'C'$, $\angle B''C'''A = \angle B'C'A'$, $\angle C'''B''A = \angle C'B'A'$.
> By construction we have $AB'' = 2 \cdot AB$, and $AC''' = 2 \cdot AC$.
> Hence (by the Midpoint Theorem): $B''C''' = 2 \cdot BC$ (so $B'C' = 2 \cdot BC$), and $BC \parallel B''C'''$ (so $\angle BCA = \angle B''C'''A$ and $\angle CBA = \angle C'''B''A$).
> $\therefore \angle B''C'''A = \angle B'C'A' = \angle BCA$,
> and $\angle C'''B''A = \angle C'B'A' = \angle CBA$. QED

The SAS-similarity (×2) interpretation of the Midpoint Theorem is like the SAS-congruence criterion in that one pair of corresponding angles in $\triangle BAC$ and $\triangle B'A'C'$ are equal, while the sides on either side of this angle in the two triangles are related; but instead of the two pairs of corresponding sides being equal, the sides of $\triangle B'A'C'$ are double the corresponding sides of $\triangle BAC$.

In general we say that

> $\triangle ABC$ is *similar* to $\triangle A'B'C'$ (written as $\triangle ABC \sim \triangle A'B'C'$) with scale-factor m if each angle of $\triangle ABC$ is equal to the corresponding angle of $\triangle A'B'C'$, and if corresponding sides are all in the same ratio:
>
> $$A'B' : AB = B'C' : BC = C'A' : CA = m : 1.$$

If two triangles $\triangle A'B'C'$ and $\triangle ABC$ are similar, with (linear) scale factor $A'B' : AB = m : 1$, then the ratio between their areas is

$$\text{area}(\triangle A'B'C') : \text{area}(\triangle ABC) = m^2 : 1.$$

Two similar triangles $\triangle ABC$ and $\triangle A'B'C'$ give rise to six matching pairs:

- the three pairs of corresponding angles (which are equal in pairs), and
- the three pairs of corresponding sides (which are in the same ratio).

In the case of congruence, the **congruence criteria** tell us that we do not need to check all six pairs to guarantee that two triangles are congruent: these criteria guarantee that certain *triples* suffice. The **similarity criteria** guarantee much the same for similarity.

Suppose we are given triangles $\triangle ABC$, $\triangle A'B'C'$.

AAA-similarity: If

$$\angle ABC = \angle A'B'C', \ \angle BCA = \angle B'C'A', \ \angle CAB = \angle C'A'B',$$

then

$$A'B' : AB = B'C' : BC = C'A' : CA,$$

so the two triangles are similar.

SSS-similarity: If

$$A'B' : AB = B'C' : BC = C'A' : CA,$$

then

$$\angle ABC = \angle A'B'C', \ \angle BCA = \angle B'C'A', \ \angle CAB = \angle C'A'B',$$

so the two triangles are similar.

SAS-similarity: If

$$A'B' : AB = A'C' : AC = m : 1$$

and

$$\angle B'A'C' = \angle BAC,$$

then

and
$$B'C' : BC = A'B' : AB = A'C' : AC$$
$$\angle A'B'C' = \angle ABC, \quad \angle B'C'A' = \angle BCA,$$
so the two triangles are similar.

Our rewording of the Midpoint Theorem gave rise to a version of the third of these criteria, with $m = 2$.

AAA-similarity in right angled triangles is what makes trigonometry possible. Suppose that two triangles $\triangle ABC$, $\triangle A'B'C'$ have right angles at A and at A'. If $\angle ABC = \angle A'B'C'$, then (since the angles in each triangle add to two right angles) we also have $\angle BCA = \angle B'C'A'$. It then follows (from AAA-similarity) that
$$A'B' : AB = B'C' : BC = C'A' : CA,$$
so the trig ratio in $\triangle ABC$
$$\sin B = \frac{AC}{BC}$$
has the same value as the corresponding ratio in $\triangle A'B'C'$
$$\sin B' = \frac{A'C'}{B'C'}.$$

Hence this ratio *depends only on the angle B*, and not on the triangle in which it occurs. The same holds for $\cos \angle B$ and for $\tan \angle B$.

The art of solving geometry problems often depends on looking for, and identifying, similar triangles hidden in a complicated configuration. As an introduction to this, we focus on three classic properties involving circles, where the figures are sufficiently simple that similar triangles should be fairly easy to find.

Problem 167 The point P lies outside a circle. The tangent from P touches the circle at T, and a secant from P cuts the circle at A and at B. Prove that $PA \times PB = PT^2$. △

Problem 168 The point P lies outside a circle. Two secants from P meet the circle at A, B and at C, D respectively. Prove in two different ways that
$$PA \times PB = PC \times PD.$$
△

Problem 169 The point P lies inside a circle. Two secants from P meet the circle at A, B and at C, D respectively. Prove that

$$\underline{PA} \times \underline{PB} = \underline{PC} \times \underline{PD}. \qquad \triangle$$

We end our summary of the foundations of Euclidean geometry by deriving the familiar formula for the area of a trapezium and its 3-dimensional analogue, and a formulation of the similarity criteria which is often attributed to Thales (Greek 6th century BC).

Problem 170 Let $ABCD$ be a trapezium with $AB \parallel DC$, in which AB has length a and DC has length b.

(a) Let M be the midpoint of $\underline{AD}$ and let N be the midpoint of $\underline{BC}$. Prove that $MN \parallel AB$ and find the length of $\underline{MN}$.

(b) If the perpendicular distance between AB and DC is d, find the area of the trapezium $ABCD$. $\triangle$

Problem 171 A pyramid $ABCDE$, with apex A and square base $BCDE$ of side length b, is cut parallel to the base at height d above the base, leaving a frustum of a pyramid, with square upper face of side length a. Find a formula for the volume of the resulting solid (in terms of a, b, and d). $\triangle$

The following general result allows us to use "equality of ratios of line segments" whenever we have three parallel lines (without first having to conjure up similar triangles).

Problem 172 (Thales' Theorem) The lines AA' and BB' are parallel. The point C lies on the line AB, and C' lies on the line $A'B'$ such that $CC' \parallel BB'$. Prove that $\underline{AB} : \underline{BC} = \underline{A'B'} : \underline{B'C'}$. $\triangle$

Under certain conditions, the similarity criteria guarantee the equality of ratios of sides of two triangles. Thales' Theorem extends this "equality of ratios" to line segments which arise whenever two lines cross three parallel lines. One of the simplest, but most far-reaching, applications of this result is the tie-up between geometry and algebra which lies behind ruler and compasses constructions, and which underpins Descartes' (1596–1650) re-formulation of geometry in terms of coordinates (see Problem **173**).

Thales (c. 620–c. 546 BC) was part of the flowering of Greek thought having its roots in Milesia (in the south west of Asia Minor, or modern Turkey). Thales seems to have been interested in almost everything – philosophy, astronomy, politics, and also geometry. In Britain his name is usually attached to the fact that the angle subtended by a diameter is a right angle. On the continent, his name is more strongly attached to the result in Problem **172**. His precise contribution to geometry is unclear – but he seems to have played a significant role in kick-starting what became (300 years later) the polished version of Greek mathematics that we know today.

Thales' contributions in other spheres were perhaps even more significant than in geometry. He seems to have been among the first to try to "explain" phenomena in reductionist terms – identifying "water" as the single "element", or first principle, from which all substances are derived. Anaximenes (c. 586–c. 526 BC) later argued in favour of "air" as the first principle. These two elements, together with "fire" and "earth", were generally accepted as the four Greek "elements" – each of which was supposed to contribute to the construction of observed matter and change in different ways.

Problem 173 To define "length", we must first decide which line segment is deemed to have unit length. So suppose we are given line segments $\underline{XY}$ of length 1, $\underline{AB}$ of length a, (i.e. $\underline{AB} : \underline{XY} = a : 1$), and $\underline{CD}$ of length b.

(a) Use Problem **137** to construct a segment of length $a + b$, and if $a \geqslant b$, a segment of length $a - b$.

(b) Show how to construct a line segment of length ab and a segment of length $\frac{a}{b}$.

(c) Show how to construct a line segment of length $\sqrt{a}$. △

5.3. Areas, lengths and angles

Problem 174 A rectangular piece of fruitcake has a layer of icing on top and down one side to form a larger rectangular slab of cake (as shown in Figure 3).

Figure 3: Icing on the cake

Describe how to make a single straight cut so as to divide both the fruitcake and the icing exactly in half. (The thickness of the icing on top is not necessarily the same as the thickness down the side.) △

Problem 175

(a) What is the angle between the two hands of a clock at 1:35? Can you find another time when the angle between the two hands is the same as this?

(b) How many times each day do the two hands of a clock 'coincide'? And at what times do they coincide?

(c) If we add a second hand, how many times each day do the three hands coincide? △

Problem 176 The twelve hour marks for a clock are marked on the circumference of a unit circle to form the vertices of a regular dodecagon $ABCDEFGHIJKL$. Calculate exactly (i.e. using Pythagoras' Theorem rather than trigonometry) the lengths of all the possible line segments joining two vertices of the dodecagon. △

Problem 177 Consider the lattice of all points (k, m, n) in 3-dimensions with integer coordinates k, m, n. Which of the following distances can be realised between lattice points?

$$\sqrt{1}, \sqrt{2}, \sqrt{3}, \sqrt{4}, \sqrt{5}, \sqrt{6}, \sqrt{7}, \sqrt{8}, \sqrt{9}, \sqrt{10}, \sqrt{11}, \sqrt{12}, \sqrt{13}, \sqrt{14}, \sqrt{15}, \sqrt{16}, \sqrt{17}.$$
△

Problem 178

(a) Five vertices A, B, C, D, E are arranged in cyclic order. However instead of joining each vertex to its two immediate neighbours to form a convex pentagon, we join each vertex to the *next but one* vertex to form a pentagonal star, or pentagram $ACEBD$. Calculate the sum of the five "angles" in any such pentagonal star.

(b) There are two types of 7-gonal stars. Calculate the sum of the angles at the seven vertices for each type.

(c) Try to extend the previous two results (and the proofs) to arbitrary n-gonal stars. △

Problem 179

(a) A regular pentagon $ABCDE$ with edges of length 1 is surrounded in the plane by five new regular pentagons – $ABLMN$ joined to $\underline{AB}$, $BCOPQ$ joined to $\underline{BC}$, and so on.

(i) Prove that M, N, X, Y lie on a line.

(ii) Prove that $MPSVY$ is a regular pentagon.

(iii) Find the edge length of this larger surrounding regular pentagon.

(b) Given a regular pentagon $MPSVY$, with edge length 1, draw the five diagonals to form the pentagram $MSYPV$. Let PY meet MV at A, and MS at B; let PV meet MS at C and SY at D; and let SY meet VM at E.

(i) Prove that $ABCDE$ is a regular pentagon.

(ii) Prove that A, B, and M are three vertices of a regular pentagon $ABLMN$, where L lies on MP and N lies on MY.

(iii) Find the edge length of the regular pentagon $ABCDE$. △

5.4. Regular and semi-regular tilings in the plane

In Problem **36** we saw that a regular n-gon has a *circumcentre* O. If we join each vertex to the point O, we get n triangles, each with angle sum π. Hence the total angle sum in all n triangles is πn. Since the n angles around the point O add to 2π, the angles of the regular n-gon itself have sum $(n-2)\pi$. Hence each angle of the regular n-gon has size $\left(1 - \frac{2}{n}\right)\pi$. (In the next chapter you will prove the general result that the sum of the angles in **any** n-gon is equal to $(n-2)\pi$ radians.)

Problem 180 A *regular tiling* of the plane is an arrangement of identical regular polygons, which fit together edge-to-edge so as to cover the plane with no overlaps.

(a) Prove that if a regular tiling of the plane with p-gons is possible, then $p = 3, 4,$ or 6.

(b) Prove that a regular tiling of the plane exists for each of the values in (a). △

We refer to the arrangement of tiles around a vertex as the *vertex figure*. In a regular tiling all vertex figures are automatically identical, so it is natural to refer to the tiling in terms of its vertex figure. When $p = 3$, exactly $q = 6$ tiles fit together at each vertex, and we abbreviate "six equilateral triangles" as 3^6. In the same way we denote the tiling whose vertex figure consists of "four squares" as 4^4, and the tiling whose vertex figure consists of "three regular hexagons" as 6^3.

The natural approach in part (a) of Problem **180** is first to identify which *vertex figures* have no gaps or overlaps – giving a *necessary* condition for a regular tiling to exist. It is tempting to stop there, and to **assume** that this obvious *necessary* condition is also *sufficient*. The temptation arises in part because 2-dimensional regular tilings are so familiar. But it is important to recognize the distinction between a necessary and a sufficient condition; so the temptation should be resisted, and a construction given.

The procedure hidden in the solution to Problem **180** illustrates a key strategy, which dates back to the ancient Greeks, and which is called the *method of analysis*.

- First, we *imagine* that we have a typical solution to the problem – perhaps by giving it a name (even though we do not yet know anything about such a solution).

- We then use the given conditions to deduce features which any such solution must *necessarily* have.

- And we continue deriving more and more necessary conditions until we believe our list of derived conditions may also be *sufficient*.

- Finally we show that any configuration which satisfies our final derived list of necessary conditions is in fact a solution to the original problem, so that the list of necessary conditions is in fact *sufficient*, and we have effectively pinned down all possible solutions.

This is what we did in a very simple way in the solution to Problem **180**: the condition on vertex figures gave an evident necessary condition, which turned out to be sufficient to guarantee that such a tiling exists. The same general strategy guided our classification of *primitive Pythagorean triples* back in Problem **23**.

In the seventeenth century, this ancient Greek strategy was further developed by Fermat (1601–1665), and by Descartes (1596–1650). For example, Fermat left very few proofs; but his proof that the equation

$$x^4 + y^4 = z^4$$

has no solutions in positive integers x, y, z illustrated the method:

- Fermat started by supposing that a solution exists, and concluded that (x^2, y^2, z^2) would then be a Pythagorean triple.

- The known formula for such Pythagorean triples then allowed him to derive even stronger necessary conditions on x, y, z.

- These conditions were so strong they could never be satisfied!

Descartes developed a "method", whereby hard geometry problems could be solved by translating them into algebra – essentially using the *method of analysis*.

- Faced with a hard problem, Descartes first imagined that he had a point, or a locus, or a curve of the kind required for a solution.
- Then he introduced coordinates "x" and "y" to denote unknowns that were linked in the problem to be solved, and interpreted the given conditions as *equations* which the unknowns x and y would have to satisfy (i.e. as *necessary* constraints).
- The solutions to these equations then corresponded to possible solutions of the original problem.
- Sometimes the algebra did not quite generate a *sufficient* condition, giving rise to "pseudo-solutions" (values of x that satisfy the necessary conditions, but which did not correspond to actual solutions). So it was important to check each apparent solution – exactly as we did in Problem **180**(b), where we checked that we can construct tilings for each of the vertex figures that arise in part (a).

The importance of the final step in this process (checking that the list of necessary constraints is also sufficient) is underlined in the next problem where we try to classify certain "almost regular" tilings.

Problem 181 A *semi-regular tiling* of the plane is an arrangement of regular polygons (not necessarily all identical), which fit together edge-to-edge so as to cover the plane without overlaps, and such that the arrangements of tiles around any two vertices are congruent.

(a)(i) Refine your argument in Problem **180**(a) to list all possible vertex figures in a semi-regular tiling.

 (ii) Try to find additional necessary conditions to eliminate vertex figures which cannot be realized, until your list of necessary conditions seems likely to be sufficient.

(b) The necessary conditions in part (a) give rise to a finite list of possible vertex figures. Construct all possible tilings corresponding to this list of possible vertex figures. △

Semi-regular tilings are often called *Archimedean* tilings. The reason for this name remains unclear. Pappus (c. 290–c. 350 AD), writing more than 500 years after the death of Archimedes (d. 212 BC), stated that Archimedes classified the semi-regular *polyhedra*. Now the classification of semi-regular polyhedra (Problem **190**) uses a similar approach to the classification of planar tilings, except that the sum of the angles at each vertex has sum *less than* (rather than exactly equal to) 360°. So it may be that the semi-regular tilings are named after Archimedes simply because he did something similar for polyhedra; or it may be that, since inequalities are harder to control than equalities, someone inferred (perhaps dodgily) that Archimedes must have known about semi-regular tilings as well as about semi-regular polyhedra.

Whatever the reason, tilings and polyhedra have fascinated mathematicians, artists and craftsmen for all sorts of unexpected reasons – as indicated by:

- the fact that the classification and construction of the five regular polyhedra appear as the culmination of the thirteen books of *Elements* by Euclid (flourished c. 300 BC);
- the ancient Greek attempt to link the five regular polyhedra with the four elements (earth, air, fire, and water) and the cosmos;
- the ceramic tilings to be found in Islamic art - for example, on the walls of the *Alhambra* in Grenada;
- the book *De Divina Proportione* by Luca Pacioli (c. 1445–1509), and the continuing fascination with the Golden Ratio;
- the geometric sketches of Leonardo da Vinci (1452–1519);
- the work of Kepler (1571–1630) who used the regular polyhedra to explain his bold theoretical cosmology in the *Astronomia Nova* (1609).

5.5. Ruler and compasses constructions for regular polygons

Euclid's *Elements* include methods for constructing the regular polygons that are required for the construction of the regular polyhedra (see Section 5.6). In one sense, Euclid is thoroughly modern: he is reluctant to work with entities that cannot be *constructed*. And for him, geometrical construction means construction "using ruler and compasses" only.

For each regular polygon, there are two related (and sometimes very different) construction problems:

- given two points A and B, construct the regular n-gon with $\underline{AB}$ as an *edge* of the regular polygon;
- given two points O and A, construct the regular n-gon $ABCD\cdots$ inscribed in the circle with centre O and passing through A, that is with *circumradius* $\underline{OA}$.

Before Problem **137** we saw how to construct an equilateral triangle ABC given the points A, B. And in Problem **36** we saw that every regular polygon has a circumcentre O.

Problem 182 Given points O, A, show how to construct the regular 3-gon ABC with circumcentre O. △

Problem 183

(a) Given two points O, A, show how to construct a regular 4-gon $ABCD$ with circumcentre O.

(b) Given points A, B, show how to construct a regular 4-gon $ABCD$. △

Problem 184

(a)(i) Given two points O, A, show how to construct a regular 6-gon $ABCDEF$ with circumcentre O.

 (ii) Given two points O, A, show how to construct a regular 8-gon $ABCDEFGH$ with circumcentre O.

(b)(i) Given points A, B, show how to construct a regular 6-gon $ABCDEF$.

 (ii) Given points A, B, show how to construct a regular 8-gon $ABCDEFGH$. △

Problem 185

(a)(i) Given two points O, A, show how to construct a regular 5-gon $ABCDE$ with circumcentre O.

 (ii) Given points O, A, show how to construct a regular 10-gon $ABCDEFGHIJ$ with circumcentre O.

(b)(i) Given points A, B, show how to construct a regular 5-gon $ABCDE$.

 (ii) Given points A, B, show how to construct a regular 10-gon $ABCDEFGHIJ$. △

We shall not prove it here, but it is impossible to construct a regular 7-gon, or a regular 9-gon, or a regular 11-gon using ruler and compasses. All constructions with ruler and compasses come down to two moves:

- if a is a known length, then $\sqrt{a}$ can be constructed (see Problem **173**(c));

- if an n-gon can be constructed, then the sides can be bisected to produce a $2n$-gon.

Put slightly differently, all ruler and compasses constructions involve solving linear or quadratic equations, so the only new points, or lengths we can construct are those which involve iterated square roots of expressions or lengths which were previously known.

This iterated extraction of square roots is linked to a fact first proved by Gauss (1777–1855), namely that the only regular p-gons (where p is a prime) that can be constructed are those where p is a Fermat prime – that is, a prime of the form $p = 2^k + 1$ (in which case k has to be a power of 2: see Problem **118**). Gauss proved (as a teenager, though it was first published in his book *Disquisitiones arithmeticae* in 1801):

> a regular n-gon can be constructed with ruler and compasses if and only if n has the form
> $$2^m \cdot p_1 \cdot p_2 \cdot p_3 \cdots p_k,$$
> where $p_1, p_2, p_3, \ldots, p_k$ are distinct Fermat primes.

As we noted in Chapter 2, the only known Fermat primes are the five discovered by Fermat himself, namely 3, 5, 17, 257, and 65 537.

5.6. Regular and semi-regular polyhedra

We have seen how regular polygons sometimes fit together edge-to-edge in the plane to create tilings of the whole plane. When tiling the plane, the angles of polygons meeting edge-to-edge around each vertex must add to 360°, or two straight angles. If the angles at a vertex add to *less than* 360°, then we are left with an empty gap and two free edges; and when these two free edges are joined, or glued together, the vertex figure rises out of the plane and becomes a 3-dimensional corner, or *solid angle*.

To form such a corner we need at least three polygons, or faces – and hence at least three edges and three faces meet around each vertex. For example, three squares fit nicely together in the plane, but leave a 90° gap. When the two spare edges are glued together, the result is to form a corner of a cube, where we have a vertex figure consisting of three regular 4-gons: so we refer to this vertex figure as 4^3.

Given a 3-dimensional corner, it may be possible to extend the construction, repeating the same vertex figure at every vertex. The resulting shape may then 'close up' to form a *convex* polyhedron. The assumption that in each vertex figure, the angles meeting at that vertex add to less than 360°, means that all the corners then project outwards – which is roughly what we mean when we say that the polyhedron is "convex".

A *regular polygon* is an arrangement of finitely many congruent line segments, with two line segments meeting at each vertex (and never crossing, or meeting internally), and with all vertices alike; a regular polygon can be inscribed in a circle (Problem **36**), and so encloses a convex subset of the plane. In the same spirit, a *regular polyhedron* is an arrangement of finitely many congruent regular polygons, with two polygons meeting at each edge, and with the same number of polygons in a single cycle around every vertex, enclosing a convex subset of 3-dimensional space (i.e. the polyhedron separates the remaining points of 3D into those that lie 'inside' and those that lie 'outside', and the line segment joining any two points of the polyhedral surface contains no points lying outside the polyhedron).

The important constraints here are the assumptions: that the polygons meet edge-to-edge with exactly two polygons meeting at each edge; that the same number of polygons meet around every vertex; and that the overall number of polygons, or faces, is *finite*. The assumption that the figure is convex should be seen as a temporary additional constraint, which means that the angles in polygons meeting at each vertex have sum less than $360°$.

Problem 186 A vertex figure is to be formed by fitting regular p-gons together, edge-to-edge, for a fixed p. If there are q of these p-gons at a vertex, we denote the vertex figure by p^q. If the angles at each vertex add to less than $360°$, prove that the only possible vertex figures are 3^3, 3^4, 3^5, 4^3, 5^3. △

The vertex figure 4^3 is realized by the way the positive axes meet at the vertex $(0, 0, 0)$, where

- the unit square $(0, 0, 0)$, $(1, 0, 0)$, $(1, 1, 0)$, $(0, 1, 0)$ in the xy-plane (with equation $z = 0$) meets

- the unit square $(0, 0, 0)$, $(1, 0, 0)$, $(1, 0, 1)$, $(0, 0, 1)$ in the xz-plane (with equation $y = 0$), and

- the unit square $(0, 0, 0)$, $(0, 1, 0)$, $(0, 1, 1)$, $(0, 0, 1)$ in the yz-plane (with equation $x = 0$).

If we include an eighth vertex $(1, 1, 1)$, and

- the unit square $(0, 0, 1)$, $(1, 0, 1)$, $(1, 1, 1)$, $(0, 1, 1)$ in the plane with equation $z = 1$,

- the unit square $(0, 1, 0)$, $(1, 1, 0)$, $(1, 1, 1)$, $(0, 1, 1)$ in the plane with equation $y = 1$,

- the unit square $(1, 0, 0)$, $(1, 1, 0)$, $(1, 1, 1)$, $(1, 0, 1)$ in the plane with equation $x = 1$, we see that all eight vertices have the same vertex figure 4^3. Hence the possible vertex figure 4^3 in Problem **186** arises as the vertex figure of a regular polyhedron – namely the *cube*.

If we select the four vertices whose coordinates have odd sum $A = (1,0,0)$, $B = (0,1,0)$, $C = (0,0,1)$, $D = (1,1,1)$, then the distance between any two of these vertices is equal to $\sqrt{2}$, so each triple of vertices (such as $(1,0,0)$, $(0,1,0)$, $(0,0,1)$) defines a regular 3-gon ABC, with three such 3-gons meeting at each vertex of $ABCD$. Hence the possible vertex figure 3^3 in Problem **186** arises as the vertex figure of a regular polyhedron – namely the regular *tetrahedron* (*tetra* = four; *hedra* = faces).

Problem 187 With $A = (1,0,0)$ etc. as above, write down the coordinates of the six midpoints of the edges of the regular tetrahedron $ABCD$ (or equivalently, the six centres of the faces of the original cube). Each edge of the regular tetrahedron meets four other edges of the regular tetrahedron (e.g. $\underline{AB}$ meets $\underline{AC}$ and $\underline{AD}$ at one end, and $\underline{BC}$ and $\underline{BD}$ at the other end). Choose an edge $\underline{AB}$ and its midpoint P. Calculate the distance from P to the midpoints Q, R, S, T of the four edges which $\underline{AB}$ meets (namely the midpoints of $\underline{AC}$, $\underline{AD}$, $\underline{BD}$, $\underline{BC}$ respectively). Confirm that the triangles $\triangle PQR$, $\triangle PRS$, $\triangle PST$, $\triangle PTQ$ are all regular 3-gons, and that the vertex figure at P is of type 3^4. Conclude that the possible vertex figure 3^4 in Problem **186** arises as the vertex figure of a regular polyhedron $PQRSTU$ – namely the regular *octahedron* (*octa* = eight; *hedra* = faces). △

Problem 188

(a) A *regular tetrahedron* $ABCD$ has edges of length 2, and sits with its base BCD on the table. Find the height of A above the base.

(b) A *regular octahedron* $ABCDEF$ has four triangles meeting at each vertex.

  (i) Let the four triangles which meet at A be ABC, ACD, ADE, AEB. Prove that $BCDE$ must be a square.

  (ii) Suppose that all the triangles have edges of length 2, and that the octahedron sits with one face BCF on the table – next to the regular tetrahedron from part (a). Which of these two solids is the taller? △

Problem 189 Let $O = (0,0,0)$, $A = (1,0,0)$, $B = (0,1,0)$, $C = (0,0,1)$ be four vertices of the cube as described after Problem **186** above. Draw equal and parallel line segments (initially of unknown length $1 - 2a$) through the centres of each pair of opposite faces – running in the three directions parallel to OA, or to OB, or to OC,

- from $N = (a, \frac{1}{2}, 0)$ to $P = (1-a, \frac{1}{2}, 0)$ and from $Q = (a, \frac{1}{2}, 1)$ to $R = (1-a, \frac{1}{2}, 1)$

- from $S = (\frac{1}{2}, 0, a)$ to $T = (\frac{1}{2}, 0, 1-a)$ and from $U = (\frac{1}{2}, 1, a)$ to $V = (\frac{1}{2}, 1, 1-a)$

- from $W = (0, a, \frac{1}{2})$ to $X = (0, 1-a, \frac{1}{2})$ and from $Y = (1, a, \frac{1}{2})$ to $Z = (1, 1-a, \frac{1}{2})$.

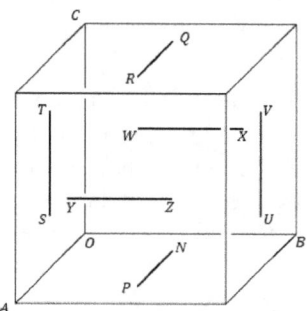 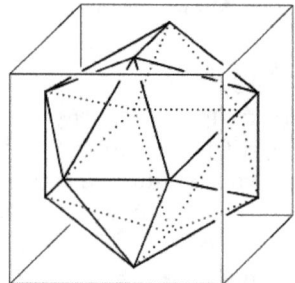

Figure 4: Construction of the regular icosahedron.

These are to form all 12 vertices and six of the 30 edges (of length $1 - 2a$) of a polyhedron, see Figure 4. The other 24 edges join each of these 12 vertices to its four natural neighbours on adjacent faces of the cube – to form the 20 triangular faces of the polyhedron: for example,

N joins: to S; to W; to X; and to U.

(i) Prove that $\underline{NS} = \underline{NW} = \underline{NX} = \underline{NU}$ and calculate the length of $\underline{NS}$.

(ii) Choose the value of the parameter a to guarantee that $\underline{NP} = \underline{NS}$, so that the five triangular faces meeting at the vertex N are all equilateral triangles, and each vertex figure of the resulting polyhedron then has vertex figure 3^5. △

The polyhedron is called the *regular icosahedron* (*icosa* = twenty, *hedra* = faces).

In the paragraph before Problem **187** we constructed the *dual* of the cube by marking the circumcentre of each of the six square faces of the cube, and then joining each circumcentre to its four natural neighbours. We now construct the *dual* of the regular icosahedron in exactly the same way. Each of the 20 circumcentres of the 20 triangular faces of a regular icosahedron has three natural neighbours (namely the circumcentres of the three neighbouring triangular faces). If we construct the 30 edges joining these 20 circumcentres, the five circumcentres of the five triangles in each vertex figure of the regular icosahedron form a regular pentagon, which becomes a face of the dual polyhedron – so we get 12 regular pentagons (one for each vertex of the regular icosahedron), with three pentagons meeting at each vertex of the dual polyhedron to give a vertex figure 5^3 at each of the 20 vertices, which form a *regular dodecahedron*.

Hence each of the five possible vertex figures in Problem **186** can be realised by a regular polyhedron. These are sometimes called the *Platonic solids* because Plato (c. 428–347 BC) often used them as illustrative examples in his writings on philosophy.

Constructing the five regular polyhedra is part of the essence of mathematics for everyone. In contrast, what comes next (in Problem **190**) may be viewed as "optional" at this stage. The ideas are worth noting, but the details may be best postponed for a rainy day.

Just as you classified semi-regular *tilings* in Section 5.4, so one can look for semi-regular *polyhedra*. A polyhedron is semi-regular if all of its faces are regular polygons (possibly with differing numbers of edges), fitting together edge-to-edge, with exactly the same ring of polygons around each vertex – the *vertex figure* of the polyhedron. Problem **190** uses "the method of analysis" - combining simple arithmetic, inequalities, and a little geometric insight – to achieve a remarkable complete classification of semi-regular polyhedra. There are usually said to be thirteen individual semi-regular polyhedra (excluding the five regular polyhedra); but one of these has a vertex figure that extends to a polyhedron in two different ways – each being the reflection of the other. There are in addition two infinite families – namely

- the n-gonal *prisms*, which consist of two parallel regular n-gons, with the top one positioned exactly above the bottom one, the two being joined by a belt of n squares (so with vertex figure $n \cdot 4^2$); and

- the n-gonal *antiprisms*, which consist of two parallel regular n-gons, but with the top n-gon turned through an angle of $\frac{\pi}{n}$ radians relative to the bottom one, the two being joined by a belt of $2n$ equilateral triangles (so with a vertex figure $n \cdot 3^3$).

Notice that the cube can also be interpreted as being a "4-gonal prism", and the regular octahedron can be interpreted as being a "3-gonal antiprism". Those interested in regular and semi-regular polyhedra are referred to the classic book *Mathematical models* by H.M. Cundy and A.P. Rollett.

Problem 190 Find possible combinations of three or more regular polygons whose angles add to less than 360°, and hence derive a complete list of possible vertex figures for a (convex) semi-regular polyhedron. Try to eliminate those putative vertex figures that cannot be extended to a semi-regular polyhedron. △

5.7. The Sine Rule and the Cosine Rule

Where given information, or a specified geometrical construction, determines an angle or length uniquely, it is sometimes – but not always – possible to find this angle or length using simple-minded angle-chasing and congruence.

Problem 191

(a) In the quadrilateral $ABCD$ the two diagonals $\underline{AC}$ and $\underline{BD}$ cross at X. Suppose $\underline{AB} = \underline{BC}$, $\angle BAC = 60°$, $\angle DAC = 40°$, $\angle BXC = 100°$.

  (i) Calculate (exactly) $\angle ADB$ and $\angle CBD$.

  (ii) Calculate $\angle BDC$ and $\angle ACD$.

(b) In the quadrilateral $ABCD$ the two diagonals $\underline{AC}$ and $\underline{BD}$ cross at X. Suppose $\underline{AB} = \underline{BC}$, $\angle BAC = 70°$, $\angle DAC = 40°$, $\angle BXC = 100°$.

  (i) Calculate (exactly) the size of $\angle BDC + \angle ACD$.

  (ii) Explain how we can be sure that $\angle BDC$ and $\angle ACD$ are uniquely determined, even though we cannot calculate them immediately. △

If it turns out that the simplest tools do not allow us to determine angles and lengths, this is usually because we are only using the most basic properties: the congruence criteria, and the parallel criterion. The general art of 'solving triangles' depends on the similarity criterion (usually via trigonometry). And the two standard techniques for 'solving triangles' that go beyond "angle-chasing" and congruence are the *Sine Rule*, which was established back in Problem **32** (and its consequences, such as the area formula $\frac{1}{2}ab \sin C$ – see Problem **33**), and the *Cosine Rule*.

The next problem invites you to use Pythagoras' Theorem to prove the Cosine Rule – an extension of Pythagoras' Theorem which applies to *all* triangles ABC (including those where the angle at C may not be a right angle).

Problem 192 (The Cosine Rule) Given $\triangle ABC$, let the perpendicular from A to BC meet BC at P. If $P = C$, then we know (by Pythagoras' Theorem) that $c^2 = a^2 + b^2$. Suppose $P \neq C$.

(i) Suppose first that P lies on the line segment $\underline{CB}$, or on $\underline{CB}$ extended beyond B. Express the lengths of $\underline{PC}$ and $\underline{AP}$ in terms of b and $\angle C$. Then apply Pythagoras' Theorem to $\triangle APB$ to conclude that

$$c^2 = a^2 + b^2 - 2ab\cos C.$$

(ii) Suppose next that P lies on the line segment $\underline{BC}$ extended beyond C. Prove once again that

$$c^2 = a^2 + b^2 - 2ab\cos C. \qquad \triangle$$

Problem 193 Go back to the configuration in Problem **191**(b). The required angles are unaffected by scaling, so we may choose $AB = BC = 1$. Devise a strategy using the Sine Rule and the Cosine Rule to calculate $\angle BDC$ and $\angle ACD$ exactly. $\triangle$

It is worth reflecting on what the Cosine Rule really tells us:

(i) if in a triangle, we know any two sides (a and b) and the included angle (C), then we can calculate the third side (c); and

(ii) if we know all three sides (a, b, c), then we can calculate any angle (say C).

Hence if we know three sides, or two sides and the angle between them, we can work out all of the angles. The Sine Rule then complements this by ensuring that:

(iii) if we know any side and two angles (in which case we also know the third angle), then we can calculate the other two sides; and

(iv) if we know any angle A, and two sides – one of which is the side a opposite A, then we can calculate (one and hence) both the other angles (and hence the third side).

The upshot is that once a triangle is uniquely determined by the given data, we can "solve" to find all three sides and all three angles.

Trigonometry has a long and very interesting history (which is not at all easy to unravel). Euclid (flourished c. 300 BC) understood that corresponding sides in similar figures were "proportional". And he stated and proved the generalization of Pythagoras' Theorem, which we now call the Cosine Rule; but he did this in a theoretical form, without introducing cosines. Euclid's versions for acute-angled and obtuse-angled triangles involved correction terms with opposite signs, so he proved them separately (*Elements*, Book II, Propositions 12 and 13).

However, the development of trigonometry as an effective theoretical and practical tool seems to have been due to Hipparchus (died c. 125 BC), to Menelaus (c. 70–130 AD), and to Ptolemy (died 168 AD). Once trigonometry moved beyond the purely theoretical, the combination of

- the (exact) language of trigonometry, together with the Sine Rule and the Cosine Rule, and

- (approximate) "tables of trigonometric ratios" (nowadays replaced by calculators)

liberated astronomers, and later engineers, to calculate lengths and angles efficiently, and as accurately as they required.

In mathematics we either work with *exact* values, or we have to control errors precisely. But trigonometry can still be a valuable *exact* tool, provided we remember the lessons of working with fractions such as $\frac{2}{3}$, or with surds such as $\sqrt{2}$, or with constants such as π, and resist the temptation to replace them by some unenlightening approximate decimal. We can replace $\cos^{-1}\left(\frac{1}{2}\right)$ $\left(=\frac{\pi}{3}\right)$ and $\cos^{-1}\left(-\frac{1}{2}\right)$ $\left(=\frac{2\pi}{3}\right)$ by their exact values; but in general we need to be willing to work with, and to think about, *exact forms* such as "$\cos^{-1}\left(\frac{1}{3}\right)$" and "$\cos^{-1}\left(-\frac{1}{3}\right)$", without switching to some approximate evaluation.

Problem 194

(a) Let $ABCD$ be a regular tetrahedron with edges of length 2. Calculate the (exact) angle between the two faces ABC and DBC.

(b) We know that in 2D five equilateral triangles fit together *at a point* leaving just enough of an angle to allow a sixth triangle to fit. How many identical regular tetrahedra can one fit together, without overlaps *around an edge*, so that they all share the edge BC (say)? △

Problem 195

(a) Let $ABCDEF$ be a regular octahedron with vertices B, C, D, E adjacent to A forming a square $BCDE$, and with edges of length 2. Calculate the (exact) angle between the two faces ABC and FBC.

(b) How many identical regular octahedra can one fit together *around an edge*, without overlaps, so that they all share the edge BC (say)? △

5.7. The Sine Rule and the Cosine Rule

Problem 196 Go back to the scenario of Problem **188**, with a regular tetrahedron and a regular octahedron both having edges of length 2, and both having one face flat on the table. Suppose we slide the tetrahedron across the table towards the octahedron. What unexpected phenomenon is guaranteed by Problems **194**(a) and **195**(a)? △

Problem 197 Consider the cube with edges of length 2 running parallel to the coordinate axes, with its centre at the origin $(0,0,0)$, and with opposite corners at $(1,1,1)$ and $(-1,-1,-1)$. The x-, y-, and z-axes, and the xy-, yz-, and zx-planes cut this cube into eight unit cubes – one sitting in each octant.

(i) Let $A = (0,0,1)$, $B = (1,0,0)$, $C = (0,1,0)$, $W = (1,1,1)$. Describe the solid $ABCW$.

(ii) Let $D = (-1,0,0)$, $X = (-1,1,1)$. Describe the solid $ACDX$.

(iii) Let $E = (0,-1,0)$, $Y = (-1,-1,1)$. Describe the solid $ADEY$.

(iv) Let $Z = (1,-1,1)$. Describe the solid $AEBZ$.

(v) Let $F = (0,0,-1)$ and repeat steps (i)–(iv) to obtain the four mirror image solids which lie beneath the xy-plane.

(vi) Describe the solid $ABCDEF$ which is surrounded by the eight identical solids in (i)–(v). △

Problem 198 Consider a single face $ABCDE$ of the regular dodecahedron, with edges of length 1, together with the five pentagons adjacent to it – so that each of the vertices A, B, C, D, E has vertex figure 5^3. Each vertex figure is rigid, so the whole arrangement of six regular pentagons is also rigid. Let V, W, X, Y, Z be the five vertices adjacent to A, B, C, D, E respectively. Calculate the dihedral angle between the two pentagonal faces that meet at the edge $\underline{AB}$. △

Problem 199 Suppose a regular icosahedron (Problem **189**) has edges of length 2. Position vertex A at the 'North pole', and let $BCDEF$ be the regular pentagon formed by its five neighbours.

(a)(i) Calculate the exact angle between the two faces ABC and ACD.

(ii) How many identical regular icosahedra can one fit together, without overlaps, around a single edge?

(b) Let $\mathcal{C}$ be the circumcircle of $BCDEF$, and let O be the circumcentre of this regular pentagon.

   (i) Prove that the three edge lengths of the right-angled triangle $\triangle BOA$ are the edge lengths of the regular hexagon inscribed in the circle $\mathcal{C}$, of the regular 10-gon inscribed in the circle $\mathcal{C}$, and of the regular 5-gon inscribed in the circle $\mathcal{C}$.

   (ii) Calculate the distance separating the plane of the regular pentagon $BCDEF$, and the plane of the corresponding regular pentagon joined to the 'South pole'. △

Notice that Problem **199**(b) shows that the regular icosahedron can be 'constructed' in the Euclidean spirit: part (b)(i) is essentially Proposition 10 of Book XIII of Euclid's *Elements*, and part (b)(ii) is implicit in Proposition 16 of the Book XIII. Once we are given the radius $\underline{OB}$, we can:

- construct the regular pentagon $BCDEF$ in the circle $\mathcal{C}$;

- bisect the sides of the regular pentagon and hence construct the regular 10-gon $BVC\cdots$ in the same circle;

- construct the vertical perpendicular at O, and transfer the length $\underline{BV}$ to the point O to determine the vertex A directly above O;

- transfer the radius $\underline{OB}$ to the vertical perpendicular at O to determine the plane directly below O, and hence construct the lower regular pentagon; etc..

It may be worth commenting on a common confusion concerning the regular icosahedron. Each regular polyhedron has a circumcentre, with all vertices lying on a corresponding sphere. If we join any triangular face of the regular icosahedron to the circumcentre O, we get a tetrahedron. These 20 tetrahedra are all congruent and fit together exactly at the point O "without gaps or overlaps". But they are **not** *regular* tetrahedra: the circumradius is *less than* the edge length of the regular icosahedron.

Problem 200 Prove that the only regular polyhedron that tiles 3D (without gaps or overlaps) is the cube. △

In one sense the result in Problem **200** is disappointing. However, since we know that there are all sorts of interesting 3-dimensional arrangements related to crystals and the way atoms fit together, the message is really that we need to look *beyond* regular tilings. For example, the construction in Problem **197** shows how the familiar *regular* tiling of space with cubes incorporates a *semi-regular* tiling of space with eight regular tetrahedra and two regular octahedra at each vertex.

5.8. Circular arcs and circular sectors

Length is defined for *straight* line segments, and area is defined in terms of *rectangles*; neither measure is defined for shapes with curved boundaries – unless, that is, they can be cunningly dissected and the pieces rearranged to make a straight line, or a rectangle.

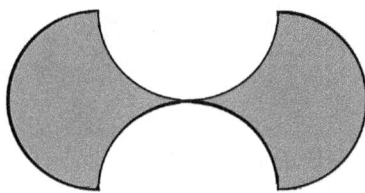

Figure 5: Dumbbell.

Problem 201 Four identical semicircles of radius 1 fit together to make the dumbbell shape shown in Figure 5. Find the exact area enclosed without using the formula for the area of a circle. △

In general, making sense of length and area for shapes with curved boundaries requires us to combine a little imagination with what we know about straight line segments and polygons. Our goal here is to lead up to the familiar results for the length of circular arcs and the area of circular sectors. But first we need to explore the perimeter and area of regular polygons, and the surface area of prisms and pyramids.

As so often in mathematics, to make sense of the perimeter and area of regular polygons we need to look beyond their actual values (which will vary according to the size of the polygon), and instead interpret these values as a function of some normalizing parameter – such as the *radius*. The calculations will be simpler if you first prove a general result.

Problem 202

(a) A regular n-gon and a regular $2n$-gon are inscribed in a circle of radius 1. The regular n-gon has edges of length $s_n = s$, while the regular $2n$-gon has edges of length $s_{2n} = t$. Prove that

$$t^2 = 2 - \sqrt{4 - s^2}.$$

(b) A regular "2-gon" inscribed in the unit circle is just a diameter (repeated twice), so has two identical edges of length $s_2 = 2$. Use the result in part (a) to calculate the edge length s_4 of a regular 4-gon, and the edge length s_8 of a regular 8-gon inscribed in the same circle.

(c) A regular 6-gon inscribed in the unit circle has edge length s_6 equal to the radius 1. Use the result in part (a) to calculate the edge length s_3 of a regular 3-gon inscribed in the unit circle, and the edge length s_{12} of a regular 12-gon inscribed in the unit circle.

(d) In Problem **185** we saw that a regular 5-gon inscribed in the unit circle has edge length
$$s_5 = \frac{\sqrt{10 - 2\sqrt{5}}}{2}.$$
Use the result in part (a) to calculate the edge length s_{10} of a regular 10-gon inscribed in the same circle. △

Problem 203

(a) A regular n-gon is inscribed in a circle of radius r.

  (i) Find the exact perimeter p_n (in surd form): when $n = 3$; when $n = 4$; when $n = 5$; when $n = 6$; when $n = 8$; when $n = 10$; when $n = 12$.

  (ii) Check that, for each n:
  $$p_n = c_n \times r$$
  for some constant c_n, where
  $$c_3 < c_4 < c_5 < c_6 < c_8 < c_{10} < c_{12} \cdots$$

(b) A regular n-gon is circumscribed about a circle of radius r.

  (i) Find the exact perimeter P_n (in surd form): when $n = 3$; when $n = 4$; when $n = 5$; when $n = 6$; when $n = 8$; when $n = 10$; when $n = 12$.

  (ii) Check that, for each n:
  $$P_n = C_n \times r$$
  for some constant C_n, where
  $$C_3 > C_4 > C_5 > C_6 > C_8 > C_{10} > C_{12} \cdots$$

(c) Explain why $c_{12} < C_{12}$. △

It follows from Problem **203** that

5.8. Circular arcs and circular sectors

- the perimeters p_n and P_n of regular n-gons inscribed in, or circumscribed about, a circle of radius r all have the same form:

(inscribed) $p_n = c_n \times r$; (circumscribed) $P_n = C_n \times r$.

- The perimeters of inscribed regular n-gons all increase with n, but remain less than the perimeter of the circle, while

- the perimeters of the circumscribed regular n-gons all decrease with n, but remain greater than the perimeter of the circle.

Hence

- the perimeter P of the circle appears to have the form $P = K \times r$, where the ratio

$$K = \frac{\text{perimeter}}{\text{radius}}$$

satisfies

$$c_3 < c_4 < c_5 < c_6 < c_8 < c_{10} < \cdots < K < \cdots < C_{10} < C_8 < C_6 < C_5 < C_4 < C_3.$$

In particular, the value of the constant K lies somewhere between $c_{12} = 6.21\cdots$ and $C_{12} = 6.43\cdots$. If we now define the quotient K to be equal to "2π", we see that

(perimeter of circle of radius r) $= 2\pi r$,

where π denotes some constant lying between 3.1 and 3.22

In this spirit one might reinterpret the first two bullet points as defining two sequences of constants "π_n" and "Π_n" for $n \geqslant 3$, such that

- (perimeter of a regular n-gon with circumradius r) $= 2\pi_n r$, where

$$\pi_3 = \frac{3\sqrt{3}}{2} = 2.59\cdots, \pi_4 = 2\sqrt{2} = 2.82\cdots, \pi_5 = \frac{5\sqrt{10 - 2\sqrt{5}}}{4} = 2.93\cdots, \pi_6 = 3,$$

etc.,

and

- (perimeter of a regular n-gon with inradius r) $= 2\Pi_n r$, where

$$\Pi_3 = 3\sqrt{3} = 5.19\cdots, \Pi_4 = 4, \Pi_5 = 5\sqrt{5 - 2\sqrt{5}} = 3.63\cdots, \Pi_6 = 2\sqrt{3} = 3.46\cdots,$$

etc..

Moreover

- $\pi_3 < \pi_4 < \pi_5 < \pi_6 < \pi_8 < \cdots < \pi < \cdots < \Pi_8 < \Pi_6 < \Pi_5 < \Pi_4 < \Pi_3.$

Geometry

Problem 204 Find the exact length (in terms of π)

(i) of a semicircle of radius r;

(ii) of a quarter circle of radius r;

(iii) of the length of an arc of a circle of radius r that subtends an angle θ radians at the centre. △

In the next problem we follow a similar sequence of steps to conclude that the quotient
$$L = \frac{\text{area of circle of radius } r}{r^2}$$
is also constant. The surprise lies in the fact that this different constant is so closely related to the previous constant K.

Problem 205

(a) A regular n-gon is inscribed in a circle of radius r.

   (i) Find the exact area a_n (in surd form): when $n = 3$; when $n = 4$; when $n = 5$; when $n = 6$; when $n = 8$; when $n = 10$; when $n = 12$.

   (ii) Check that, for each n:
   $$a_n = d_n \times r^2$$
   for some constant d_n, where
   $$d_3 < d_4 < d_5 < d_6 < d_8 < d_{10} < d_{12} \cdots$$

(b) A regular n-gon is circumscribed about a circle of radius r.

   (i) Find the exact area A_n (in surd form): when $n = 3$; when $n = 4$; when $n = 5$; when $n = 6$; when $n = 8$; when $n = 10$; when $n = 12$.

   (ii) Check that, for each n:
   $$A_n = D_n \times r^2$$
   for some constant D_n, where
   $$D_3 > D_4 > D_5 > D_6 > D_8 > D_{10} > D_{12} \cdots$$

(c) Explain why $d_{12} < D_{12}$. △

It follows from Problem **205** that

- the areas a_n and A_n of regular n-gons inscribed in, or circumscribed about, a circle of radius r all have the same form:

(inscribed) $a_n = d_n \times r^2$; (circumscribed) $A_n = D_n \times r^2$.

- The areas of inscribed regular n-gons all increase with n, but remain less than the area of the circle, while

- the areas of the circumscribed regular n-gons all decrease with n, but remain greater than the area of the circle, whence

- the area A of the circle appears to have the form $A = L \times r^2$, where the ratio
$$L = \frac{\text{area of circle of radius } r}{\text{radius squared}}$$
satisfies
$$d_3 < d_4 < d_5 < d_6 < d_8 < d_{10} \cdots < L < \cdots < D_{10} < D_8 < D_6 < D_5 < D_4 < D_3.$$

In particular, the value of L lies somewhere between $d_{12} = 3$ and $D_{12} = 12(2 - \sqrt{3}) = 3.21 \cdots$. The surprise lies in the fact that the constant L is exactly half of the constant K – that is, $L = \pi$, so

$$(\text{area of circle of radius } r) = \pi r^2.$$

The next problem offers a heuristic explanation for this surprise.

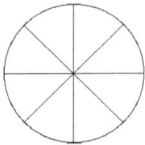

Figure 6: Circle cut into 8 slices.

Problem 206 A regular $2n$-gon $ABCDE \cdots$ is inscribed in a circle of radius r. The $2n$ radii $OA, OB, \ldots$ joining the centre O to the $2n$ vertices cut the circle into $2n$ sectors, each with angle $\frac{\pi}{n}$ (Figure 6).

These $2n$ sectors can be re-arranged to form an "almost rectangle", by orienting them alternately to point "up" and "down". In what sense does this "almost rectangle" have "height $= r$" and "width $= \pi r$"? △

Problem 207

(a) Find a formula for the surface area of a right cylinder with height h and with circular base of radius r.

(b) Find a similar formula for the surface area of a right prism with height h, whose base is a regular n-gon with inradius r. △

Problem 208

(a) Find the exact area (in terms of π)
  (i) of a semicircle of radius r;
  (ii) of a quarter circle of radius r;
  (iii) of a sector of a circle of radius r that subtends an angle θ radians at the centre.

(b) Find the area of a sector of a circle of radius 1, whose total perimeter (including the two radii) is exactly half that of the circle itself. △

Problem 209

(a) Find a formula for the surface area of a right circular cone with base of radius r and slant height l.

(b) Find a similar formula for the surface area of a right pyramid with apex A whose base $BCDE\cdots$ is a regular n-gon with inradius r. △

Problem 210

(a) Find an expression involving "$\sin \frac{\pi}{n}$" for the ratio
$$\frac{\text{perimeter of inscribed regular } n\text{-gon}}{\text{perimeter of circumscribed circle}}.$$

(b) Find an expression involving "$\tan \frac{\pi}{n}$" for the ratio
$$\frac{\text{perimeter of circumscribed regular } n\text{-gon}}{\text{perimeter of inscribed circle}}.$$ △

Problem 211

(a) Find an expression involving "$\sin \frac{2\pi}{n}$" for the ratio
$$\frac{\text{area of inscribed regular } n\text{-gon}}{\text{area of circumscribed circle}}.$$

(b) Find an expression involving "$\tan \frac{\pi}{n}$" for the ratio

$$\frac{\text{area of circumscribed regular } n\text{-gon}}{\text{area of inscribed circle}}.$$
△

5.9. Convexity

This short section presents a simple result which to some extent justifies the assumptions made in the previous section – namely that the perimeter (or area) of a regular n-gon inscribed in a circle is less than the perimeter (or area) of the circle, and of the circumscribed regular n-gon.

Problem 212 A convex polygon P_1 is drawn in the interior of another convex polygon P_2.

(a) Explain why the area of P_1 must be less than the area of P_2.

(b) Prove that the perimeter of P_1 must be less than the perimeter of P_2. △

5.10. Pythagoras' Theorem in three dimensions

Pythagoras' Theorem belongs in 2-dimensions. But does it generalise to 3-dimensions? The usual answer is to interpret the result in terms of coordinates.

Problem 213

(a) Construct a right angled triangle that explains the standard formula for the distance from $P = (a, b)$ to $Q = (d, e)$.

(b) Use part (a) to derive the standard formula for the distance from $P = (a, b, c)$ to $Q = (d, e, f)$. △

This extension of Pythagoras' Theorem to 3-dimensions is extremely useful, but not very profound. In contrast, the next result is more intriguing, but seems to be a complete fluke of limited relevance. In 2D, a right angled triangle is obtained by

- taking one corner A of a rectangle $ABCD$, together with its two neighbours B and D;

- then "cutting off the corner" to get the triangle ABD.

This suggests that a corresponding figure in 3D might be obtained by

218   Geometry

- taking one corner A of a cuboid, together with its three neighbours B, C, D;

- then cutting off the corner to get a pyramid $ABCD$, with the right angled triangle $\triangle ABC$ as base, and with apex D.

The obvious candidate for the "3D-hypotenuse" is then the sloping face BCD, and the three right angled triangles $\triangle ABC$, $\triangle ACD$, $\triangle ADB$ presumably correspond to the 'legs' (the shorter sides) of the right triangle in 2D.

Problem 214 You are given a pyramid $ABCD$ with all three faces meeting at A being right angled triangles with right angles at A. Suppose $\underline{AB} = b$, $\underline{AC} = c$, $\underline{AD} = d$.

(a) Calculate the areas of $\triangle ABC$, $\triangle ACD$, $\triangle ADB$ in terms of b, c, d.

(b) Calculate the area of $\triangle BCD$ in terms of b, c, d.

(c) Compare your answer in part (b) with the sum of the squares of the three areas you found in part (a). △

More significant (e.g. for navigation on the surface of the Earth) and more interesting than Problem **214** is to ask what form Pythagoras' Theorem takes for "lines on a sphere".

For simplicity we work on a unit sphere. We discovered in the run-up to Problem **34** that lines, or shortest paths, on a sphere are arcs of great circles. So, if the triangle $\triangle ABC$ on the unit sphere is right angled at A, we may rotate the sphere so that the arc $\underline{AB}$ lies along the equator and the arc $\underline{AC}$ runs up a circle of longitude. It is then clear that, once the lengths c, b of $\underline{AB}$ and $\underline{AC}$ are known, the locations of B and C are essentially determined, and hence the length of the arc $\underline{BC}$ on the sphere is determined. So we would like to have a simple formula that would allow us to calculate the length of the arc $\underline{BC}$ directly in terms of c and b.

Problem 215 Given a spherical triangle $\triangle ABC$ on the unit sphere with centre O, such that $\angle BAC$ is a right angle, and such that $\underline{AB}$ has length c, and $\underline{AC}$ has length b.

(a) We have (rightly) referred to b and c as 'lengths'. But what are they really?

(b) We want to know how the inputs b and c determine the value of the length a of the arc $\underline{BC}$; that is, we are looking for a function with inputs b and c, which will allow us to determine the value of the "output" a. Think about the answer to part (a). What kind of standard functions do we already know that could have inputs b and c?

5.10. Pythagoras' Theorem in three dimensions

(c) Suppose $c = 0 \neq b$. What should the output a be equal to? (Similarly if $b = 0 \neq c$.) Which standard function of b and of c does this suggest is involved?

(d)(i) Suppose $\angle B = \angle C = \frac{\pi}{2}$, what should the output a be equal to?

(ii) Suppose $\angle B = \frac{\pi}{2}$, but $\angle C$ (and hence c) is unconstrained. The output a is then determined – but the formula must give this fixed output for different values of c. What does this suggest as the "simplest possible" formula for a? △

The answers to Problem **215** give a pretty good idea what form Pythagoras' Theorem must take on the unit sphere. The next problem proves this result as a simple application of the familiar 2D Cosine Rule.

Problem 216 Given any triangle $\triangle ABC$ on the unit sphere with a right angle at the point A, we may position the sphere so that A lies on the equator, with $\underline{AB}$ along the equator and $\underline{AC}$ up a circle of longitude. Let O be the centre of the sphere and let $\mathbf{T}$ be the tangent plane to the sphere at the point A. Extend the radii $\underline{OB}$ and $\underline{OC}$ to meet the plane $\mathbf{T}$ at B' and C' respectively.

(a) Calculate the lengths of the line segments $\underline{AB'}$ and $\underline{AC'}$, and hence of $\underline{B'C'}$.

(b) Calculate the lengths of $\underline{OB'}$ and $\underline{OC'}$, and then apply the Cosine Rule to $\triangle B'OC'$ to find an equation linking b and c with $\angle B'OC' \, (= a)$. △

When "solving triangles" on the sphere the same principles apply as in the plane: right angled triangles hold the key – but Pythagoras' Theorem and trig in right angled triangles must be extended to obtain variations of the Sine Rule and the Cosine Rule for *spherical* triangles. The corresponding results on the sphere are both similar to, and intriguingly different from, those we are used to in the plane. For example, there are two forms of the Cosine Rule extending the result in Problem **216**.

Problem 217 Given a (not necessarily right angled) triangle $\triangle ABC$ on the unit sphere, apply the same proof as in Problem **216** to show (with the usual labelling) that:

$$\cos a = \cos b \cdot \cos c + \sin b \cdot \sin c \cdot \cos A \qquad △$$

The other form of the Cosine Rule is "dual" to that in Problem **217** (with arcs and angles interchanged, and with an unexpected change of sign) – namely:

$$\cos A = -\cos B \cdot \cos C + \sin B \cdot \sin C \cdot \cos a.$$

The next two problems derive a version of the Sine Rule for spherical triangles.

Problem 218 Let $\triangle ABC$ be a triangle on the unit sphere with a right angle at A. Let A' lie on the arc BA produced, and C' lie on the arc BC produced so that $\triangle A'BC'$ is right angled at A'. With the usual labelling (so that x denotes the length of the side of a triangle opposite vertex X, with arc $AC = b$, arc $BC = a$, arc $BC' = a'$, and arc $A'C' = b'$, prove that:

$$\frac{\sin b}{\sin a} = \frac{\sin b'}{\sin a'}. \qquad \triangle$$

Problem 219 Let $\triangle ABC$ be a general triangle on the unit sphere with the usual labelling (so that x denotes the length of the side of a triangle opposite vertex X, and X is used both to label the vertex and to denote the size of the angle at X). Prove that:

$$\frac{\sin a}{\sin A} = \frac{\sin b}{\sin B} = \frac{\sin c}{\sin C}. \qquad \triangle$$

It is natural to ask (cf Problem **32**):

"If the three ratios in Problem **219** are all equal, what is it that they are all equal to?"

The answer may not at first seem quite as nice as in the Euclidean 2-dimensional case: one answer is that they are all equal to

$$\frac{\sin a \cdot \sin b \cdot \sin c}{\text{volume of the tetrahedron } OABC}.$$

Notice that this echoes the result in the Euclidean plane, where the three ratios in the Sine Rule are all equal to $2R$, and

$$2R = \frac{abc}{2(\text{area of } \triangle ABC)}.$$

5.11. Loci and conic sections

This section offers a brief introduction to certain classically important *loci* in the plane. The word *locus* here refers to the set of all points satisfying some simple geometrical condition; and all the examples in this section are based on the notion of distance from a point and from a line.

Given a point O and a positive real r, the locus of points at distance r from O is precisely the circle of radius r with centre O. If $r < 0$, then the locus is empty; while if $r = 0$, the locus consists of the point O alone.

Given a line m and a positive real r, the locus of all points at distance r from the line m consists of a pair of parallel lines – one either side of the line m. Given a circle of radius r, and a positive real number $d < r$; the locus of points at distance d from the circle consists of two circles, each concentric with the given circle (one inside the given circle and one outside). If $d > r$, the locus consists of a single circle outside the given circle.

Given two points A and B, the locus of points which are equidistant from A and from B is precisely the perpendicular bisector of the line segment AB. And given two lines m, n the locus of points which are equidistant from m and from n takes different forms according as m and n are, or are not, parallel.

- If m and n are parallel, then the locus consists of a single line parallel to m and n and half way between them.

- If m and n meet at X (say), then the locus consists of the pair of perpendicular lines through X, that bisect the four angles at X.

Problem 220 Given a point F and a line m, choose m as the x-axis and the line through F perpendicular to m as the y-axis. Let F have coordinates $(0, 2a)$.

(i) Find the equation that defines the locus of points which are equidistant from F and from m.

(ii) Does the equation suggest a more natural choice of axes – and hence a simpler equation for the locus? △

The locus, or curve, in Problem **220** is called a *parabola*; the point F is called the *focus* of the parabola, and the line m is called the *directrix*. In general, the ratio

"the distance from X to F" : "the distance from X to m"

is called the *eccentricity* of the curve. Hence the parabola has eccentricity $e = 1$.

The parabola has many wonderful properties: for example, it is the path followed by a projectile under the force of gravity; if viewed as the surface of a mirror, a parabola reflects the sun's rays (or any parallel beam) to a single point – the focus F. Since the only variable in the construction of the parabola is the distance "$2a$" between the focus and the directrix, we can scale distances to see that any two different-looking parabolas must in fact be *similar to one another* – just as with any two circles. (It is hard not to infer from the graphs that $y = 10x^2$ is a "thin" parabola, and that

$y = \left(\frac{1}{10}\right) x^2$ is a "fat" parabola. But the first can be rewritten in the form $10y = (10x)^2$, and the second can be rewritten in the form $\left(\frac{y}{10}\right) = \left(\frac{x}{10}\right)^2$, so each is a re-scaled version of $Y = X^2$.)

So far we have considered loci defined by some pair of distances being equal, or in the ratio $1 : 1$. More interesting things begin to happen when we consider conditions in which two distances are in a fixed ratio other than $1 : 1$.

Problem 221

(a) Given two points A, B, with $\underline{AB} = 6$. Find the locus of all points X such that $\underline{AX} : \underline{BX} = 2 : 1$.

(b) Given points A, B, with $\underline{AB} = 2b$ and a positive real number f. Find the locus of all points X such that $\underline{AX} : \underline{BX} = f : 1$. △

Problem 222

(a) Given points A, B, with $\underline{AB} = 2c$ and a real number $a > c$. Find the locus of all points X such that $\underline{AX} + \underline{BX} = 2a$.

(b) Given a point F and a line m, find the locus of all points X such that the ratio

distance from X to the point F : distance from X to the line m

is a positive constant $e < 1$.

(c) Prove that parts (a) and (b) give different ways of specifying the same curve, or locus. △

Problem 223

(a) Given points A, B, with $\underline{AB} = 2c$, and a positive real number a. Find the locus of all points X such that $|\underline{AX} - \underline{BX}| = 2a$.

(b) Given a point F and a line m, find the locus of all points X such that the ratio

distance from X to the point F : distance from X to the line m

is a constant $e > 1$.

(c) Prove that parts (a) and (b) give different ways of specifying the same curve, or locus. △

Problem **221** is sometimes presented in the form of a mild joke.

> Two dragons are sleeping, one at A and one at B. Dragon A can run twice as fast as dragon B. A specimen of *homo sapiens* is positioned on the line segment AB, twice as far from A as from B, and cunningly decides to crawl quietly away, while maintaining the ratio of his distances from A and from B (so as to make it equally difficult for either dragon to catch him should they wake).

The locus that emerges generally comes as a surprise: if the man sticks to his imposed restriction, by moving so that his position X satisfies $XA = 2 \cdot XB$, then he follows a circle and lands back where he started! The circle is called the *circle of Apollonius*, and the points A and B are sometimes referred to as its foci.

The locus in Problem **222** is an *ellipse* – with *foci* A (or $F = (-ae, 0)$) and B ($= (ae, 0)$), and with directrix m (the line $y = -\frac{a}{e}$; the line $y = \frac{a}{e}$ is the second directrix of the ellipse). The "focus-focus" description in part (a) is symmetrical under reflection in both the line AB and the perpendicular bisector of AB. The "focus-directrix" description in (b) is clearly symmetrical in the line through F perpendicular to m; but it is a surprise to find that the equation

$$\frac{x^2}{a^2} + \frac{y^2}{a^2(1-e^2)} = 1$$

is also symmetrical under reflection in the y-axis. If we set $b^2 = a^2(1 - e^2)$, the equation takes the form

$$\frac{x^2}{a^2} + \frac{y^2}{b^2} = 1,$$

which crosses the x-axis when $x = \pm a$, and crosses the y-axis when $y = \pm b$. In its standard form, we usually choose coordinates so that $b < a$: the line segment from $(-a, 0)$ to $(a, 0)$ is then called the *major axis*, and half of it (say from $(0, 0)$ to $(a, 0)$) – of length a – is called the *semi-major axis*; the line segment from $(0, -b)$ to $(0, b)$ is called the *minor axis*, and half of it (say from $(0, 0)$ to $(0, b)$) – of length b – is called the *semi-minor axis*.

The form of the equation shows that an ellipse is obtained from a unit circle by stretching by a factor "a" in the x-direction, and by a factor "b" in the y-direction. This implies that the area of an ellipse is equal to πab (since each small s by s square that arises in the definition of the "area" of the unit circle gets stretched into an "as by bs rectangle"). However, the equation tells us nothing about the *perimeter* of an ellipse. Attempts to pin down the perimeter of an ellipse gave rise in the 18th century to the subject of "elliptic integrals".

Like parabolas, ellipses arise naturally in many important settings. For example, Kepler (1571–1630) discovered that the planetary orbits are not circular (as had previously been believed), but are ellipses – with the Sun at one focus (a conjecture which was later explained by Isaac Newton (1642–1727)). Moreover, the tangent to an ellipse at any point X is equally inclined to the two lines XA and XB, so that a beam emerging from one focus is reflected at every point of the ellipse so that all the reflected rays pass through the other focus.

The curve in Problem **223** is a *hyperbola* – with *foci* A (or $F = (-ae, 0)$) and $B (= (ae, 0))$, and with *directrix* m (the line $y = -\frac{a}{e}$; the line $y = \frac{a}{e}$ is the second directrix of the hyperbola). The "focus-focus" description in part (a) is symmetrical under reflection in both the line AB and the perpendicular bisector of $\underline{AB}$. The "focus-directrix" description in (b) is clearly symmetrical in the line through F perpendicular to m; but it is a surprise to find that the equation

$$\frac{x^2}{a^2} - \frac{y^2}{a^2(e^2 - 1)} = 1$$

is also symmetrical under reflection in the y-axis. Like parabolas and ellipses, hyperbolas arise naturally in many important settings – in mathematics and in the natural sciences.

All these loci were introduced and studied by the ancient Greeks without the benefit of coordinate geometry and equations. They were introduced as planar *cross-sections of a cone* – that is, as natural extensions of straight lines and circles (since the doubly infinite cone is the surface traced out when one rotates a line about an axis through a point on that line). The equivalence of the focus-directrix definition in Problems **220**, **222**, and **223** and cross sections of a cone follows from the next problem. All five constructions in Problem **224** work with the doubly-infinite cone, which we may represent as $x^2 + y^2 = (rz)^2$ – although this representation is not strictly needed for the derivations. The surface of the double cone extends to infinity in both directions, and is obtained by taking the line $y = rz$ in the yz-plane (where $r > 0$ is constant), and rotating it about the z-axis. Images of this rotated line are called *generators* of the cone; and the point they all pass through (i.e. the origin) is called the *apex* of the cone.

Problem 224 (**Dandelin's spheres**: Dandelin (1794–1847))

(a) Describe the cross-sections obtained by cutting such a double cone by a horizontal plane (i.e. a plane perpendicular to the z-axis). What if the cutting plane is the xy-plane?

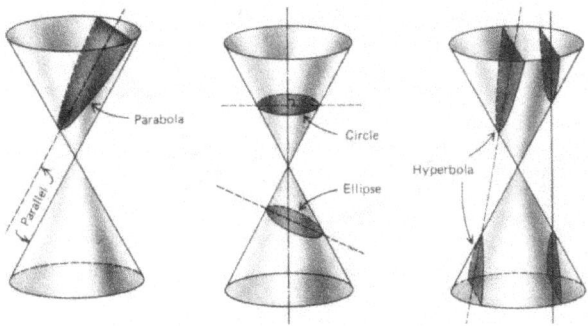

Figure 7: Conic sections.

(b)(i) Describe the cross-section obtained by cutting such a cone by a vertical plane through the origin, or apex.

(ii) What cross-section is obtained if the cutting plane passes through the apex, but is not vertical?

(c) Give a *qualitative* description of the curve obtained as a cross-section of the cone if we cut the cone by a plane which is parallel to a generator: e.g. the plane $y - rz = c$.

(i) What happens if $c = 0$?

(ii) Now assume the cutting plane is parallel to a generator, but does not pass through the apex of the cone – so we may assume that the plane cuts only the bottom half of the cone. Insert a small sphere inside the bottom half of the cone and above the cutting plane, and inflate the sphere as much as possible – until it touches the cone around a horizontal circle (the "contact circle with the cone"), and touches the plane at a single point F. Let the horizontal plane of the "contact circle with the cone" meet the cutting plane in the line m. Prove that each point of the cross-sectional curve is equidistant from the point F and from the line m – and so is a parabola.

(d)(i) Give a *qualitative* description of the curve obtained as a cross-section of the cone if we cut the cone by a plane which is *less steep* than a generator, but does not pass through the apex – and so cuts right across the cone.

(ii) We may assume that the plane cuts only the bottom half of the cone. Insert a small sphere inside the bottom half of the cone and above the cutting plane (i.e. on the same side of the cutting plane as the apex of the cone), and inflate the sphere as much as possible – until it touches the cone around a horizontal circle, and touches the plane at a single point F. Let the horizontal plane of the contact circle meet the cutting plane in the line m. Prove that, for each point X on the cross-sectional curve, the ratio

"distance from X to F" : "distance from X to m" $= e : 1$

is constant, with $e < 1$, and so is an ellipse.

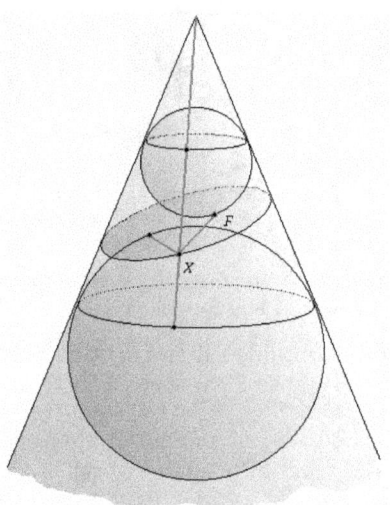

Figure 8: The conic section arising in Problem **224**(d).

(e)(i) Give a *qualitative* description of the curve obtained as a cross-section if we cut the cone by a plane which is *steeper* than a generator, but does not pass through the apex (and hence cuts both halves of the cone)?

(ii) We can be sure that the plane cuts the bottom half of the cone (as well as the top half). Insert a small sphere inside the bottom half of the cone and on the same side of the cutting plane as the apex, and inflate the sphere as much as possible – until it touches the cone around a horizontal circle, and touches the plane at a single point F. Let the horizontal plane of the contact circle meet the cutting plane in the line m. Prove that, for each point X on the cross-sectional curve, the ratio

"distance from X to F" : "distance from X to m" $= e : 1$

is constant, with $e > 1$, and so is a hyperbola. △

Problem **224** reveals a remarkable correspondence. It is not hard to show algebraically that any quadratic equation in two variables x, y represents either a point, or a pair of crossing (possibly identical) straight lines, or a parabola, or an ellipse, or a hyperbola: that is, by changing coordinates, the quadratic equation can be transformed to one of the standard forms obtained in this section. Hence, *the possible quadratic curves* are precisely the same as the possible cross-sections of a cone. This remarkable equivalence is further reinforced by the many natural contexts in which these conic sections arise.

5.12. Cubes in higher dimensions

This final section on elementary geometry seeks to explore fresh territory by going beyond three dimensions. Whenever we try to jump up to a new level, it can help to first take a step *back* and 'take a longer run up'. So please be patient if we initially take a step or two backwards.

We all know what a unit 3D-cube is. And – going backwards – it is not hard to guess what is meant by a unit "2D-cube": a unit 2D-cube is just another name for a unit *square*. It is then not hard to notice that a unit 3D-cube can be constructed from two unit 2D-cubes as follows:

- first position two unit 2D-cubes 1 unit apart in 3D space, with one directly above the other;
- make sure that each vertex of the lower 2D-cube is directly beneath a vertex of the upper 2D-cube;
- then join each vertex of the upper 2D-cube to the corresponding vertex below it.

Perhaps a unit 2D-cube can be constructed in a similar way from "unit 1D-cubes"! This idea suggests that a unit "1D-cube" is just another name for a unit *line segment*.

Take the unit 1D-cube to be the line segment from 0 to 1:

- position two such 1D-cubes in 2D (e.g. one joining $(0,0)$ to $(1,0)$, and the other joining $(0,1)$ to $(1,1)$);
- check that each vertex of the lower 1D-cube is directly beneath a vertex of the upper 1D-cube;
- then join corresponding pairs of vertices – one from the upper 1D-cube and one from the lower 1D-cube ($(0,0)$ to $(0,1)$, and $(1,0)$ to $(1,1)$) – to obtain a unit 2D-cube.

Having taken a step back, we repeat (and reformulate) the previous construction of a 3D-cube:

- position two such unit 2D-cubes in 3D: with one 2D-cube joining $(0,0,0)$ to $(1,0,0)$, then to $(1,1,0)$, then to $(0,1,0)$ and back to $(0,0,0)$, and the other 2D-cube joining $(0,0,1)$ to $(1,0,1)$, then to $(1,1,1)$, then to $(0,1,1)$, and back to $(1,0,0)$;

- with one 2D-cube directly above the other,
- then join corresponding pairs of vertices – one from the upper 2D-cube and one from the lower 2D-cube $((0,0,0)$ to $(0,0,1)$, and $(1,0,0)$ to $(1,0,1)$, and $(1,1,0)$ to $(1,1,1)$, and $(0,1,0)$ to $(0,1,1))$ – to obtain a unit 3D-cube.

To sum up: a unit cube in 1D, or in 2D, or in 3D:

- has as "vertices" all points whose coordinates are all "0s or 1s" (in 1D, or 2D, or 3D)
- has as "edges" all the unit segments (or unit 1D-cubes) joining vertices whose coordinates differ in exactly one place
- and a unit 3D-cube has as "faces" all the unit 2D-cubes spanned by vertices with a constant value (0 or 1) in one of the three coordinate places (that is, for the unit 3D-cube: the four vertices with $x = 0$, or the four vertices with $x = 1$; or the four vertices with $y = 0$, or the four vertices with $y = 1$; or the four vertices with $z = 0$, or the four vertices with $z = 1$).

A 3D-cube is surrounded by six 2D-cubes (or faces), and a 2D-cube is surrounded by four 1D-cubes (or faces). So it is natural to interpret the two end vertices of a 1D-cube as being '0D-cubes'. We can then see that a cube in any dimension is made up from cubes of smaller dimensions. We can also begin to make a reasonable guess as to *what we might expect to find in a '4D-cube'*.

Problem 225

(a)(i) How many vertices (i.e. 0D-cubes) are there in a 1D-cube?

(ii) How many edges (i.e. 1D-cubes) are there in a 1D-cube?

(b)(i) How many vertices (or 0D-cubes) are there in a 2D-cube?

(ii) How many "faces" (i.e. 2D-cubes) are there in a 2D-cube?

(iii) How many edges (i.e. 1D-cubes) are there in a 2D-cube?

(c)(i) How many vertices (or 0D-cubes) are there in a 3D-cube?

(ii) How many 3D-cubes are there in a 3D-cube?

(iii) How many edges (i.e. 1D-cubes) are there in a 3D-cube?

(iv) How many "faces" (i.e. 2D-cubes) are there in a 3D-cube?

(d)(i) How many vertices (or 0D-cubes) do you expect to find in a 4D-cube?

(ii) How many 4D-cubes do you expect to find in a 4D-cube?
(iii) How many edges (i.e. 1D-cubes) do you expect to find in a 4D-cube?
(iv) How many "faces" (i.e. 2D-cubes) do you expect to find in a 4D-cube?
(v) How many 3D-cubes do you expect to find in a 4D-cube? △

Problem 226

(a)(i) Sketch a unit 2D-cube as follows. Starting with two unit 1D-cubes – one directly above the other. Then join up each vertex in the upper 1D-cube to the vertex it corresponds to in the lower 1D-cube (directly beneath it).

(ii) Label each vertex of your sketch with coordinates (x, y) ($x, y = 0$ or 1) so that the lower 2D-cube has the equation "$y = 0$" and the upper 2D-cube has the equation "$y = 1$".

(b)(i) Sketch a unit 3D-cube, starting with two unit 2D-cubes – one directly above the other. Then join up each vertex in the upper 2D-cube to the vertex it corresponds to in the lower 2D-cube (directly beneath it).

(ii) Label each vertex of your sketch with coordinates (x, y, z) (where each $x, y, z = 0$ or 1) so that the lower 2D-cube has the equation "$z = 0$" and the upper 2D-cube has the equation "$z = 1$".

(c)(i) Now sketch a unit 4D-cube in the same way – starting with two unit 3D-cubes, one "directly above" the other.

[**Hint:** In part (b) your sketch was a projection of a 3D-cube onto 2D paper, and this forced you to represent the lower and upper 2D-cubes as rhombuses rather than genuine 2D-cubes (unit squares). In part (c) you face the even more difficult task of representing a 4D-cube on 2D paper; so you must be prepared for other "distortions". In particular, it is almost impossible to see what is going on if you try to physically position one 3D-cube "directly above" the other on 2D paper. So start with the "upper" unit 3D-cube towards the top right of your paper, and then position the "lower" unit 3D-cube not directly below it on the paper, but below and slightly to the left, before pairing off and joining up each vertex of the upper 3D-cube with the corresponding vertex in the lower 3D-cube.]

(ii) Label each vertex of your sketch with coordinates (w, x, y, z) (where each $w, x, y, z = 0$ or 1) so that the lower 3D-cube has the equation "$z = 0$" and the upper 3D-cube has the equation "$z = 1$". △

Problem 227 The only possible path along the edges of a 2D-cube uses each vertex once and returns to the start after visiting all four vertices.

(a)(i) Draw a path along the edges of a 3D-cube that visits each vertex exactly once and returns to the start.

(ii) Look at the sequence of coordinate triples as you follow your path. What do you notice?

(b)(i) Draw a path along the edges of a 4D-cube that visits each vertex exactly once and returns to the start.

(ii) Look at the sequence of coordinate 4-tuples as you follow your path. What do you notice? △

5.13. Chapter 5: Comments and solutions

137. Note: The spirit of constructions restricts us to:

- drawing the line joining two known points
- drawing the circle with centre a known point and passing through a known point.

The whole thrust of this first problem is to find some way to "jump" from A (or B) to C. So the problem leaves us with very little choice; $\underline{AB}$ is given, and A and B are more-or-less indistinguishable, so there are only two possible 'first moves' – both of which work with the line segment $\underline{AC}$ (or $\underline{BC}$).

Join AC.

Then construct the point X such that $\triangle ACX$ is equilateral (i.e. use Euclid's *Elements*, Book I, Proposition 1). Construct the circle with centre A which passes through B; let this circle meet the line AX at the point Y, where either

(i) Y lies on the segment $\underline{AX}$ (if $\underline{AB} \leqslant \underline{AC}$), or

(ii) Y lies on $\underline{AX}$ produced (i.e. beyond X, if $\underline{AB} > \underline{AC}$).

In each case, $\underline{AY} = \underline{AB}$. Finally construct the circle with centre X which passes through Y. In case (i), let the circle meet the line segment $\underline{XC}$ at D; in case (ii), let the circle meet $\underline{CX}$ produced (beyond X) at D.

In case (i), $\underline{CX} = \underline{CD} + \underline{DX}$; therefore

$$\begin{aligned}\underline{CD} &= \underline{CX} - \underline{DX} \\ &= \underline{AX} - \underline{YX} \\ &= \underline{AY} = \underline{AB}.\end{aligned}$$

In case (ii),

$$\begin{aligned}\underline{CD} &= \underline{CX} + \underline{XD} \\ &= \underline{AX} + \underline{XY} \\ &= \underline{AY} = \underline{AB}.\end{aligned}$$

QED

138.

$$\angle AXC = \angle AXB - \angle CXB$$
$$= \angle CXD - \angle CXB \quad \text{(since the two straight angles}$$
$$\angle AXB \text{ and } \angle CXD \text{ are equal)}$$
$$= \angle BXD.$$

QED

139.

$AM = AM$
$MB = MC$ (by construction of the midpoint M)
$BA = CA$ (given).
$\therefore \triangle AMB \equiv \triangle AMC$ (by SSS-congruence)
$\therefore \angle AMB = \angle AMC$, so each angle is exactly half the straight angle $\angle BMC$.
Hence AM is perpendicular to BC. QED

140. Let $ABCDEF$ be a regular hexagon with sides of length 1. Then $\triangle ABC$ (formed by the first three vertices) satisfies the given constraints, with $\angle ABC = 120°$.

Let $\triangle B'C'D$ be an equilateral triangle with sides of length 2, and with A' the midpoint of $B'D$. Then $\triangle A'B'C'$ satisfies the given constraints with angle $\angle A'B'C' = 60°$.

141.

(i) Join CO. Then $\triangle ACO$ is isosceles (since $OA = OC$) and $\triangle BCO$ is isosceles (since $OB = OC$).
Hence $\angle OAC = \angle OCA = x$ (say), and $\angle OBC = \angle OCB = y$ (say).
So $\angle C = x + y$ (and $\angle A + \angle B + \angle C = x + (x+y) + y = 2(x+y)$). QED

(ii) If $\angle A + \angle B + \angle C = 2(x+y)$ is equal to a straight angle, then $\angle C = x + y$ is half a straight angle, and hence a right angle. QED

142.

(i) Draw the circle with centre A and passing through B, and the circle with centre B passing through A. Let these two circles meet at C and D.
Wherever the midpoint M of AB may be, we know from Euclid Book I, Proposition 1 and Problem **139**:

> that $\triangle ABC$ is equilateral, and that CM is perpendicular to AB, and that $\triangle ABD$ is equilateral, and that DM is perpendicular to AB.

Hence CMD is a straight line.
So if we join CD, then this line cuts AB at its midpoint M. QED

(ii) We may suppose that $BA \leqslant BC$.

Then the circle with centre B, passing through A meets BC internally at A' (say).

Let the circle with centre A and passing through B meet the circle with centre A' and passing through B at the point D.

Claim BD bisects $\angle ABC$.
Proof

$$BA = BA' \text{ (radii of the same circle with centre } B)$$
$$AD = AB \text{ (radii of the same circle with centre } A)$$
$$= A'B = A'D \text{ (radii of same circle with centre } A')$$
$$BD = BD$$

Hence $\triangle BAD \equiv \triangle BA'D$ (by SSS-congruence).
$\therefore \angle ABD = \angle A'BD$. $\hfill$ QED

(iii) Suppose first that $PA = PB$. Then the circle with centre P and passing through A meets the line AB again at B. If we construct the midpoint M of AB as in part (i), then PM will be perpendicular to AB.

Now suppose that one of PA and PB is longer than the other. We may suppose that $PA > PB$, so B lies inside the circle with centre P and passing through A. Hence this circle meets the line AB again at A' where B lies between A and A'. If we now construct the midpoint M of AA' as in part (i), then PM will be perpendicular to AA', and hence to AB. $\hfill$ QED

143. Let M be the midpoint of AB.

(a) Let X lie on the perpendicular bisector of AB.
$\therefore \triangle XMA \equiv \triangle XMB$ (by SAS congruence, since $XM = XM$, $\angle XMA = \angle XMB$, $MA = MB$)
$\therefore XA = XB$.

(b) If X is equidistant from A and from B, then $\triangle XMA \equiv \triangle XMB$ (by SSS-congruence, since $XM = XM$, $MA = MB$, $AX = BX$).
$\therefore \angle XMA = \angle XMB$, so each must be exactly half a straight angle.
$\therefore X$ lies on the perpendicular bisector of AB. $\hfill$ QED

144. Let X lie on the plane perpendicular to NS, through the midpoint M.
$\therefore \triangle XMN \equiv \triangle XMS$ (by SAS congruence, since $XM = XM$, $\angle XMN = \angle XMS$, $MN = MS$)
$\therefore XN = XS$.

Let X be equidistant from N and from S, then $\triangle XMN \equiv \triangle XMS$ (by SSS-congruence, since $XM = XM$, $MN = MS$, $NX = SX$).
$\therefore \angle XMN = \angle XMS$, so each must be exactly half a straight angle.
$\therefore X$ lies on the perpendicular bisector of NS. $\hfill$ QED

145. Let M be the midpoint of AC. Join BM and extend the line beyond M to the point D such that $MB = MD$. Join CD. Then $\triangle AMB \equiv \triangle CMD$ (by SAS-congruence, since

$AM = CM$ (by construction of the midpoint M)
$\angle AMB = \angle CMD$ (vertically opposite angles)
$MB = MD$ (by construction)).

$\therefore \angle DCM = \angle BAM$.
Now
$$\begin{aligned}\angle ACX &= \angle DCM + \angle DCX \\ &> \angle DCM \\ &= \angle BAM \\ &= \angle A.\end{aligned}$$

Hence $\angle ACX > \angle A$.

Similarly, we can extend AC beyond C to a point Y. Let N be the midpoint of BC. Join AN and extend the line beyond N to the point E such that $NA = NE$. Join CE.
Then $\triangle BNA \equiv \triangle CNE$ (again by SAS-congruence).
$\therefore \angle BCY > \angle BCE = \angle CBA = \angle B$. QED

146.

(a) Suppose that $AB > AC$.
Let the circle with centre A, passing through C, meet AB (internally) at X.
Then $\triangle ACX$ is isosceles, so $\angle ACX = \angle AXC$.
By Problem **145**, $\angle AXC > \angle ABC$.
$\therefore \angle ACB\ (> \angle ACX = \angle AXC) > \angle ABC$. QED

(b) Suppose the conclusion does not hold. Then either

 (i) $AB = AC$, or

 (ii) $AC > AB$.

 (i) If $AB = AC$, then $\triangle ABC$ is isosceles, so $\angle ACB = \angle ABC$ – contrary to assumption.

 (ii) If $AC > AB$, then $\angle ABC > \angle ACB$ (by part (a)) – again contrary to assumption.

 Hence, if $\angle ACB > \angle ABC$, it follows that $AB > AC$. QED

(c) Extend AB beyond B to the point D, such that $BD = BC$.
Then $\triangle BDC$ is isosceles with apex B, so $\angle BDC = \angle BCD$.
Now
$$\angle ACD = \angle ACB + \angle BCD > \angle BCD = \angle BDC.$$

Hence, by part (b), $AD > AC$.
By construction,
$$AD = AB + BD = AB + BC,$$
so $AB + BC > AC$. QED

147. Suppose $\angle C = \angle A + \angle B$, but that C does not lie on the circle with diameter AB. Then C lies either inside, or outside the circle. Let O be the midpoint of AB.

(i) If C lies outside the circle, then $OC > OA = OB$.
∴ $\angle OAC > \angle OCA$ and $\angle OBC > \angle OCB$ (by Problem **146**(a)).
∴ $\angle C = \angle OCA + \angle OCB < \angle A + \angle B$ – contrary to assumption.
Hence C does not lie outside the circle.

(ii) If C lies inside the circle, then $OC < OA = OB$.
∴ $\angle OAC < \angle OCA$ and $\angle OBC < \angle OCB$ (by Problem **146**(a)).
∴ $\angle C = \angle OCA + \angle OCB > \angle A + \angle B$ – contrary to assumption.
Hence C does not lie inside the circle.
Hence C lies on the circle with diameter AB. QED

148. Suppose, to the contrary, that OP is **not** perpendicular to the tangent at P.

Drop a perpendicular from O to the tangent at P to meet the tangent at Q. Extend PQ beyond Q to some point X. Then $\angle OQP$ and $\angle OQX$ are both right angles.

Since Q ($\neq P$) lies on the tangent, Q lies outside the circle, so $OQ > OP$. Hence (by Problem **146**(a)) $\angle OPQ > \angle OQP = \angle OQX$ – contrary to the fact that $\angle OQX > \angle OPQ$ (by Problem **145**). QED

149. Let Q lie on the line m such that PQ is perpendicular to m.
Let X be any other point on the line m, and let Y be a point on m such that Q lies between X and Y.
Then $\angle PQX$ and $\angle PQY$ are both right angles.
Suppose that $PX < PQ$.
Then $\angle PQY = \angle PQX < \angle PXQ$ (by Problem **146**(a)), which contradicts Problem **145** (since $\angle PQY$ is an exterior angle of $\triangle PQX$).
Hence $PX \geqslant PQ$ as required. QED

150. $\angle A + \angle B + \angle C$, and $\angle XCA + \angle C$ are both equal to a straight angle. So $\angle A + \angle B = \angle XCA$.

151. Join OA, OB, OC. Since these radii are all equal, this produces three isosceles triangles. There are **five** cases to consider.

(i) Suppose first that O lies on AB. Then AB is a diameter, so $\angle ACB$ is a right angle (by Problem **141**), $\angle AOB$ is a straight angle, and the result holds.

(ii) Suppose O lies on AC, or on BC. These are similar, so we may assume that O lies on AC.
Then $\triangle OBC$ is isosceles, so $\angle OBC = \angle OCB$.
$\angle AOB$ is the exterior angle of $\triangle OBC$, so
$$\angle AOB = \angle OBC + \angle OCB = 2 \cdot \angle ACB$$
(by Problem **150**).

(iii) Suppose O lies inside $\triangle ABC$.
$\triangle OAB$, $\triangle OBC$, $\triangle OCA$ are isosceles, so let $\angle OAB = \angle OBA = x$, $\angle OBC = \angle OCB = y$, $\angle OCA = \angle OAC = z$.
Then $\angle ACB = y+z$, $\angle ABC = x+y$, $\angle BAC = x+z$.
The three angles of $\triangle ABC$ add to a straight angle, so $2(x+y+z)$ equals a straight angle.
Hence, in $\triangle OBA$,
$$\angle AOB = 2(x+y+z) - (\angle OAB + \angle OBA) = 2(y+z) = 2 \cdot \angle ACB.$$

(iv) Suppose O lies outside $\triangle ABC$ with O and B on opposite sides of AC.
$\triangle OAB$, $\triangle OBC$, $\triangle OCA$ are isosceles, so let $\angle OAB = \angle OBA = x$, $\angle OBC = \angle OCB = y$, $\angle OCA = \angle OAC = z$.
Then $\angle ACB = y-z$, $\angle ABC = x+y$, $\angle BAC = x-z$.
The three angles of $\triangle ABC$ add to a straight angle, so $2x+2y-2z$ equals a straight angle.
Hence
$$2x + 2y - 2z = \angle AOB + \angle OAB + \angle OBA = \angle AOB + 2x,$$
so $\angle AOB = 2y - 2z = 2 \cdot \angle ACB$.

(v) Suppose O lies outside $\triangle ABC$ with O and B on the same side of AC.
$\triangle OAB$, $\triangle OBC$, $\triangle OCA$ are isosceles, so let $\angle OAB = \angle OBA = x$, $\angle OBC = \angle OCB = y$, $\angle OCA = \angle OAC = z$.
Then $\angle ACB = y+z$, $\angle ABC = y-x$, $\angle BAC = z-x$.
The three angles of ABC add to a straight angle, so $2y+2z-2x$ equals a straight angle.
Since C lies on the minor arc relative to the chord $\underline{AB}$, we need to interpret "the angle subtended by the chord $\underline{AB}$ at the centre O" as the *reflex* angle **outside** the triangle $\triangle AOB$ – which is equal to "$2x$ more than a straight angle", so $\angle AOB = 2y + 2z = 2 \cdot \angle ACB$. QED

152. The chord $\underline{AB}$ subtends angles at C and at D on the same arc. Similarly $\underline{BC}$ subtends angles at A and at D on the same arc. Hence (by Problem **151**) $\angle ACB = \angle ADB$, and $\angle BAC = \angle BDC$.
Hence
$$\angle ADC = \angle ADB + \angle BDC = \angle ACB + \angle BAC,$$
so $\angle ADC + \angle ABC$ equals the sum of the three angles in $\triangle ABC$. QED

153. Let O be the circumcentre of $\triangle ABC$, and let $\angle XAB = x$.
Then $\angle XAO$ is a right angle, and $\triangle OAB$ is isosceles. There are two cases.

(i) If X and O lie on opposite sides of AB, then $\angle OBA = \angle OAB = 90° - x$.
∴ $\angle AOB = 2x$.
∴ $\angle ACB = x = \angle XAB$.

(ii) If X and O lie on the same side of AB, then Y and O lie on opposite sides of AB and of AC. Hence we can apply case (i) (with Y in place of X, AC in place of AB, $\angle YAC$ in place of $\angle XAB$, and $\angle ABC$ in place of $\angle ACB$) to conclude that $\angle YAC = \angle ABC$. Hence

$$\angle XAB + \angle BAC + \angle YAC = \angle XAB + \angle BAC + \angle ABC$$

are both straight angles, so $\angle XAB = \angle ACB$ (since the three angles of $\triangle ABC$ also add to a straight angle). QED

154.

(a)(i) Extend AD to meet the circumcircle at X. Then (applying Problem **145** to $\triangle DXB$), the exterior angle $\angle ADB > \angle DXB = \angle AXB = \angle ACB$.

(ii) We are told that the points C, D lie "on the same side of the line AB". This "side" of the line AB (or "half-plane") is split into three parts by the half-lines "AC produced beyond C" and "BC produced beyond C". There are two very different possibilities.
Suppose first that D lies in one of the two overlapping wedge-shaped regions "between AB produced and AC produced" or "between BA produced and BC produced". Then either DA or DB cuts the arc AB at a point X (say), and $\angle ACB = \angle AXB > \angle ADB$ (by Problem **145** applied to $\triangle AXD$ or to $\triangle BXD$).
The only alternative is that D lies in the wedge shaped region outside the circle at the point C, lying "between AC produced and BC produced". Then C lies inside $\triangle ADB$, so C lies inside the circumcircle of $\triangle ADB$. Hence part (i) implies that $\angle ACB < \angle ADB$ as required. QED

(b) If D does not lie on the circumcircle of $\triangle ABC$, then it lies either inside, or outside the circle – in which case $\angle ADB \neq \angle ACB$ (by part (a)).

(c) $\angle D$ is less than a straight angle, so D must lie outside $\triangle ABC$. Moreover, B, D lie on opposite sides of the line AC (since the edges BC, DA do not cross internally, and neither do the edges CD, AB). Consider the circumcircle of $\triangle ABC$, and let X be any point on the circle such that X and D lie on the same side of the line AC. Then $ABCX$ is a cyclic quadrilateral, so $\angle ABC$ and $\angle CXA$ are supplementary. Hence $\angle CXA = \angle CDA = \angle D$. But then D lies on the circle (by part (b)).

155. $\triangle OPQ \equiv \triangle OP'Q$ (by RHS-congruence: $\angle OPQ = \angle OP'Q$ are both right angles, $\underline{OP} = \underline{OP'}$, $OQ = OQ$)
∴ $\underline{QP} = \underline{QP'}$, and $\angle QOP = \angle QOP'$. QED

156.

(a) Let X be any point on the bisector of $\angle ABC$. Drop the perpendiculars XY from X to AB and XZ from X to CB. Then: $\angle XYB = \angle XZB$ are both right angles by construction; and $\angle XBY = \angle XBZ$ since X lies on the bisector of $\angle YBZ$; hence $\angle BXY = \angle BXZ$ (since the three angles in each triangle have the same sum; so as soon as two of the angles are equal in pairs, the third pair must also be equal). Hence $\triangle BXY \equiv \triangle BXZ$ (by ASA-congruence: $\angle YBX = \angle ZBX$, $\underline{BX} = \underline{BX}$, $\angle BXY = \angle BXZ$.)
$\therefore \underline{XY} = \underline{XZ}$. QED

(b) Suppose X is equidistant from m and from n. Drop the perpendiculars XY from X to m, and XZ from X to n.
Then $\triangle BXY \equiv \triangle BXZ$ (by RHS-congruence:

$\angle BYX = \angle BZX$ are both right angles
$\underline{XY} = \underline{XZ}$, since we are assuming X is equidistant from m and from n
$\underline{BX} = \underline{BX}$).

Hence $\angle XBY = \angle XBZ$, so X lies on the bisector of $\angle YBZ$. QED

157.

(i) Join $\underline{AC}$. Then $\triangle ABC \equiv \triangle CDA$ by ASA-congruence:

$\angle BAC = \angle DCA$ (alternate angles, since $AB \parallel DC$)
$\underline{AC} = \underline{CA}$
$\angle ACB = \angle CAD$ (alternate angles, since $CB \parallel DA$).

In particular, $\triangle ABC$ and $\triangle CDA$ must have equal area, and so each is exactly half of $ABCD$. QED

Note: Once we prove (Problem **161** below) that a parallelogram has the same area as the rectangle on the same base and lying between the same pair of parallels (whose area is equal to "base $\times$ height"), the result in part (i) will immediately translate into the familiar formula for the area of the triangle

$$\frac{1}{2}(\text{base} \times \text{height}).$$

(ii) $\triangle ABC \equiv \triangle CDA$ by part (i). Hence $\underline{AB} = \underline{CD}$, $\underline{BC} = \underline{DA}$; and $\angle B = \angle ABC = \angle CDA = \angle D$.
To show that $\angle A = \angle C$, we can either copy part (i) after joining BD to prove that $\triangle BCD \equiv \triangle DAB$, or we can use part (i) directly to note that $\angle A = \angle BAC + \angle DAC = \angle DCA + \angle BCA = \angle C$. QED

(iii) $\triangle AXB \equiv \triangle CXD$ by ASA-congruence:

$\angle XAB = \angle XCD$ (alternate angles, since $AB \parallel DC$)
$\underline{AB} = \underline{CD}$ (by part (ii))
$\angle XBA = \angle XDC$ (alternate angles, since $AB \parallel DC$).

Hence $\underline{XA}$ (in $\triangle AXB$) = $\underline{XC}$ (in $\triangle CXD$), and $\underline{XB}$ (in $\triangle AXB$) = $\underline{XD}$ (in $\triangle CXD$). QED

158. We may assume that m cuts the opposite sides AB at Y and DC at Z. $\triangle XYB \equiv \triangle XZD$ by ASA-congruence:

$\angle YXB = \angle ZXD$ (vertically opposite angles)
$\underline{XB} = \underline{XD}$ (by Problem **157**(iii))
$\angle XBY = \angle XDZ$ (alternate angles).

Therefore

$$\begin{aligned} \text{area}(YZCB) &= \text{area}(\triangle BCD) - \text{area}(\triangle XZD) + \text{area}(\triangle XYB) \\ &= \text{area}(\triangle BCD) \\ &= \frac{1}{2}\text{area}(ABCD). \end{aligned}$$

QED

159.

(a) Join AC. Then $\triangle ABC \equiv \triangle CDA$ by SAS-congruence:

$\underline{BA} = \underline{DC}$ (given)
$\angle BAC = \angle DCA$ (alternate angles, since $AB \parallel DC$)
$\underline{AC} = \underline{CA}$.

Hence $\angle BCA = \angle DAC$, so $AD \parallel BC$ as required. QED

(b) Join AC. Then $\triangle ABC \equiv \triangle CDA$ by SSS-congruence:

$\underline{AB} = \underline{CD}$ (given)
$\underline{BC} = \underline{DA}$ (given)
$\underline{CA} = \underline{AC}$.

Hence $\angle BAC = \angle DCA$, so $AB \parallel DC$; and $\angle BCA = \angle DAC$, so $BC \parallel AD$. QED

(c) $\angle A + \angle B + \angle C + \angle D = 2\angle A + 2\angle B = 2\angle A + 2\angle D$ are each equal to two straight angles.
$\therefore \angle A + \angle B$ is equal to a straight angle, so $AD \parallel BC$; and $\angle A + \angle D$ is equal to a straight angle, so $AB \parallel DC$. QED

Note: The fact that the angles in a quadrilateral add to two straight angles is proved in the next chapter. However, if preferred, it can be proved here directly. If we imagine pins located at A, B, C, D, then a string tied around the four points defines their "convex hull" – which is either a 4-gon (if the string touches all four pins), or a 3-gon (if one vertex is inside the triangle formed by the other three). In the first case, either diagonal (AC or BD) will split the quadrilateral internally into two triangles; in the second case, one of the three 'edges' joining vertices of the convex hull to the internal vertex cannot be an edge of the quadrilateral, and so must be a diagonal, which splits the quadrilateral internally into two triangles.

160. $AM = MD$ (by construction of M as the midpoint), and $BN = NC$.
∴ $AM = BN$ (since $AD = BC$ by Problem **157**(ii)).
∴ $ABNM$ is a parallelogram (by Problem **159**(a)), so $MN \parallel AB$.
Let AC cross MN at Y.
Then $\triangle AYM \equiv \triangle CYN$ (by ASA-congruence, since

$\angle YAM = \angle YCN$ (alternate angles, since $AD \parallel BC$)
$AM = CN$
$\angle AMY = \angle CNY$ (alternate angles, since $AD \parallel BC$).

Hence $AY = CY$, so Y is the midpoint of AC – the centre of the parallelogram (where the two diagonals meet (by Problem **157**(iii))). QED

161.

Note: In the easy case, where the perpendicular from A to the line DC meets the side DC internally at X, it is natural to see the parallelogram $ABCD$ as the "sum" of a trapezium $ABCX$ and a right angled triangle $\triangle AXD$. If the perpendicular from B to DC meets DC at Y, then $\triangle AXD \equiv \triangle BYC$. Hence we can rearrange the two parts of the parallelogram $DCBA$ to form a rectangle $XYBA$.

However, a general proof cannot assume that the perpendicular from A (or from B) to DC meets DC internally. Hence we are obliged to think of the parallelogram in terms of **differences**. This is a strategy that is often useful, but which can be surprisingly elusive.

Draw the perpendiculars from A and B to the line CD, and from C and D to the line AB. Choose the two perpendiculars which, together with the lines AB and CD define a rectangle that completely contains the parallelogram $ABCD$ (that is, if AB runs from left to right, take the left-most, and the right-most perpendiculars). These will be either the perpendiculars from B and from D, or the perpendiculars from A and from C (depending on which way the sides AC and BD slope).

Suppose the chosen perpendiculars are the one from B – meeting the line DC at P, and the one from D – meeting the line AB at Q.

Then $BP \parallel QD$ (by Problem **159**(c)), so $BQDP$ is a parallelogram with a right angle – and hence a rectangle. Hence $BQ = PD$, and $BP = QD$ (by Problem **157**(ii)).

$\triangle QAD \equiv \triangle PCB$ (by RHS-congruence), so each is equal to half the rectangle on base PC and height PB. Hence

$$\begin{aligned}\text{area}(ABCD) &= \text{area(rectangle } BQDP) \\ &\quad - \text{area(rectangle on base } PC \text{ with height } PB) \\ &= \text{area(rectangle on base } CD \text{ with height } DQ).\end{aligned}$$

QED

162.

(a) $ABCB'$ is a parallelogram, so $AB' \parallel CB$ and $\underline{AB'} = \underline{BC}$.
Similarly, $BCAC'$ is a parallelogram, so $AC' \parallel CB$ and $\underline{AC'} = \underline{CB}$.
Hence $\underline{B'A} = \underline{AC'}$, so A is the midpoint of $\underline{B'C'}$.
Similarly B is the midpoint of $\underline{C'A'}$, and C is the midpoint of $\underline{A'B'}$.

(b) Let H be the circumcentre of $\triangle A'B'C'$ – that is, the common point of the perpendicular bisectors of $\underline{A'B'}$, $\underline{B'C'}$, and $\underline{C'A'}$. Then H is a common point of the three perpendiculars from A to BC, from B to CA, and from C to AB.

163. Consider any path from H to V. Suppose this reaches the river at X. The shortest route from H to X is a straight line segment; and the shortest route from X to V is a straight line segment.

If we reflect the point H in the line of the river, we get a point H', where HH' is perpendicular to the river and meets the river at Y (say).

Then $\triangle HXY \equiv \triangle H'XY$ (by SAS-congruence, since $\underline{HY} = \underline{H'Y}$, $\angle HYX = \angle H'YX$, and $\underline{YX} = \underline{YX}$). Hence $\underline{HX} = \underline{H'X}$, so the distance from H to V via X is equal to $\underline{HX} + \underline{XV} = \underline{H'X} + \underline{XV}$, and this is shortest when H', X, and V are collinear. (So to find the shortest route, reflect H in the line of the river to H', then draw $H'V$ to cross the line of the river at X, and travel from H to V via X.)

164.

(a) Let $\triangle PQR$ be any triangle inscribed in $\triangle ABC$, with P on BC, Q on CA, R on AB (not necessarily the orthic triangle). Let P' be the reflection of P in the side AC, and let P'' be the reflection of P in the side AB. Then $\underline{PQ} = \underline{P'Q}$, and $\underline{PR} = \underline{P''R}$ (as in Problem **163**).
$\therefore \underline{PQ} + \underline{QR} + \underline{RP} = \underline{P'Q} + \underline{QR} + \underline{RP''}$.

Each choice of the point P on $\underline{AB}$ determines the positions of P' and P''. Hence the shortest possible perimeter of $\triangle PQR$ arises when $P'QRP''$ is a straight line. That is, given a choice of the point P, choose Q and R by:

– constructing the reflections P' of P in AC, and P'' of P in AB;

– join $\underline{P'P''}$ and let Q, R be the points where this line segment crosses AC, AB respectively.

It remains to decide how to choose P on $\underline{BC}$ so that $\underline{P'P''}$ is as short as possible. The key here is to notice that A lies on both AC and on AB.
$\therefore \underline{AP} = \underline{AP'}$, and $\underline{AP} = \underline{AP''}$, so $\triangle AP'P''$ is isosceles.
Also $\angle PAC = \angle P'AC$, and $\angle PAB = \angle P''AB$.
$\therefore \angle P'AP'' = 2 \cdot \angle A$.

Hence, for each position of the point P, $\triangle AP'P''$ is isosceles with apex angle equal to $2 \cdot \angle A$. Any two such triangles are similar (by SAS-similarity). Hence the triangle $\triangle AP'P''$ with the shortest "base" $\underline{P'P''}$ occurs when the legs $\underline{AP'}$ and $\underline{AP''}$ are as short as possible. But $\underline{AP} = \underline{AP'} = \underline{AP''}$, so this occurs when $\underline{AP}$ is as short as possible – namely when $\underline{AP}$ is perpendicular to BC.

Since the same reasoning applies to Q and to R, it follows that the required triangle $\triangle PQR$ must be the orthic triangle of $\triangle ABC$. QED

(b) Let $\triangle PQR$ be the orthic triangle of $\triangle ABC$, with P on BC, Q on CA, R on AB. Let H be the orthocenter of $\triangle ABC$.
$\angle BPH$ and $\angle BRH$ are both right angles, so add to a straight angle. Hence (by Problem **154**(c)), $BPHR$ is a cyclic quadrilateral. Similarly $CPHQ$ and $AQHR$ are cyclic quadrilaterals.
In the circumcircle of $CPHQ$, we see that the initial "angle of incidence" $\angle CQP = \angle CHP$. Also $\angle CHP = \angle AHR$ (vertically opposite angles); and in the circumcircle of $AQHR$, $\angle AHR = \angle AQR$.
Hence $\angle CQP = \angle AQR$, so a ray of light which traverses $\underline{PQ}$ will reflect at Q along the line $\underline{QR}$. Similarly one can show that $\angle ARQ = \overline{\angle BRP}$, so that the ray will then reflect at R along $\underline{RP}$; and $\angle BPR = \angle CPQ$, so the ray will then reflect at P along $\underline{PQ}$. QED

165.

(a)(i) Triangles $\triangle ABL$ and $\triangle ACL$ have equal bases $\underline{BL} = \underline{CL}$, and the same apex A – so lie between the same parallels. Hence they have equal area (by Problems **157** and **161**).
Similarly, $\triangle GBL$ and $\triangle GCL$ have equal bases $\underline{BL} = \underline{CL}$, and the same apex G – so have equal area.
Hence the differences $\triangle ABG = \triangle ABL - \triangle GBL$ and $\triangle ACG = \triangle ACL - \triangle GCL$ have equal area.

(ii) [Repeat the solution for part (i) replacing A, B, C, L, G by B, C, A, M, G.]

(b) $\triangle ABL$ and $\triangle ACL$ have equal area (as in (a)(i)). Similarly $\triangle GBL$ and $\triangle GCL$ have the same area – say x (since $\underline{BL} = \underline{CL}$). Hence $\triangle ABG$ and $\triangle ACG$ have equal area.
In the same way $\triangle BCM$ and $\triangle BAM$ have equal area; and $\triangle GCM$ and $\triangle GAM$ have the same area – say y (since $CM = AM$). Hence $\triangle BCG$ and $\triangle BAG$ have equal area.
But then $\triangle ABG = \triangle ACG = \triangle BCG$ and $\triangle ACG = \triangle AMG + \triangle CMG = 2y$, $\triangle BCG = \triangle BLG + \triangle CLG = 2x$. Hence $x = y$, so $\triangle AMG$, $\triangle CMG$, $\triangle CLG$, $\triangle BLG$ all have the same area x, and $\triangle ABG$ has area $2x$.
The segment $\underline{GN}$ divides $\triangle ABG$ into two equal parts ($\triangle ANG$ and $\triangle BNG$), so each part has area x.
Hence $\triangle CAG + \triangle ANG$ has the same area ($3x$) as $\triangle CAN$. Hence $\angle CGN$ is a straight angle, and the three medians AL, BM, CN all pass through the point G.

Note: At first sight, the 'proof' of the result in (b) using vectors seems considerably easier. (If A, B, C have position vectors $\mathbf{a}$, $\mathbf{b}$, $\mathbf{c}$ respectively, then L has the position vector $\frac{1}{2}(\mathbf{b} + \mathbf{c})$, and M has position vector $\frac{1}{2}(\mathbf{c} + \mathbf{a})$, and it is easy to see that AL and BM meet at G with position vector $\frac{1}{3}(\mathbf{a} + \mathbf{b} + \mathbf{c})$. One can then check directly that G lies on CN, or notice that the symmetry of the expression $\frac{1}{3}(\mathbf{a}+\mathbf{b}+\mathbf{c})$ guarantees that G is also the point where BM and CN meet.

The inscrutable aspect of this 'proof' lies in the fact that all the geometry has been silently hidden in the algebraic assumptions which underpin the unstated axioms of the 2-dimensional vector space, and the underlying field of real numbers. Hence, although the vector 'proof' may seem simpler, the two different approaches cannot really be compared.

166.

(a) $\underline{AN} = \underline{BN}$ (by construction of N as the midpoint of $\underline{AB}$)
$\angle ANM = \angle BNM'$ (vertically opposite angles)
$\underline{NM} = \underline{NM'}$ (by construction).
$\therefore \triangle ANM \equiv \triangle BNM'$ (by SAS-congruence). QED

(b) $\underline{BM'} = \underline{AM} = \underline{CM}$.
$\angle NAM = \angle NBM'$ (since $\triangle ANM \equiv \triangle BNM'$)
$\therefore BM' \parallel MA$ (i.e. $BM' \parallel CM$). QED

(c) $BM'MC$ is a quadrilateral with opposite sides $\underline{CM}$, $\underline{BM'}$ equal and parallel.
$\therefore BM'MC$ is a (by Problem **158**(a)). QED

167. Since A and B are interchangeable in the result to be proved, we may assume that A is the point on the secant that lies between P and B.

In order to make deductions, we have to create triangles – so join AT and BT. This creates two triangles: $\triangle PAT$ and $\triangle PTB$, in which we see that:
$\angle TPA = \angle BPT$,
$\angle PTA = \angle PBT$ (by Problem **153**),
$\therefore \angle PAT = \angle PTB$ (since the three angles in each triangle add to a straight angle).
Hence $\triangle PAT \sim \triangle PTB$ (by AAA-similarity).
$\therefore \underline{PT} : \underline{PB} = \underline{PA} : \underline{PT}$, or $\underline{PA} \times \underline{PB} = \underline{PT}^2$. QED

168. Since A, B are interchangeable in the result to be proved, and C, D are interchangeable, we may assume that A lies between P and B, and that C lies between P and D.

(i) Let the tangent from P to the circle touch the circle at T.
Then Problem **167** guarantees that $\underline{PA} \times \underline{PB} = \underline{PT}^2$.
Replacing the secant PAB by PCD shows similarly that $\underline{PC} \times \underline{PD} = \underline{PT}^2$.
Hence $\underline{PA} \times \underline{PB} = \underline{PC} \times \underline{PD}$. QED

(ii) **Note:** The first proof is so easy, one may wonder why anyone would ask for a second proof. The reason lies in Problem **169** – which looks very much like Problem **168**, but with P *inside* the circle. Hence, when we come to the next problem, the easy approach in (i) will not be available, so it is worth looking for another proof of **168** which has a chance of generalizing.

Notice that $\underline{AC}$ is a chord which links the two secants PAB and PCD.
So join $\underline{AD}$ and $\underline{CB}$.

5.13. Chapter 5: Comments and solutions   243

Then $\triangle PAD \sim \triangle PCB$ (by AAA-similarity: since $\angle APD = \angle CPB$, and $\angle PDA = \angle PBC$).
$\therefore \underline{PA} : \underline{PC} = \underline{PD} : \underline{PB}$, or $\underline{PA} \times \underline{PB} = \underline{PC} \times \underline{PD}$.   QED

169. Join $\underline{AD}$ and $\underline{CB}$.
Then $\triangle PAD \sim \triangle PCB$ (by AAA-similarity: since
$\angle APD = \angle CPB$ (vertically opposite angles)
$\angle PDA = \angle PBC$ (angles subtended by chord $\underline{AC}$ on the same arc)
$\angle PAD = \angle PCB$ (angles subtended by a chord $\underline{BD}$ on the same arc)).
$\therefore \underline{PA} : \underline{PC} = \underline{PD} : \underline{PB}$, or $\underline{PA} \times \underline{PB} = \underline{PC} \times \underline{PD}$.   QED

170.

(a) If $a = b$, then $ABCD$ is a parallelogram (by Problem **159**(a)).
$\therefore \underline{AD} = \underline{BC}$ (by Problem **157**(ii)).
$\therefore \underline{AM} = \underline{BN}$, so $ABNM$ is a parallelogram (by Problem **159**(a)).
$\therefore \underline{MN} = \underline{AB}$ has length a, and $MN \parallel AB$ (by Problem **160**).

If $a \neq b$, then $a < b$, or $b < a$. We may assume that $a < b$.
Construct the line through B parallel to AD, and let this line meet $\underline{DC}$ at Q.
Then $ABQD$ is a parallelogram, so $\underline{DQ} = \underline{AB}$, and $\underline{AD} = \underline{BQ}$. Hence $\underline{QC}$ has length $b - a$.
Construct the line through M parallel to QC (and hence parallel to BA), and let this meet BQ at P, and BC at N'.
Then $ABPM$ and $MPQD$ are both parallelograms.
$\therefore \underline{MP} = \underline{AB}$ has length a, and
$$\underline{BP} = \underline{AM} = \underline{MD} = \underline{PQ}.$$
Now $\triangle BPN' \sim \triangle BQC$ (by AAA-similarity, since $PN' \parallel QC$); so
$$\underline{BP} : \underline{BQ} = \underline{BN'} : \underline{BC} = 1 : 2.$$
Hence $N' = N$ is the midpoint of $\underline{BC}$, $MN \parallel BC$, and $\underline{MN}$ has length
$$a + \frac{b-a}{2} = \frac{a+b}{2}.$$

(b) Suppose first that $a = b$. Then $ABCD$ is a parallelogram (by Problem **159**(a)), so the area of $ABCD$ is given by $a \times d$ ("length of base" $\times$ "height"). (by Problem **161**). Hence we may suppose that $a < b$.

Solution 1: Extend $\underline{AB}$ beyond B to a point X such that $\underline{BX} = \underline{DC}$ (so $\underline{AX}$ has length $a + b$).
Extend $\underline{DC}$ beyond C to a point Y such that $\underline{CY} = \underline{AB}$ (so $\underline{DY}$ has length $a + b$).
Clearly $ABCD$ and $YCBX$ are congruent, so each has area one half of area($AXYD$).

Now $AX \parallel DY$, and $\underline{AX} = \underline{DY}$, so $AXYD$ is a parallelogram (by Problem **159**(a)).
Hence $AXYD$ has area "(length of base) × height" (by Problem **161**), so $ABCD$ has area $\frac{a+b}{2} \times d$.

Solution 2: [We give a second solution as preparation for Problem **171**.]
Now $a < b$ implies that $\angle BAD + \angle ABC$ is greater than a straight angle.

[**Proof.** The line through B parallel to AD meets DC at Q, and $ABQD$ is a parallelogram.
Hence $\underline{DQ} = \underline{AB}$, so Q lies between D and C, and
$$\angle ABC = \angle ABQ + \angle QBC > \angle ABQ.$$
$\therefore \angle BAD + \angle ABC > \angle BAD + \angle ABQ$, which is equal to a straight angle.]

So if we extend $\underline{DA}$ beyond A, and $\underline{CB}$ beyond B, the lines meet at X, where X, D are on opposite sides of AB. Then $\triangle XAB \sim \triangle XDC$ (by AAA-similarity), whence corresponding lengths in the two triangles are in the ratio $\underline{AB} : \underline{DC} = a : b$. In particular, if the perpendicular from X to AB has length h, then $\frac{h}{h+d} = \frac{a}{b}$, so $h(b-a) = ad$.
Now

$$\begin{aligned}
\text{area}(ABCD) &= \text{area}(\triangle XDC) - \text{area}(\triangle XAB) \\
&= \frac{1}{2}b(h+d) - \frac{1}{2}ah \\
&= \frac{1}{2}bd + \frac{1}{2}h(b-a) \\
&= \frac{1}{2}bd + \frac{1}{2}ad \\
&= \left(\frac{a+b}{2}\right)d.
\end{aligned}$$

171. Note: The volume of a pyramid or cone is equal to

$$\frac{1}{3} \times (\text{area of base}) \times \text{height}.$$

There is no elementary general proof of this fact. The initial coefficient of $\frac{1}{3}$ arises because we are "adding up", or integrating, cross-sections parallel to the base, whose area involves a square x^2, where x is the distance from the apex (just as the coefficient $\frac{1}{2}$ in the formula for the area of a triangle arises because we are integrating linear cross-sections whose size is a multiple of x – the distance from the apex). Special cases of this formula can be checked – for example, by noticing that a cube $ABCDEFGH$ of side s, with base $ABCD$, and upper surface $EFGH$, with E above D, F above A, and so on, can be dissected into three identical pyramids

– all with apex E: one with base $ABCD$, one with base $BCHG$, and one with base $ABGF$. Hence each pyramid has volume $\frac{1}{3}s^3$, which may be interpreted as

$$\frac{1}{3} \times (\text{area of base} = s^2) \times (\text{height} = s).$$

To obtain the frustum of height d, a pyramid with height h (say) is cut off a pyramid with height $h + d$.

$$\begin{aligned}
\therefore \text{volume(frustum)} &= \left[\frac{1}{3} \times b^2 \times (h+d)\right] - \left[\frac{1}{3} \times a^2 \times h\right] \\
&= \left[\frac{1}{3} \times b^2 \times d\right] + \left[\frac{1}{3} \times (b^2 - a^2) \times h\right].
\end{aligned}$$

Let N be the midpoint of $\underline{BC}$, and let the line AN meet the upper square face of the frustum at M.
Let the perpendicular from the apex A to the base $BCDE$ meet the upper face of the frustum at Y and the base $BCDE$ at Z.
Then $\triangle AYM \sim \triangle AZN$ (by AAA-similarity), so $\underline{AY} : \underline{AZ} = \underline{YM} : \underline{ZN}$.
$\therefore \frac{h}{h+d} = \frac{a}{b}$, so $h(b-a) = ad$.

$$\begin{aligned}
\therefore \text{volume(frustum)} &= \frac{1}{3}b^2 d + \frac{1}{3}(b^2 - a^2)h \\
&= \frac{1}{3}b^2 d + \frac{1}{3}(b+a)ad \\
&= \frac{1}{3}(b^2 + ab + a^2)d.
\end{aligned}$$

172. Construct the line through A which is parallel to $A'B'$, and let it meet the line BB' at P.
Similarly, construct the line through B which is parallel to $B'C'$ and let it meet the line CC' at Q.
Then $\triangle ABP \sim \triangle BCQ$ (by AAA-similarity), so $\underline{AB} : \underline{BC} = \underline{AP} : \underline{BQ}$.
Now $AA'B'P$ is a parallelogram, so $\underline{AP} = \underline{A'B'}$, and $BB'C'Q$ is a parallelogram, so $\underline{BQ} = \underline{B'C'}$.
$\therefore \underline{AB} : \underline{BC} = \underline{A'B'} : \underline{B'C'}$. QED

173.

(a) Suppose $\underline{AB}$ has length a and $\underline{CD}$ has length b. Problem **137** allows us to construct a point X such that $\underline{AX} = \underline{CD}$, so $\underline{AX}$ has length b. Now construct the point Y where the circle with centre A and radius $\underline{AX}$ meets $\underline{BA}$ produced (beyond A). Then $\underline{YB}$ has length $a + b$.
If $a > b$, let Z be the point where the circle with centre A and passing through X meets the segment $\underline{AB}$ internally Then $\underline{ZB}$ has length $a - b$.

(b) Use Problem **137** to construct a point U (not on the line CD) such that $DU = XY$, and a point V on DU (possibly extended beyond U) such that $DV = AB$. Hence DU has length 1 and DV has length a.

Construct the line through V parallel to UC and let it meet the line DC at W. Then $\triangle DVW \sim \triangle DUC$, with scale factor $DW : DC = DV : DU = a : 1$. Hence DW has length ab.

To construct a line segment of length $\frac{a}{b}$ construct U, V as above. Then let the circle with centre D and radius DU meet CD at U'. Let the line through U' parallel to CV meet DV at X. The $\triangle DU'X \sim \triangle DCV$, with scale factor $DX : DV = DU' : DC = 1 : b$, so DX has length $\frac{a}{b}$.

(c) Use Problem **137** to construct a point G so that $AG = XY = 1$. Then draw the circle with centre A passing through G to meet BA extended at H. Hence $HA = 1$, $AB = a$.

Construct the midpoint M of HB; draw the circle with centre M and passing through H and B.

Construct the perpendicular to HB at the point A to meet the circle at K. Then $\triangle HAK \sim \triangle KAB$, so $AB : AK = AK : AH$. Hence $AK = \sqrt{a}$.

174. The key to this problem is to use Problem **158**: a parallelogram (and hence any rectangle) is divided into two congruent pieces by any straight cut through the centre. If A is the centre of the rectangular piece of fruitcake, and B is the centre of the combined rectangle consisting of "fruitcake plus icing", then the line AB gives a straight cut that divides both the fruitcake and the icing exactly in two.

175.

(a) The minute hand is pointing exactly at "7", but the hour hand has moved $\frac{7}{12}$ of the way from "1" to "2": that is, $\frac{7}{12}$ of $30°$, or $17\frac{1}{2}°$. Hence the angle between the hands is $162\frac{1}{2}°$.

The same angle arises whenever the hands are trying to point in opposite directions, but are off-set by $\frac{7}{12}$ of $30°$, or $17\frac{1}{2}°$. This suggests that instead of "35 minutes after 1" we should consider "35 minutes before 11", or 10:25.

(b) The two hands coincide at midnight. The minute hand then races ahead, and the hands do not coincide again until shortly after 1:00 – and indeed after 1:05. More precisely, in 60 minutes, the minute hand turns through $360°$, so in x minutes, it turns through $6x°$. In 60 minutes the hour hand turns through $30°$, so in x minutes, the hour hand turns through $\frac{1}{2}x°$.

The hands overlap when

$$\frac{1}{2}x \equiv 6x \pmod{360},$$

or when $\frac{11}{2}x$ is a multiple of 360.

This occurs when $x = 0$ (i.e. at midnight); then not until

$$x = \frac{720}{11} = 65\frac{5}{11},$$

and the time is $5\frac{5}{11}$ minutes past 1: that is, after $1\frac{1}{11}$ hours.
It occurs again $1\frac{1}{11}$ hours, or $65\frac{5}{11}$ minutes, later – namely at $10\frac{10}{11}$ minutes past 2, and so on.

Hence the phenomenon occurs at midnight, and then 11 more times until noon (with noon as the 12th time; and then 11 more times until midnight – and hence 23 times in all (including both midnight occurrences).

(c) If we add a third hand (the 'second hand'), all three hands coincide at midnight. In x minutes, the second hand turns through $360x°$.

We now know exactly when the hour hand and minute hand coincide, so we can check where the second hand is at these times. For example, at $5\frac{5}{11}$ minutes past 1, the second hand has turned through $(360 \times 65\frac{5}{11})°$, and

$$360 \times 65\frac{5}{11} \equiv 360 \times \frac{5}{11} \pmod{360},$$

so the second hand is nowhere near the other two hands.

The kth occasion when the hour hand and minute hands coincide occurs at $k \times 1\frac{1}{11}$ hours after midnight, when the two hands point in a direction $\left(\frac{360k}{11}\right)°$ clockwise beyond "12". At the same time, the second hand has turned through $(360 \times 65\frac{5}{11} \times k)°$, and

$$360 \times 65\frac{5}{11} \times k \equiv 360 \times \frac{5}{11} \times k \pmod{360},$$

or five times as far round, and these two are never equal (mod 360).

176.

Note: One of the things that makes it possible to calculate distances exactly here is that the angles are all known exactly, and give rise to lots of right angled triangles.

The rotational symmetry of the clockface means we only have to consider segments with one endpoint at 12 o'clock (say A). The reflectional symmetry in the line AG means that we only have to find $\underline{AB}$, $\underline{AC}$, $\underline{AD}$, $\underline{AE}$, $\underline{AF}$, and $\underline{AG}$.

Clearly $\underline{AG} = 2$. If O denotes the centre of the clockface, then $\underline{AD}$ is the hypotenuse of an isosceles triangle $\triangle OAD$ with legs of length 1, so $\underline{AD} = \sqrt{2}$.

$\triangle ACO$ is isosceles ($\underline{OA} = OC = 1$) with apex angle $\angle AOC = 60°$, so the triangle is equilateral. Hence $\underline{AC} = OA = 1$.

It follows that $ACEGIK$ is a regular hexagon, so $\underline{AE} = \sqrt{3}$ (if OC meets AE at X, then $\triangle ACX$ is a 30-60-90 triangle, and so is half of an equilateral triangle, whence $\underline{AX} = \frac{\sqrt{3}}{2}$).

It remains to find $\underline{AB}$ and $\underline{AF}$.

Let OB meet AC at Y. Then $OY = \frac{\sqrt{3}}{2}$ (since $\triangle OYA$ is a 30-60-90 triangle).
$\therefore BY = 1 - \frac{\sqrt{3}}{2}$, $AY = \frac{1}{2}$, so $AB = \sqrt{2 - \sqrt{3}}$.
Finally, AB subtends $\angle AOB = 30°$ at the centre, whence $\angle OAB = \angle OBA = 75°$ and $\angle AGB = 15°$.
$\therefore \angle ABG = 90°$ so we may apply Pythagoras' Theorem to $\triangle ABG$ to find $BG = AF = \sqrt{2 + \sqrt{3}}$.

177.

$\sqrt{1}$ is the distance from $(0,0,0)$ to $(1,0,0)$.
$\sqrt{2}$ is the distance from $(0,0,0)$ to $(1,1,0)$.
$\sqrt{3}$ is the distance from $(0,0,0)$ to $(1,1,1)$.
$\sqrt{4}$ is the distance from $(0,0,0)$ to $(2,0,0)$.
$\sqrt{5}$ is the distance from $(0,0,0)$ to $(2,1,0)$.
$\sqrt{6}$ is the distance from $(0,0,0)$ to $(2,1,1)$.
$\sqrt{7}$ cannot be realized as a distance between integer lattice points in 3D.
$\sqrt{8}$ is the distance from $(0,0,0)$ to $(2,2,0)$.
$\sqrt{9}$ is the distance from $(0,0,0)$ to $(3,0,0)$.
$\sqrt{10}$ is the distance from $(0,0,0)$ to $(3,1,0)$.
$\sqrt{11}$ is the distance from $(0,0,0)$ to $(3,1,1)$.
$\sqrt{12}$ is the distance from $(0,0,0)$ to $(2,2,2)$.
$\sqrt{13}$ is the distance from $(0,0,0)$ to $(3,2,0)$.
$\sqrt{14}$ is the distance from $(0,0,0)$ to $(3,2,1)$.
$\sqrt{15}$ cannot be realized as a distance between integer lattice points in 3D.
$\sqrt{16}$ is the distance from $(0,0,0)$ to $(4,0,0)$.
$\sqrt{17}$ is the distance from $(0,0,0)$ to $(4,1,0)$.

Note: The underlying question extends Problem **32**:

Which integers can be represented as a sum of three squares?

This question was answered by Legendre (1752–1833):

> **Theorem.** A positive integer can be represented as a sum of three squares if and only if it is not of the form $4^a(8b+7)$.

178.

(a) Let AD and CE cross at X. Join DE. Then $\angle AXC = \angle EXD$ (vertically opposite angles). Hence $\angle CAD + \angle ACE = \angle ADE + \angle CED$, so the five angles of the pentagonal star have the same sum as the angles of $\triangle BED$. Hence the five angles have sum π radians.

(b) Start with seven vertices A, B, C, D, E, F, G arranged in cyclic order.

(i) Consider first the 7-gonal star $ADGCFBE$. Let GD and BE cross at X and let GC and BF cross at Y. Join DE, BG. As in part (a),

$$\angle BGC + \angle GBF = \angle BFC + \angle GCF,$$

and
$$\angle BGD + \angle GBE = \angle BED + \angle GDE,$$
so the angles in the 7-gonal star have the same sum as the angles in $\triangle ADE$. Hence the seven angles have sum π radians.

(ii) Similar considerations with the 7-gonal star $ACEGBDF$ show that its seven angles have sum 3π radians.

Note: Notice that the three possible "stars" (including the polygon $ABCDEFG$) have angle sums π radians, 3π radians, and 5π radians.

(c) if $n = 2k + 1$ is prime, we may join A to its immediate neighbour B (1-step), or to its second neighbor C (2-step), ..., or to its kth neighbour (k-step), so there are k different stars, with angle sums
$$(n-2)\pi, (n-4)\pi, (n-6)\pi, \ldots, 3\pi, \pi$$
respectively.

If n is not prime, the situation is slightly more complicated, since, for each divisor m of n, the km-step stars break up into separate components.

179.

(a)(i) Let the other three pentagons be $CDRST$, $DEUVW$, $EAXYZ$.
At A we have three angles of $108°$, so $\angle NAX = 36°$.
$\triangle ANX$ is isosceles ($\underline{AN} = \underline{AB} = \underline{AE} = \underline{AX}$), so $\angle AXN = 72°$.
Hence $\angle AXY + \angle AXN = 180°$, so Y, X, N lie in a straight line.
Similarly M, N, X lie in a straight line.
Hence M, N, X, Y lie on a straight line segment $\underline{MY}$ of length $1 + \underline{YN} = 2 + \underline{NX}$.

(ii) In the same way it follows that MP passes through L and Q, that PS passes through O and T, etc. so that the figure fits snugly inside the pentagon $MPSVY$, whose angles are all equal to $108°$. Moreover, $\triangle ANX \equiv \triangle BQL$ (by SAS-congruence), so $\underline{XN} = \underline{LQ}$, whence $\underline{MP} = \underline{YM}$. Similarly $\underline{MP} = \underline{PS} = \underline{SV} = \underline{VY}$, so $MPSVY$ is a regular pentagon.

(iii) In the regular pentagon $EAXYZ$ the diagonal $EY \parallel AX$.
Moreover XAC is a straight line, and $\angle ACB + \angle NBC = 180°$, so $AC \parallel NB$.
Hence $YE \parallel NB$, and $\underline{YE} = \underline{NB} = \tau$ (the Golden Ratio $\frac{1+\sqrt{5}}{2}$), so $YEBN$ is a parallelogram.
Hence $\underline{YN} = \underline{EB} = \tau$, so $\underline{YM} = 1 + \tau = \tau^2$.

(b)(i)
$$\triangle MPY \equiv \triangle PSM \equiv \triangle SVP \equiv \triangle VYS \equiv \triangle YMV$$
(by SAS-congruence), so
$$\underline{YP} = \underline{MS} = \underline{PV} = \underline{SY} = \underline{VM}$$

250   *Geometry*

Also $\angle PMS = \angle MPY = 36°$;
$\therefore \triangle BMP$ has equal base angles, and so is isosceles.
Hence $\underline{BM} = \underline{BP}$, and $\angle MBP = 108° = \angle ABC$.
Similarly $\angle AMY = \angle AYM = 36°$.
$\therefore \triangle AMY \equiv \triangle BPM$ (by ASA-congruence), so

$$\underline{AY} = \underline{AM} = \underline{BM} = \underline{BP}.$$

$\triangle MAB \equiv \triangle PBC$ – by ASA-congruence:

$$\begin{aligned}\angle BPC &= 108° - \angle BPM - \angle CPS = 36° = \angle AMB,\\ MA &= PB, \text{ and}\\ \angle PBC &= \angle MBA = \angle MAB.\end{aligned}$$

Hence $\underline{AB} = \underline{BC}$.

Continuing round the figure we see that

$$\underline{AB} = \underline{BC} = \underline{CD} = \underline{DE} = \underline{EA}$$

and that

$$\angle A = \angle B = \angle C = \angle D = \angle E.$$

(ii) Extend DB to meet MP at L, and extend DA to meet MY at N.
Then $36° = \angle DBC = \angle LBM$ (vertically opposite angles). Hence $\triangle LBM$ has equal base angles and so is isosceles: $\underline{LM} = \underline{LB}$. Similarly $\underline{NM} = \underline{NA}$.
Now $\triangle LBM \equiv \triangle NAM$ (by ASA-congruence, since $\underline{MA} = \underline{MB}$), so $\underline{LM} = \underline{LB} = \underline{NA} = \underline{NM}$.
In the regular pentagon $ABCDE$ we know that $\angle DBC = 36°$; and in $\triangle LBM$, $\angle LBM = \angle DBC$ (vertically opposite angles), so $\angle MLB = 108°$. Hence $\angle BLP = 72° = \angle LBP$, so $\triangle PLB$ is isosceles: $\underline{PL} = \underline{PB}$.
In the regular pentagon $MPSVY$, $\triangle PMA$ is isosceles, so $\underline{PM} = \underline{PA}$.
$\therefore \underline{LM} = \underline{PM} - \underline{PL} = \underline{PA} - \underline{PB} = \underline{BA}$.
Hence $ABLMN$ is a pentagon with five equal sides. It is easy to check that the five angles are all equal.

(iii) We saw in (i) that the five diagonals of $MPSVY$ are equal. We showed in Problem **3** that each has length τ, and that $MPDY$ is a rhombus, so $\underline{DY} = \underline{PM} = 1$. Similarly $\underline{SE} = 1$.
Hence $\underline{SD} = \tau - 1$, and $\underline{DE} = \underline{SE} - \underline{SD} = 2 - \tau = (\tau^2)^{-1}$.

180.

(a) Since the tiles fit together "edge-to-edge", all tiles have the same edge length. The number k of tiles meeting at each vertex must be at least 3 (since the angle at each vertex of a regular n-gon $= \left(1 - \frac{2}{n}\right)\pi < \pi$), and can be at most 6 (since the smallest possible angle in a regular n-gon occurs when $n = 3$, and is then $\frac{\pi}{3}$).
We consider each possible value of k in turn.

- If $k = 6$, then $n = 3$ and we have six equilateral triangles at each vertex.
- If $k = 5$, then we would have $\left(1 - \frac{2}{n}\right)\pi = \frac{2\pi}{5}$, so $n = \frac{10}{3}$ is not an integer.
- If $k = 4$, then $n = 4$ and we have four squares at each vertex.
- If $k = 3$, then $n = 6$ and we have three regular hexagons at each vertex.

Hence $n = 3$, or 4, or 6.

(b)(i) If $k = n = 4$, it is easy to form the vertex figure. It may seem 'obvious' that it continues "to infinity"; but if we think it is obvious, then we should explain why: (choose the scale so that the common edge length is "1"; then let the integer lattice points be the vertices of the tiling, with the tiles as translations of the unit square formed by $(0, 0)$, $(1, 0)$, $(1, 1)$, $(0, 1)$).

(ii) If $k = 6$, $n = 3$, let the vertices correspond to the complex numbers $p + q\omega$, where p, q are integers, and where ω is a complex cube root of 1 (that is, a solution of the equation $\omega^3 = 1 \neq \omega$, or $\omega^2 + \omega + 1 = 0$), with the edges being the line segments of length 1 joining nearest neighbours (at distance 1).

(iii) If $k = 3$, $n = 6$, take the same vertices as in (ii), but eliminate all those for which $p + q \equiv 0 \pmod{3}$, then let the edges be the line segments of length 1 joining nearest neighbours.

181.

(a)(i) As before, the number k of tiles at each vertex lies between 3 and 6. However, this time k does not determine the shape of the tiles. Hence we introduce a new parameter: the number of t of triangles at each vertex, which can range from 0 up to 6. The derivations are based on elementary arithmetic, for which it is easier to work with angles in degrees.

* If $t = 6$, then the vertex figure must be $\mathbf{3^6}$.
* If $t = 5$, the remaining gap of $60°$ could only take a sixth triangle, so this case cannot occur.
* If $t = 4$, we are left with angle of $120°$, so the only possible vertex figure is $\mathbf{3^4.6}$.
* If $t = 3$, then we are left with an angle of $180°$, so the only possible configurations are $\mathbf{3^3.4^2}$ (with the two squares together), or $\mathbf{3^2.4.3.4}$ (with the two squares separated by a triangle).
* If $t = 2$, we are left with an angle of $240°$, which cannot be filled with 3 or more tiles (since the average angle size would then be at most $80°$, and no more triangles are allowed), so the only possible vertex figures are $\mathbf{3^2.4.12}$, $\mathbf{3.4.3.12}$, $\mathbf{3^2.6^2}$, $\mathbf{3.6.3.6}$ (since $\mathbf{3^2.5.n}$ or $\mathbf{3.5.3.n}$ would require a regular n-gon with an angle of $132°$, which is impossible).

Note: The compactness of the argument based on the parameter t is about to end. We continue the same approach, with the focus shifting from the parameter t to a new parameter s – namely the number of squares in each vertex figure.

* Suppose $t = 1$. We are left with an angle of 300°.
  If $s = 0$, the 300° cannot be filled with 3 or more tiles (since then the average angle size would be at most 100°, and no squares can be used), so there are exactly two additional tiles. Since each tile has angle $\leqslant 180°$, we cannot use a hexagon, so the smallest tile has at least 7 sides; and since the average of the two remaining angle sizes is 150°, the smallest tile has at most 12 sides. It is now easy to check that the only possible vertex figures are **3.7.42, 3.8.24, 3.9.20, 3.10.15, 3.12^2**.
  If $s = 1$, we would be left with an angle of 210°, which would require two larger tiles with average angle size 105°, which is impossible.
  If $s = 2$, the only possible vertex figures are **3.4^2.6**, or **3.4.6.4**.
  Clearly we cannot have $s = 3$ (or we would be left with a gap of 30°); and $t = 1$, $s > 3$ is also impossible.

* Hence we may assume that $t = 0$, in which case s is at most 4.
  If $s = 4$, then the vertex figure is **4^4**.
  If $s = 3$, then the remaining gap could only take a fourth square, so this case does not occur.
  If $s = 2$, we are left with an angle of 180°, which cannot be filled.
  If $s = 1$, we are left with an angle of 270°, so there must be exactly two additional tiles and the only possible vertex figures are **4.5.20, 4.6.12**, or **4.8^2** (since a regular 7-gon would leave an angle of $141\frac{3}{7}°$).

* Hence we may assume that $t = s = 0$, and proceed using the parameter f – namely the number of regular pentagons. Clearly f is at most 3, and cannot equal 3 (or we would leave an angle of 36°).
  If $f = 2$, we are left with an angle of 144°, so the only vertex figure is **5^2.10**.
  If $f = 1$, we are left with an angle of 252°, which requires exactly two further tiles, whose average angle is 126°; but this forces us to use a hexagon – leaving an angle of 132°, which is impossible.

* Hence we may assume that $t = s = f = 0$. So the smallest possible tile is a hexagon, and since we need at least 3 tiles at each vertex, the only possible vertex figure is **6^3**.

  Hence, the simple minded necessary condition (namely that the vertex figure should have no gaps) gives rise to a list of twenty-one possible vertex figures:

  3^6, 3^4.6, 3^3.4^2, 3^2.4.3.4, 3^2.4.12, 3.4.3.12, 3^2.6^2, 3.6.3.6, 3.7.42, 3.8.24, 3.9.20, 3.10.15, 3.12^2, 3.4^2.6, 3.4.6.4, 4^4, 4.5.20, 4.6.12, 4.8^2, 5^2.10, 6^3.

(ii) **Lemma.** The vertex figures **3^2.4.12, 3.4.3.12, 3^2.6^2, 3.7.42, 3.8.24, 3.9.20, 3.10.15, 3.4^2.6, 4.5.20, 5^2.10** do not extend to semi-regular tilings of the plane.

Proof. Suppose to the contrary that any of these vertex figures could be realized by a semi-regular tiling of the plane. Choose a vertex B and consider the tiles around vertex B.

In the first eight of the listed vertex figures we may choose a triangle $T = ABC$, which is adjacent to polygons of different sizes on the edges $\underline{BA}$ (say) and $\underline{BC}$.

In the two remaining vertex figures, there is a face $T = ABC \cdots$ with an **odd** number of edges, which has the same property – namely that of being adjacent to an a-gon on the edge $\underline{BA}$ (say) and a b-gon on the edge $\underline{BC}$ with $a \neq b$. (For example, in the vertex figure $\mathbf{3^2.4.12}$, $T = ABC$ is a triangle, and the faces on $\underline{BA}$ and on $\underline{BC}$ are – in some order – a 3-gon and a 4-gon, or a 3-gon and a 12-gon.)

In each case let the face T be a p-gon.

If the face adjacent to T on the edge $\underline{BA}$ is an a-gon, and that on edge $\underline{BC}$ is a b-gon, then the vertex figure symbol must include the sequence "$\ldots a.p.b \ldots$".

If we now switch attention from vertex B to the vertex A, then we know that A has the same vertex figure, so must include the sequence "$\ldots a.p.b \ldots$", so the face adjacent to the other edge of T at A must be a b-gon. As one traces round the edges of the face T, the faces adjacent to T are alternately a-gons and b-gons – contradicting the fact that T has an **odd** number of edges.

Hence none of the listed vertex figures extends to a semi-regular tiling of the plane. QED

(b) It transpires that the remaining eleven vertex figures

$$3^6,\ 3^4.6,\ 3^3.4^2,\ 3^2.4.3.4,\ 3.6.3.6,\ 3.12^2,\ 3.4.6.4,\ 4^4,\ 4.6.12,\ 4.8^2,\ 6^3$$

can all be realized as semi-regular tilings (and one – namely $\mathbf{3^4.6}$ – can be realized in two different ways, one being a reflection of the other).

In the spirit of Problem **180**(b), one should want to do better than to produce plausible pictures of such tilings, by specifying each one in some canonical way. We leave this challenge to the reader.

182. [We construct a regular hexagon, and take alternate vertices.]

Draw the circle with centre O passing through A. The circle with centre A passing through O meets the circle again at X and Y.

The circle with centre X and passing through A and O meets the circle again at B; and the circle with centre Y and passing through A and O, meets the circle again at C.

Then $\triangle AOX$, $\triangle AOY$, $\triangle XOB$, $\triangle YOC$ are equilateral, so $\angle AXB = 120° = \angle AYC$, and $\angle XAB = 30° = \angle YAC$.

Hence $\triangle AXB \equiv \triangle AYC$, so $\underline{AB} = \underline{AC}$. $\triangle ABC$ is isosceles so $\angle B = \angle C$, with apex angle $\angle BAC = \angle XAY - \angle XAB - \angle YAC = 60°$; hence $\triangle ABC$ is equiangular and so equilateral.

183.

(a) Draw the circle with centre O and passing through A.

Extend AO beyond O to meet the circle again at C.

Construct the perpendicular bisector of AC, and let this meet the circle at B and at D.

Then $BA = BC$ (since the perpendicular bisector of AC is the locus of points equidistant from A and from C); similarly $DA = DC$.

$\triangle OAB$ and $\triangle OCB$ are both isosceles right angled triangles, so $\angle ABC$ is a right angle (or appeal to "the angle subtended on the circle by the diameter AC"). Similarly $\angle A$, $\angle C$, $\angle D$ are right angles, so $ABCD$ is a rectangle with $BA = BC$, and hence a square.

Note: This construction starts with the regular 2-gon AC inscribed in its circumcircle, and doubles it to get a regular 4-gon, by constructing the perpendicular bisectors of the "two sides" to meet the circumcircle at B and at D.

(b) Erect the perpendiculars to AB at A and at B.

Then draw the circles with centre A and passing through B, and with centre B and passing through A.

Let these circles meet the perpendiculars to AB (on the same side of AB) at D and at C.

Then $AD = BC$, and $AD \parallel BC$, so $ABCD$ is a parallelogram (by Problem **159**(a)), and hence a (being a parallelogram with a right angle), and so a square (since $AB = AD$).

184.

(a)(i) First construct the regular 3-gon ACE with circumcentre O. Then construct the perpendicular bisectors of the three sides AC, CE, EA, and let these meet the circumcircle at B, D, F.

Note: Here we emphasise the general step from inscribed regular n-gon to inscribed regular $2n$-gon – even though this may seem perverse in the case of a regular 3-gon (since we constructed the inscribed regular 3-gon in Problem **182** by first constructing the regular 6-gon and then taking alternate vertices).

(ii) First construct the regular 4-gon $ACEG$ with circumcentre O. Then construct the perpendicular bisectors of the four sides AC, CE, EG, GA, and let these meet the circumcircle at B, D, F, H.

(b)(i) Construct an equilateral triangle ABO.

Then draw the circle with centre O and passing through A and B. Then proceed as in **182**.

(ii) Extend AB beyond B, and let this line meet the circle with centre B and passing through A at X.

Now construct a square $BXYZ$ on the side BX as in **183**(a), and let the diagonal BY meet the circle at C.

Construct the circumcentre O of $\triangle ABC$ (the point where the perpendicular bisectors of AB, BC meet).

Construct the next vertex D as the point where the circle with centre O and passing through A meets the circle with centre C and passing through B. The remaining points E, F, G, H can be found in a similar way.

185.

(a)(i) There are various ways of doing this – none of a kind that most of us might stumble upon. Draw the circumcircle with centre O and passing through A. Extend the line AO beyond O to meet the circle again at X.

Construct the perpendicular bisector of AX, and let this meet the circle at Y and at Z. Construct the midpoint M of OZ, and join MA.

Let the circle with centre M and passing through A meet the line segment OY at the point F.

Finally let the circle with centre A and passing through F meet the circumcircle at B. Then AB is a side of the required regular 5-gon. (The vertex C on the circumcircle is then obtained as the second meeting point of the circumcircle with the circle having centre B and passing through A. The points D, E can be found in a similar way.)

The proof that this construction does what is claimed is most easily accomplished by calculating lengths.

Let $OA = 2$. Then $OF = \sqrt{5} - 1$, so $AF = \sqrt{10 - 2\sqrt{5}} = AB$. It remains to prove that this is the correct length for the side of a regular pentagon inscribed in a circle of radius 2. Fortunately the work has already been done, since $\triangle OAB$ is isosceles with apex angle equal to $72°$. If we drop a perpendicular from O to AB, then we need to check whether it is true that

$$\sin 36° = \frac{\sqrt{10 - 2\sqrt{5}}}{4}.$$

But this was already shown in Problem **3**(c).

(ii) To construct a regular 10-gon $ABCDEFGHIJ$, first construct a regular 5-gon $ACEGI$ with circumcentre O; then construct the perpendicular bisectors of the five sides, and so find B, D, F, H, J as the points where these bisectors meet the circumcircle.

(b)(i) The first move is to construct a line BX through B such that $\angle ABX = 108°$. Fortunately this can be done using part (a), by temporarily treating B as the circumcentre, drawing the circle with centre B and passing through A, and beginning the construction of a regular 5-gon $AP\cdots$ inscribed in this circle.

256 *Geometry*

Then $\angle ABP = 72°$; so if we extend the line PB beyond B to X, then $\angle ABX = 108°$. Let BX meet the circle with centre B through A at the point C. Then $\underline{BA} = \underline{BC}$ and $\angle ABC = 108°$, so we are up and running.

If we let the perpendicular bisectors of $\underline{AB}$ and $\underline{BC}$ meet at O, then the circle with centre O and passing through A also passes through B and C (and the yet to be located points D and E). The circle with centre C and passing through B meets this circle again at D; and the circle with centre A and passing through B meets the circle again at E.

(ii) To construct the regular 10-gon $ABCDEFGHIJ$, treat B as the point O in (a)(i) and construct a regular 5-gon $AXCYZ$ inscribed in the circle with centre B and passing through A.

Then $\angle ABC = 144°$, and $\underline{BA} = \underline{BC}$, so C is the next vertex of the required regular 10-gon. We may now proceed as in (a)(ii) to first construct the circumcentre O of the required regular 10-gon as the point where the perpendicular bisectors of $\underline{AB}$ and $\underline{BC}$ meet, then draw the circumcircle, and finally step off successive vertices D, E, ... of the 10-gon around the circumcircle.

186. The number k of faces meeting at each vertex can be at most five (since more would produce an angle sum that is too large). And $k \geq 3$ (in order to create a genuine corner.

- If $k = 5$, then the vertex figure must be $\mathbf{3^5}$ (or the angle sum would be $> 360°$).
- If $k = 4$, then the vertex figure must be $\mathbf{3^4}$ (or the angle sum would be too large).
- If $k = 3$, the angle in each of the regular polygons must be $< 120°$, so the only possible vertex figures are $\mathbf{5^3}$, $\mathbf{4^3}$, and $\mathbf{3^3}$.

187. The respective midpoints have coordinates:
of $\underline{AB}$: $\left(\frac{1}{2}, \frac{1}{2}, 0\right)$; of $\underline{AC}$: $\left(\frac{1}{2}, 0, \frac{1}{2}\right)$; of $\underline{AD}$: $\left(1, \frac{1}{2}, \frac{1}{2}\right)$; of $\underline{BC}$: $\left(0, \frac{1}{2}, \frac{1}{2}\right)$;
of $\underline{BD}$: $\left(\frac{1}{2}, 1, \frac{1}{2}\right)$; of $\underline{CD}$: $\left(\frac{1}{2}, \frac{1}{2}, 1\right)$.
$\therefore PQ = \frac{1}{\sqrt{2}} = PR = PS = PT = QR = RS = ST = TQ$.

188.

(a) There are infinitely many planes through the apex A and the base vertex B. Among these planes, the one perpendicular to the base BCD is the one that passes through the midpoint M of $\underline{CD}$. Let the perpendicular from A to the base, meet the base BCD at the point X, which must lie on $\underline{BM}$. Let $\underline{AX}$ have length h. To find h we calculate the area of $\triangle ABM$ in two ways.
First, $\underline{BM}$ has length $\sqrt{3}$, so area$(\triangle ABM) = \frac{1}{2}\left(\sqrt{3} \times h\right)$
Second, $\triangle ABM$ is isosceles with base $\underline{AB}$ and apex M, so has height $\sqrt{2}$.
$\therefore$ area$(\triangle ABM) = \frac{1}{2}(2 \times \sqrt{2}) = \sqrt{2}$.
If we now equate the two expressions for area$(\triangle ABM)$, we see that $h = \sqrt{\frac{8}{3}}$.

(b) (i) When constructing a regular octahedron (whether using card, or tiles such as Polydron®) one begins by arranging four equilateral triangles around a vertex such as A. This 'vertex figure' is not rigid: though we know that $\underline{BC} = \underline{CD} = \underline{DE} = \underline{EB}$, there is no *a priori* reason why the four neighbours B, C, D, E of A should form a square, or a rhombus, or a planar quadrilateral. We show that these four neighbours lie in a single plane: B, C, D, E are all distance 2 from A and distance 2 from F, so (by Problem **144**) they must all lie in the plane perpendicular to the line AF and passing through the midpoint X of $\underline{AF}$. Moreover, $\triangle ABX \equiv \triangle ACX$ (by RHS-congruence, since $\underline{AB} = \underline{AC} = 2$, $\underline{AX} = \underline{AX}$, and $\angle AXB = \angle AXC$ are both right angles). Hence $\underline{XB} = \underline{XC}$, so B, C lie on a circle in this plane with centre X. Similarly $\triangle ABX \equiv \triangle ADX \equiv \triangle AEX$, so $BCDE$ is a cyclic quadrilateral (and a rhombus), and hence a square – with X as the midpoint of both $\underline{BD}$ and $\underline{CE}$.

(ii) Let M be the midpoint of $\underline{BC}$ and N the midpoint of $\underline{DE}$.
Then NM and AF cross at X and so define a single plane. In this plane, $\triangle ANM \equiv \triangle FMN$ (by SSS-congruence, since $\underline{AN} = \underline{FM} = \sqrt{3}$, $\underline{NM} = \underline{MN}$, $\underline{MA} = \underline{NF} = \sqrt{3}$); hence $\angle ANM = \angle FMN$, so $AN \parallel MF$.
Similarly, if P is the midpoint of $\underline{AE}$ and Q is the midpoint of $\underline{FC}$, then $\triangle DPQ \equiv \triangle BQP$, so $DP \parallel QB$. Hence the top face DEA is parallel to the bottom face BCF, so the height of the octahedron sitting on the table is equal to the height of $\triangle FMN$. But this triangle has sides of lengths $2, \sqrt{3}, \sqrt{3}$, so this height is exactly the same as the height h in part (a).

189.

(i) Let $L = (\frac{1}{2}, 0, 0)$ and $M = (\frac{1}{2}, \frac{1}{2}, 0)$. Then L lies on the line ST and M is the midpoint of $\underline{NP}$.
$\triangle LSM$ is a right angled triangle with legs of length $\underline{LS} = a$, $\underline{LM} = \frac{1}{2}$, so $MS = \sqrt{a^2 + \frac{1}{4}}$.
$\triangle MNS$ is a right angled triangle with legs of length $\underline{MN} = \frac{1}{2} - a$, $\underline{MS} = \sqrt{a^2 + \frac{1}{4}}$. Hence $\underline{NS} = \sqrt{2a^2 - a + \frac{1}{2}}$.
Similarly $\underline{NU} = \sqrt{2a^2 - a + \frac{1}{2}}$.
Let $L' = (0, \frac{1}{2}, 0)$ and $M' = (0, \frac{1}{2}, \frac{1}{2})$. Then M' is the midpoint of $\underline{WX}$ and L' lies immediately below M' on the line joining $(0, 0, 0)$ to $(0, 1, 0)$. In the right angled triangle $\triangle L'NM'$ we find $\underline{NM'} = \sqrt{a^2 + \frac{1}{4}}$, and in the right angled triangle $\triangle M'WN$ we then find $\underline{NW} = \sqrt{2a^2 - a + \frac{1}{2}}$. Similarly $\underline{NX} = \sqrt{2a^2 - a + \frac{1}{2}}$.

(ii) $\underline{NP} = \underline{NS}$ precisely when $1 - 2a = \sqrt{2a^2 - a + \frac{1}{2}} > 0$; that is, when $a = \frac{3-\sqrt{5}}{4}$, so all edges of the polyhedron have length $\frac{\sqrt{5}-1}{2} = \tau - 1$.

Note: The rectangle $NPRQ$ is a "1 by $\tau - 1$" rectangle, and $1 : \tau - 1 = \tau : 1$. Hence the regular icosahedron can be constructed from three congruent, and pairwise perpendicular, copies of a "Golden rectangle".

190. We mimic the classification of possible vertex figures for semi-regular tilings. We are assuming that the angles meeting at each vertex add to $< 360°$, so the number k of faces at each vertex lies between 3 and 5. Because faces are regular, but not necessarily congruent, k does not determine the shape of the faces. Hence we let t denote the number of triangles at each vertex, which can range from 0 up to 5.

- If $t = 5$, then the vertex figure must be $\mathbf{3^5}$.
- If $t = 4$, the remaining polygons have angle sum $< 120°$, so the possible vertex figures are $\mathbf{3^4}$, $\mathbf{3^4.4}$, and $\mathbf{3^4.5}$.
- If $t = 3$, then the remaining polygons have angle sum $< 180°$, so the only possible vertex figures are $\mathbf{3^3}$, and $\mathbf{3^3.n}$ (for any $n > 3$).
- If $t = 2$, then the remaining polygons have angle sum $< 240°$, so there are at most 2 additional faces in the vertex figure (since if there were 3 or more extra faces, the average angle size would then be at most 80°, with no more triangles allowed). If there is just 1 additional face, we get the vertex figure $\mathbf{3^2.n}$ for any $n > 3$. So we may assume that there are 2 additional faces – the smallest of which must then be a 4-gon or a 5-gon.

If the next smallest face is a 4-gon, then we get the possible vertex figures

$\mathbf{3^2.4^2}$ and $\mathbf{3.4.3.4}$, $\mathbf{3^2.4.5}$ and $\mathbf{3.4.3.5}$, $\mathbf{3^2.4.6}$ and $\mathbf{3.4.3.6}$, $\mathbf{3^2.4.7}$ and $\mathbf{3.4.3.7}$, $\mathbf{3^2.4.8}$ and $\mathbf{3.4.3.8}$, $\mathbf{3^2.4.9}$ and $\mathbf{3.4.3.9}$, $\mathbf{3^2.4.10}$ and $\mathbf{3.4.3.10}$, $\mathbf{3^2.4.11}$ and $\mathbf{3.4.3.11}$.

If the next smallest face is a 5-gon, then we get the possible vertex figures

$\mathbf{3^2.5^2}$ and $\mathbf{3.5.3.5}$, $\mathbf{3^2.5.6}$ and $\mathbf{3.5.3.6}$, $\mathbf{3^2.5.7}$ and $\mathbf{3.5.3.7}$.

Note: Before proceeding further it is worth deciding which among the putative vertex figures identified so far seem to correspond to semi-regular polyhedra – and then to prove that these observations are correct.

- The vertex figure $\mathbf{3^5}$ corresponds to the regular icosahedron.
- The vertex figure $\mathbf{3^4}$ corresponds to the regular octahedron; $\mathbf{3^4.4}$ corresponds to the *snubcube*; $\mathbf{3^4.5}$ corresponds to the *snub dodecahedron* – which comes in left-handed and right-handed forms.
- The vertex figure $\mathbf{3^3}$ corresponds to the regular tetrahedron, and $\mathbf{3^3.n}$ (for any $n > 3$) corresponds to the *n-gonal antiprism*.
- The vertex figure $\mathbf{3^2.n}$ for any $n > 3$ does not seem to arise.
- The vertex figure $\mathbf{3^2.4^2}$ does not seem to arise; $\mathbf{3.4.3.4}$ corresponds to the *cuboctahedron*; $\mathbf{3^2.4.5}$ and $\mathbf{3.4.3.5}$, $\mathbf{3^2.4.6}$ and $\mathbf{3.4.3.6}$, $\mathbf{3^2.4.7}$ and $\mathbf{3.4.3.7}$, $\mathbf{3^2.4.8}$ and $\mathbf{3.4.3.8}$, $\mathbf{3^2.4.9}$ and $\mathbf{3.4.3.9}$, $\mathbf{3^2.4.10}$ and $\mathbf{3.4.3.10}$, $\mathbf{3^2.4.11}$ and $\mathbf{3.4.3.11}$ do not seem to arise.
- The vertex figure $\mathbf{3^2.5^2}$ does not seem to arise, whereas $\mathbf{3.5.3.5}$ corresponds to the *icosidodecahedron*; $\mathbf{3^2.5.6}$ and $\mathbf{3.5.3.6}$, $\mathbf{3^2.5.7}$ and $\mathbf{3.5.3.7}$ do not seem to arise.

To avoid further proliferation of spurious 'putative vertex figures' we inject a version of the **Lemma** used for tilings somewhat earlier than we did for tilings, and then apply the underlying idea to eliminate other spurious possibilities as they arise.

Lemma. The vertex figures

$3^2.4^2$, $3^2.n$ ($n > 3$), $3^2.4.5$, $3.4.3.5$, $3^2.4.6$, $3.4.3.6$, $3^2.4.7$, $3.4.3.7$, $3^2.4.8$, $3.4.3.8$, $3^2.4.9$, $3.4.3.9$, $3^2.4.10$, $3.4.3.10$, $3^2.4.11$, $3.4.3.11$, $3^2.5^2$, $3^2.5.6$, $3.5.3.6$, $3^2.5.7$, $3.5.3.7$

do not arise as vertex figures of any semi-regular polyhedron.

Proof outline. Each of these requires that the vertex figure of any vertex B includes a tile $T = ABC \cdots$ with an **odd** number of edges, for which the edge $\underline{BA}$ is adjacent to an a-gon, the edge $\underline{BC}$ is adjacent to a b-gon (where $a \neq b$), and where the subsequent faces adjacent to T are forced to alternate – a-gon, b-gon, a-gon, ... – which is impossible. QED

For the rest we introduce the additional parameter s to denote the number of 4-gons in the vertex figure.

Suppose $t = 1$. Then the remaining polygons have angle sum $< 300°$, so there are at most 3 additional faces in the vertex figure (since if there were 4 or more extra faces, the average angle size would be $< 75°$, with no more triangles allowed).

If there are 3 additional faces, the average angle size is $< 100°$, so $s > 0$.

- If $s > 1$, then the possible vertex figures are $\mathbf{3.4^3}$ (which corresponds to the *rhombicuboctahedron*), $\mathbf{3.4^2}$ (which corresponds to the 3-gonal prism), $\mathbf{3.4^2.5}$ (which is impossible as in the Lemma) and $\mathbf{3.4.5.4}$ (which corresponds to the *small rhombicosidodecahedron*).

- If $s = 1$, then the remaining faces have angle sum $< 210°$, so there can only be one additional face, and every $\mathbf{3.4.n}$ ($n > 4$) is impossible as in the Lemma.

- If $s = 0$, then the remaining faces have angle sum $< 300°$, so there are exactly two other faces and the smallest face has < 12 edges, so the only possible vertex figures are

  - $\mathbf{3.5.n}$ ($4 < n$), which is impossible as in the Lemma;
  - $\mathbf{3.6^2}$, which corresponds to the *truncated tetrahedron*;
  - $\mathbf{3.6.n}$ ($6 < n$), which is impossible as in the Lemma;
  - $\mathbf{3.7.n}$ ($6 < n$), which is impossible as in the Lemma;
  - $\mathbf{3.8^2}$, which corresponds to the *truncated cube*;
  - $\mathbf{3.8.n}$ ($8 < n$), which is impossible as in the Lemma;
  - $\mathbf{3.9.n}$ ($8 < n$), which is impossible as in the Lemma;
  - $\mathbf{3.10^2}$, which corresponds to the *truncated dodecahedron*;
  - $\mathbf{3.10.n}$ ($10 < n$), which is impossible as in the Lemma;
  - $\mathbf{3.11.n}$ ($10 < n$), which is impossible as in the Lemma.

Thus we may assume that $t = 0$ – in which case, $s < 4$.

- If $s = 3$, then the only possible vertex figure is 4^3, which corresponds to the *cube*.

- If $s = 2$, then the remaining faces have angle sum $< 180°$, so there is exactly one additional face, and every $4^2.n$ ($n > 4$) corresponds to the *n-gonal prism*.

- If $s = 1$, then the remaining faces have angle sum $< 270°$, so there are at most 2 other faces with the smallest face having < 8 edges, so the only possible vertex figures are $4.5.n$ (for $4 < n < 20$), 4.6^2, $4.6.n$ (for $6 < n < 12$), 4.7^2, $4.7.n$ (for $7 < n < 10$). Among these $4.5.n$ is impossible as in the Lemma; 4.6^2 corresponds to the *truncated octahedron*; $4.6.n$ with n odd is impossible as in the Lemma; $4.6.8$ corresponds to the *great rhombicuboctahedron*; $4.6.10$ corresponds to the *great rhombicosidodecahedron*; $4.7.n$ is impossible as in the Lemma.

- If $s = 0$, there must be exactly three faces at each vertex, and the smallest must be a 5-gon, so the only possible vertex figures are 5^3, which corresponds to the *regular dodecahedron*; $5^2.n$ (for $n > 5$), which is impossible as in the Lemma; 5.6^2, which corresponds to the *truncated icosahedron*; or $5.6.7$, which is impossible as in the Lemma.

191.

(a)(i) $\angle AXD = 100°$, so $\angle ADB = 40°$. In $\triangle ABD$ we then see that $\angle ABD = 40°$, so $\triangle ABD$ is isosceles with $\underline{AB} = \underline{AD}$. $\triangle ABC$ is isosceles with a base angle $\angle BAC = 60°$, so $\triangle ABC$ is equilateral. Hence $\angle CBD = 20°$.

(ii) $\triangle ABC$ is equilateral, so $\underline{AC} = \underline{AB} = \underline{AD}$.
Hence $\triangle ADC$ is isosceles, so $\angle ACD = 70° = \angle ADC$, whence $\angle BDC = 70° - \angle ADB = 30°$.

(b)(i) As before $\angle AXD = 100°$, so $\angle ADB = 40°$.
In $\triangle ABD$ we then see that $\angle ABD = 30°$, so $\triangle ABD$ is not isosceles (as it was in (a)).
$\triangle ABC$ is isosceles, so $\angle BCA = 70°$, whence $\angle CBD = 10°$.
In $\triangle XCD$, $\angle CXD = 80°$, so $\angle BDC + \angle ACD = 100°$, but there is no obvious way of determining the individual summands: $\angle BDC$ and $\angle ACD$.

(ii) No lengths are specified, so we may choose the length of $\underline{AC}$. The point B then lies on the perpendicular bisector of $\underline{AC}$, and $\angle CAB = 70°$ determines the location of B exactly. The line AD makes an angle of $40°$ with AC, and BD makes an angle of $80°$ with AC, so the location of D is determined. Hence, despite our failure in part (i), the angles are determined.

192.

(i) If P lies on $\underline{CB}$, then $\underline{PC} = b\cos C$, $\underline{AP} = b\sin C$, and in the right angled triangle $\triangle APB$ we have:

$$\begin{aligned} c^2 &= (b\sin C)^2 + (a - b\cos C)^2 \\ &= a^2 + b^2(\sin^2 C + \cos^2 C) - 2ab\cos C \\ &= a^2 + b^2 - 2ab\cos C. \end{aligned}$$

If P lies on $\underline{CB}$ extended beyond B, then $\underline{PC} = b\cos C$, $\underline{AP} = b\sin C$ as before, and in the right angled triangle $\triangle APB$ we have:

$$\begin{aligned} c^2 &= (b\sin C)^2 + (b\cos C - a)^2 \\ &= a^2 + b^2(\sin^2 C + \cos^2 C) - 2ab\cos C \\ &= a^2 + b^2 - 2ab\cos C. \end{aligned}$$

(ii) If P lies on $\underline{BC}$ extended beyond C, then $\underline{PC} = b\cos\angle ACP = -b\cos C$, $\underline{AP} = b\sin\angle ACP = b\sin C$, and in the right angled triangle $\triangle APB$ we have:

$$\begin{aligned} c^2 &= (b\sin C)^2 + (a + \underline{PC})^2 \\ &= (b\sin C)^2 + (a - b\cos C)^2 \\ &= a^2 + b^2(\sin^2 C + \cos^2 C) - 2ab\cos C \\ &= a^2 + b^2 - 2ab\cos C. \end{aligned}$$

193. There are many ways of doing this – once one knows the Sine Rule and Cosine Rule. If we let $\angle BDC = y$ and $\angle ACD = z$, then one route leads to the identity $\cos(z - 10°) = 2\sin 10° \cdot \sin z$, from which it follows that $z = 80°$, $y = 20°$.

194.

(a) The angle between two faces, or two planes, is the angle one sees "end-on" – as one looks along the line of intersection of the two planes. This is equal to the angle between two perpendiculars to the line of intersection – one in each plane. If M is the midpoint of $\underline{BC}$, then $\triangle ABC$ is isosceles with apex A, so the median $\underline{AM}$ is perpendicular to $\underline{BC}$; similarly $\triangle DBC$ is isosceles with apex D, so the median $\underline{DM}$ is perpendicular to $\underline{BC}$.

$\triangle MAD$ is isosceles with apex M, $\underline{MA} = \underline{MD} = \sqrt{3}$, $\underline{AD} = 2$, so we can use the Cosine Rule to conclude that $2^2 = 3 + 3 - 2 \cdot 3 \cdot \cos(\angle AMD)$, whence $\cos(\angle AMD) = \frac{1}{3}$.

(b) $\cos(\angle AMD) = \frac{1}{3}$, and $\frac{1}{3} < \frac{1}{2}$, so $\angle AMD > 60°$; hence we cannot fit six regular tetrahedra together so as to share an edge.

Since $\angle AMD$ is acute, $\angle AMD < 90°$, so we can certainly fit four regular tetrahedra together with lots of room to spare.

We can now appeal to "trigonometric tables", or a calculator, to see that

$$\arccos\left(\frac{1}{3}\right) < 1.24,$$

that is 1.24 *radians*, so five tetrahedra use up less than 6.2 radians – which is less than 2π. Hence we can fit five regular tetrahedra together around a common edge with room to spare (but not enough to fit a sixth).

195.

(a) The angle between the faces ABC and FBC is equal to the angle between two perpendiculars to the common edge BC. Since the two triangles are isosceles with the common base BC, it suffices to find the angle between the two medians AM and FM.

In Problem **188** we saw that $BCDE$ is a square, with sides of length 2. If we switch attention from the opposite pair of vertices A, F to the pair C, E, then the same proof shows that $ABFD$ is a square with sides of length 2. Hence the diagonal $AF = 2\sqrt{2}$.

Now apply the Cosine Rule to $\triangle AMF$ to conclude that:

$$\left(2\sqrt{2}\right)^2 = 3 + 3 - 2 \cdot 3 \cdot \cos(\angle AMF),$$

so it follows that $\cos(\angle AMF) = -\frac{1}{3}$.

(b) $\cos(\angle AMF) = -\frac{1}{3} < 0$, so $\angle AMF > 90°$; hence we cannot fit four regular octahedra together so as to share a common edge. Moreover, $-\frac{1}{2} < -\frac{1}{3}$, so $\angle AMD < 120°$; hence we can fit three octahedra together to share an edge with room to spare (but not enough room to fit a fourth).

196. The angle $\angle AMD = \arccos\left(\frac{1}{3}\right)$ in Problem **194** is acute, and the angle $\angle AMF = \arccos\left(-\frac{1}{3}\right)$ in Problem **195** is obtuse. Hence these angles are supplementary; so the regular tetrahedron fits *exactly* into the wedge-shaped hole between the face ABC of the regular octahedron and the table.

197.

(i) $AB = \sqrt{2} = BC = CA = AW = BW = CW$. Hence the four faces ABC, BCW, CWA, WAB are all equilateral triangles, so the solid is a regular tetrahedron (or, more correctly, the surface of the solid is a regular tetrahedron).

(ii) $AC = \sqrt{2} = CD = DA = AX = CX = DX$. Hence the four faces ACD, CDX, DXA, XAC are equilateral triangles, so the solid is a regular tetrahedron.

(iii) $AD = \sqrt{2} = DE = EA = AY = DY = EY$. Hence the four faces ADE, DEY, EYA, YAD are equilateral triangles, so the solid is a regular tetrahedron.

(iv) $AE = \sqrt{2} = EB = BA = AZ = EZ = BZ$. Hence the four faces AEB, EBZ, BZA, ZAE are equilateral triangles, so the solid is a regular tetrahedron.

(v) We get another four regular tetrahedra – such as $FBCP$, where $P = (1, 1, -1)$.

(vi) $ABCDEF$ is a regular octahedron (or, more correctly, the surface of the solid is a regular octahedron).

Note: The six vertices $(\pm 1, 0, 0)$, $(0, \pm 1, 0)$, $(0, 0, \pm 1)$ span a regular octahedron, with a regular tetrahedron fitting exactly on each face. The resulting compound star-shaped figure is called the *stellated octahedron*. Johannes Kepler (1571–1630) made an extensive study of polyhedra and this figure is sometimes referred to as Kepler's 'stella octangula'. It is worth making in order to appreciate the way it appears to consist of two interlocking tetrahedra.

198. Let the unlabeled vertex of the pentagonal face $ABW - V$ be P. In the pentagon $ABWPV$ the edge $\underline{AB}$ is parallel to the diagonal $\underline{VW}$. Hence $ABWV$ is an isosceles trapezium. The sides VA and WB (produced) meet at S in the plane of the pentagon $ABWPV$.

$\triangle SAB$ has equal base angles, so $\underline{SA} = \underline{SB}$.

Hence $\underline{SV} = \underline{SW}$.

Similarly $BC \parallel WX$, and WB and XC meet at some point S' on the line WB.

Now $\triangle S'BC \equiv \triangle SAB$, so $\underline{S'B} = \underline{SA} = \underline{SB}$. Hence $S' = S$, and the lines VA, WB, XC, YD, ZE all meet at S.

Since $VW \parallel AB$, we know that $\triangle SAB \sim \triangle SVW$, with scale factor

$$\tau : 1 = \underline{VW} : \underline{AB} = \underline{SV} : \underline{SA}.$$

If $\underline{SA} = x$, then $x + 1 : x = \tau : 1$, so $x = \tau$.

Let M be the midpoint of $\underline{AB}$ and let O denote the circumcentre of the regular pentagon $ABCDE$.

$\triangle OAB$ is isosceles, so OM is perpendicular to AB, and the required dihedral angle between the two pentagonal faces is $\angle OMP$.

It turns out to be better to find not the dihedral angle $\angle OMP$, but its supplement: namely the angle $\angle SMO$.

From $\triangle OAM$ we see that

$$\underline{OA} = \frac{1}{2\sin 36°},$$

and we know that $\underline{SA} = \tau = 2\cos 36°$. One can then check that $\underline{OS} = \cot 36°$.

Similarly, in $\triangle OAM$ we have

$$\underline{OM} = \frac{\cot 36°}{2}.$$

Hence in $\triangle OMS$ we have

$$\tan(\angle SMO) = 2.$$

Hence the required dihedral angle is equal to $\pi - \arctan 2 \approx 116.56°$.

199.

(a)(i) Let M be the midpoint of $\underline{AC}$. The angle between the two faces is equal to $\angle BMD$.

264   *Geometry*

In $\triangle BMD$, we have $\underline{BM} = \underline{DM} = \sqrt{3}$, $\underline{BD} = 2\tau = 1 + \sqrt{5}$. So the Cosine Rule in $\triangle BMD$ gives:

$$6 + 2\sqrt{5} = 3 + 3 - 2 \cdot 3 \cdot \cos(\angle BMD),$$

so $\cos(\angle BMD) = -\frac{\sqrt{5}}{3}$,

$$\angle BMD = \arccos\left(-\frac{\sqrt{5}}{3}\right) = \pi - \arccos\left(\frac{\sqrt{5}}{3}\right) \approx 138.19°.$$

(ii) $\frac{\sqrt{5}}{3} > \frac{1}{2}$; hence $\angle BMD > \frac{2\pi}{3}$, so we can fit two copies along a common edge, but not three.

(b)(i) Now let M denote the midpoint of $\underline{BC}$, and suppose that OM extended beyond M meets the circumcircle at V. Then $\underline{BV}$ is an edge of the regular 10-gon inscribed in the circumcircle. The circumradius $\underline{OB}$ of $BCDEF$ (which is equal to the edge length of the inscribed regular hexagon) is

$$\underline{OB} = \frac{1}{\sin 36°};$$

and $\angle MBV = 18°$, so

$$\underline{BV} = \frac{1}{\cos 18°}.$$

It is easiest to use the converse of Pythagoras' Theorem, and to write everything first in terms of $\cos 36°$, then (since $\tau = 2\cos 36°$, so $\cos 36° = \frac{1+\sqrt{5}}{4}$) write everything in terms of $\sqrt{5}$.

If we use $\sin^2 36° = 1 - \cos^2 36°$, and $2\cos^2 18° - 1 = \cos 36° = \frac{1+\sqrt{5}}{4}$, then

$$\begin{aligned}
\underline{BV}^2 + \underline{OB}^2 &= \left(\frac{1}{\cos 18°}\right)^2 + \left(\frac{1}{\sin 36°}\right)^2 \\
&= \frac{8}{5+\sqrt{5}} + \frac{8}{5-\sqrt{5}} \\
&= \frac{80}{20} = 2^2 \\
&= \underline{AB}^2.
\end{aligned}$$

Hence, in the right angled triangle $\triangle AOB$, we must have $\underline{OA} = \underline{BV}$ as claimed.

(ii) Miraculously no more work is needed. Let the vertex at the 'south pole' be L, and let the pentagon formed by its five neighbours be $GHIJK$. It helps if we can refer to the circumcircle of $BCDEF$ as the 'tropic of Cancer', and to the circumcircle of $GHIJK$ as the 'tropic of Capricorn'.

The pentagon $GHIJK$ is parallel to $BCDEF$, but the vertices of the southern pentagon have been rotated through $\frac{\pi}{5}$ relative to $BCDEF$, so that G (say) lies on the circumcircle of the pentagon $GHIJK$, but sits directly below the midpoint of the minor arc $\underline{BC}$. Let X denote the point on the 'tropic of Capricorn' which is directly beneath B. Then $\triangle BXG$ is a right angled triangle with $BG = AB = 2$, and $XG = BV$ is equal to the edge length of a regular 10-gon inscribed in the circumcircle of $GHIJK$. Hence, by the calculation in (i), BX is equal to the edge length of the regular hexagon inscribed in the same circle – which is also equal to the circumradius.

200. A necessary condition for copies of a regular polyhedron to "tile 3D (without gaps or overlaps)" is that an integral number of copies should fit together around an edge. That is, the dihedral angle of the polyhedron should be an exact submultiple of 2π. Only the cube satisfies this necessary condition.

Moreover, if we take as vertices the points (p, q, r) with integer coordinates p, q, r, and as our regular polyhedra all possible translations of the standard unit cube having opposite corners at $(0, 0, 0)$ and $(1, 1, 1)$, then we see that it is possible to tile 3D using just cubes.

201. The four diameters form the four edges of a square $ABCD$ of edge length 2.

The protruding semicircular segments on the left and right can be cut off and inserted to exactly fill the semicircular indentations above and below.

Hence the composite shape has area exactly equal to $2^2 = 4$ square units.

202.

(a) If the regular n-gon is $ACEG\cdots$ and the regular $2n$-gon is $ABCDEFG\cdots$, then $AC = s_n = s$ and $AB = s_{2n} = t$. If M is the midpoint of AC, $AM = \frac{s}{2}$ and
$$MB = 1 - \sqrt{\left(1 - \left(\frac{s}{2}\right)^2\right)}.$$

$$\therefore \quad t^2 = AB^2$$
$$= \left(\frac{s}{2}\right)^2 + \left[1 - \sqrt{1 - \left(\frac{s}{2}\right)^2}\right]^2 \quad (1)$$
$$= 1 + 1 - 2\sqrt{1 - \left(\frac{s}{2}\right)^2}$$
$$= 2 - \sqrt{4 - s^2}.$$

(b) Put $s = s_2 = 2$ in (1) to get $t = s_4 = \sqrt{2}$. Then put $s = s_4 = \sqrt{2}$ in (1) to get
$$t = s_8 = \sqrt{2 - \sqrt{2}}.$$

(c) Put $t = s_6 = 1$ in (1) to get $s = s_3 = \sqrt{3}$. Then put $s = s_6 = 1$ to get
$$t = s_{12} = \sqrt{2 - \sqrt{3}}.$$

(d) Put $s = s_5 = \frac{\sqrt{10-2\sqrt{5}}}{2}$ to get

$$t = s_{10} = \sqrt{\frac{3-\sqrt{5}}{2}}.$$

203.

(a)(i) $n = 3$: $p_3 = 3\sqrt{3} \times r$
$n = 4$: $p_4 = 4\sqrt{2} \times r$
$n = 5$: $p_5 = \frac{5\sqrt{10-2\sqrt{5}}}{2} \times r$
$n = 6$: $p_6 = 6 \times r$
$n = 8$: $p_8 = 8\sqrt{2-\sqrt{2}} \times r$
$n = 10$: $p_{10} = 5\sqrt{6-2\sqrt{5}} \times r$
$n = 12$: $p_{12} = 12\sqrt{2-\sqrt{3}} \times r$.

(ii)

$c_3 = 5.19\cdots$ $<$ $c_4 = 5.65\cdots$
$<$ $c_5 = 5.87\cdots$
$<$ $c_6 = 6$
$<$ $c_8 = 6.12\cdots$
$<$ $c_{10} = 6.18\cdots$
$<$ $c_{12} = 6.21\cdots$.

(b)(i) $n = 3$: $P_3 = 6\sqrt{3} \times r$
$n = 4$: $P_4 = 8 \times r$
$n = 5$: $P_5 = 10\sqrt{5-2\sqrt{5}} \times r$
$n = 6$: $P_6 = 4\sqrt{3} \times r$
$n = 8$: $P_8 = 8\left(2\sqrt{2} - 2\right) \times r$
$n = 10$: $P_{10} = 4\sqrt{25 - 10\sqrt{5}} \times r$
$n = 12$: $P_{12} = 12\left(4 - 2\sqrt{3}\right) \times r$.

(ii)

$C_3 = 10.39\cdots$ $>$ $C_4 = 8$
$>$ $C_5 = 7.26\cdots$
$>$ $C_6 = 6.92\cdots$
$>$ $C_8 = 6.62\cdots$
$>$ $C_{10} = 6.49\cdots$
$>$ $C_{12} = 6.43\cdots$.

(c) Let O be the centre of the circle of radius r. Let A, B lie on the circle with $\angle AOB = 30°$.

Let M be the midpoint of $\underline{AB}$ – so that $\triangle OAB$ is isosceles, with apex O and height $h = \underline{OM}$.

Let A' lie on $\underline{OA}$ produced, and let B' lie on $\underline{OB}$ produced, such that $\underline{OA'} = \underline{OB'}$ and $A'B'$ is tangent to the circle.

Then $\triangle OAB \sim \triangle OA'B'$ with $\triangle OA'B'$ larger than $\triangle OAB$, so the scale factor $\frac{1}{h} = 2\sqrt{2-\sqrt{3}} > 1$.

Hence $P_{12} = 2\sqrt{2-\sqrt{3}} \times p_{12} > p_{12}$, so $C_{12} > c_{12}$.

204. (i) πr (ii) $\frac{\pi}{2}r$ (iii) θr

205.

(a)(i) **Note:** This could be a long slog. However we have done much of the work before: when the radius is 1, the most of the required areas were calculated exactly back in Problem **3** and Problem **19**.

Alternatively, the area of each of the n sectors is equal to $\frac{1}{2}\sin\theta$, where θ is the angle subtended at the centre of the circle, and the exact values of the required trig functions were also calculated back in Chapter 1.

$$a_3 = \frac{3\sqrt{3}}{4} \times r^2$$
$$a_4 = 2 \times r^2$$
$$a_5 = \frac{5}{8}\sqrt{10+2\sqrt{5}} \times r^2$$
$$a_6 = \frac{3\sqrt{3}}{2} \times r^2$$
$$a_8 = 2\sqrt{2} \times r^2$$
$$a_{10} = \frac{5}{4}\sqrt{10-2\sqrt{5}} \times r^2$$
$$a_{12} = 3 \times r^2$$

(ii)

$$d_3 = 1.29\cdots < d_4 = 2$$
$$< d_5 = 2.37\cdots$$
$$< d_6 = 2.59\cdots$$
$$< d_8 = 2.82\cdots$$
$$< d_{10} = 2.93\cdots$$
$$< d_{12} = 3.$$

(b)(i) **Note:** This could also be a long slog. However we have done much of the work before. But notice that, when the radius is 1, the area of each of the n sectors is equal to half the edge length times the height ($= 1$); so if $r = 1$, then the total area is numerically equal to "half the perimeter P_n".

$$
\begin{aligned}
A_3 &= 3\sqrt{3} \times r^2 \\
A_4 &= 4 \times r^2 \\
A_5 &= 5\sqrt{5 - 2\sqrt{5}} \times r^2 \\
A_6 &= 2\sqrt{3} \times r^2 \\
A_8 &= 8\left(\sqrt{2} - 1\right) \times r^2 \\
A_{10} &= 2\sqrt{25 - 10\sqrt{5}} \times r^2 \\
A_{12} &= 12\left(2 - \sqrt{3}\right) \times r^2
\end{aligned}
$$

(ii)
$$
\begin{aligned}
D_3 = 5.19\cdots &> D_4 = 4 \\
&> D_5 = 3.63\cdots \\
&> D_6 = 3.46\cdots \\
&> D_8 = 3.31\cdots \\
&> D_{10} = 3.24\cdots \\
&> D_{12} = 3.21\cdots .
\end{aligned}
$$

(c) Let O be the centre of the circle of radius r. Let A, B lie on the circle with $\angle AOB = 30°$.

Let M be the midpoint of $\underline{AB}$ – so that $\triangle OAB$ is isosceles, with apex O and height $h = \underline{OM}$.

Let A' lie on $\underline{OA}$ produced, and let B' lie on $\underline{OB}$ produced, such that $\underline{OA'} = \underline{OB'}$ and $A'B'$ is tangent to the circle.

Then $\triangle OAB \sim \triangle OA'B'$ with $\triangle OAB$ contained in $\triangle OA'B'$, so the scale factor $\frac{1}{h} = 2\sqrt{2 - \sqrt{3}} > 1$. Hence

$$4(2 - \sqrt{3}) \cdot a_{12} = A_{12} > a_{12},$$

so $D_{12} > d_{12}$.

206. The rearranged shape (shown in Figure 9) is an "almost rectangle", where $\underline{OA}$ forms an "almost height" and $\underline{OA} = r$. Half of the $2n$ circular arcs such as AB form the upper "width", and the other half form the "lower width", so each of these "almost widths" is equal to half the perimeter of the circle – namely πr.

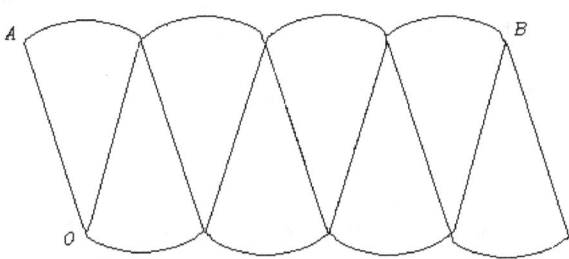

Figure 9: Rectification of a circle.

Hence the area of the rearranged shape is very close to

$$r \times \pi r = \pi r^2.$$

207.

(a) Cut along a generator, open up and lay the surface flat to obtain: a $2\pi r$ by h rectangle and two circular discs of radius r.
Hence the total surface area is

$$2\pi r^2 + 2\pi rh = \mathbf{2\pi r(r+h)}.$$

(b) The lateral surface consists of n rectangles, each with dimensions s_n by h (where s_n is the edge length of the regular n-gon), and hence has area $P_n h = 2\Pi_n rh$. Adding in the two end discs (each with area $\frac{1}{2} P_n r$) then gives total surface area $\mathbf{2\Pi_n r(r+h)}$.

208.

(a) (i) $\frac{1}{2}\pi r^2$; (ii) $\frac{1}{4}\pi r^2$; (iii) $\frac{\theta}{2} r^2$.
(b) The sector has two radii of total length $2r$. Hence the circular must have length $(\pi - 2)r$, and so subtends an angle $\pi - 2$ at the centre, so has area $\frac{\pi-2}{2} r^2$.

209.

(a) Focus first on the sloping surface. If we cut along a "generator" (a straight line segment joining the apex to a point on the circumference of the base), the surface opens up and lays flat to form a sector of a circle of radius l. The outside arc of this sector has length $2\pi r$, so the sector angle at the centre is equal to $\frac{r}{l} \cdot 2\pi$, and hence its area is $\frac{r}{l} \cdot \pi l^2 = \pi rl$.
Adding the area of the base gives the total surface area of the cone as

$$\pi r(r+l) = \frac{1}{2} \cdot \mathbf{2\pi r(r+l)}.$$

(b) Let M be the midpoint of the edge BC and let l denote the 'slant height' AM. Then the area of the n sloping faces is equal to $\frac{1}{2}P_n \cdot l$, while the area of the base is equal to $\frac{1}{2}P_n \cdot r$.

Hence the surface area is precisely $\frac{1}{2}P_n(r+l)$.

210.

(a) If AB is an edge of the inscribed regular n-gon $ABCD\cdots$, and O is the circumcentre, then $\angle AOB = \frac{2\pi}{n}$, so $AB = 2r\sin\frac{\pi}{n}$. Hence the required ratio is equal to $\frac{\sin\frac{\pi}{n}}{\frac{\pi}{n}}$, which tends to 1 as n tends to ∞.

(b) If AB is an edge of the circumscribed regular n-gon $ABCD\cdots$, and O is the circumcentre, then $\angle AOB = \frac{2\pi}{n}$, so $AB = 2r\tan\frac{\pi}{n}$. Hence the required ratio is equal to $\frac{\tan\frac{\pi}{n}}{\frac{\pi}{n}}$, which tends to 1 as n tends to ∞.

211.

(a) If AB is an edge of the inscribed regular n-gon $ABCD\cdots$, and O is the circumcentre, then $\angle AOB = \frac{2\pi}{n}$, so area$(\triangle OAB) = \frac{1}{2}r^2\sin\frac{2\pi}{n}$. Hence the required ratio is equal to $\frac{\sin\frac{2\pi}{n}}{\frac{2\pi}{n}}$, which tends to 1 as n tends to ∞.

(b) If AB is an edge of the circumscribed regular n-gon $ABCD\cdots$, and O is the circumcentre, then $\angle AOB = \frac{2\pi}{n}$, so area$(\triangle OAB) = r^2\tan\frac{\pi}{n}$. Hence the required ratio is equal to $\frac{\tan\frac{\pi}{n}}{\frac{\pi}{n}}$, which tends to 1 as n tends to ∞.

212.

(a) area(P_2) = area(P_1) + area$(P_2 - P_1)$ > area(P_1).

(b) This general result is clearly related to the considerations of the previous section. But it is not clear whether we can really expect to prove it with the tools available. So it has been included partly in the hope that readers might come to appreciate the difficulties inherent in proving such an "obvious" result.

In the end, any attempt to prove it seems to underline the need to use "proof by induction" – for example, on the number of edges of the inner polygon. This method is not formally treated until Chapter 6, but is needed here.

- Suppose the inner polygon $P_1 = ABC$ has just $n = 3$ edges, and has perimeter p_1.

Draw the line through A parallel to BC, and let it meet the (boundary of the) polygon P_2 at the points U and V. The triangle inequality (Problem **146**(c)) guarantees that the length UV is less than or equal to the length of the compound path from U to V along the perimeter of the polygon P_2 (keeping on the opposite side of the line UV from B and C). So, if we cut off the part of P_2 on the side of the line UV opposite to B and C, we obtain a new outer convex polygon P_3, which contains P_1, and whose perimeter is no larger than that of P_2.

Now draw the line through B parallel to AC, and let it meet the boundary of P_3 at points W, X. If we cut off the part of P_3 on the side of the line WX opposite to A and C, we obtain a new outer convex polygon P_4, which contains P_1, and whose perimeter is no larger than that of P_3.

If we now do the same by drawing a line through C parallel to AB, and cut off the appropriate part of P_4, we obtain a final outer convex polygon P_5, which contains the polygon P_1, and whose perimeter is no larger than that of P_4 – and hence no larger than that of the original outer polygon P_2.

All three vertices A, B, C of P_1 now lie on the boundary of the outer polygon P_5, so the triangle inequality guarantees that AB is no larger than the length of the compound path along the boundary of P_5 from A to B (staying on the opposite side of the line AB from C). Similarly BC is no larger than the length of the compound path along the boundary of P_5 from B to C; and CA is no larger than the length of the compound path along the boundary of P_5 from C to A.

Hence the perimeter p_1 of the triangle P_1 is no larger than the perimeter of the outer polygon P_5, whose perimeter was no larger than the perimeter of the original outer polygon P_2. Hence the result holds when the inner polygon is a triangle.

- Now suppose that the result has been proved when the inner polygon is a k-gon, for some $k \geqslant 3$, and suppose we are presented with a pair of polygons P_1, P_2 where the inner polygon $P_1 = ABCD \cdots$ is a convex $(k+1)$-gon.

Draw the line m through C parallel to BD. Let this line meet the outer polygon P_2 at U and V. Cut off the part of P_2 on the opposite side of the the line UCV to B and D, leaving a new outer convex polygon P with perimeter no greater than that of P_2. We prove that the perimeter of polygon P_1 is less than that of polygon P – and hence less than that of P_2. Equivalently, we may assume that UCV is an edge of P_2.

Translate the line m parallel to itself, from m to BD and beyond, until it reaches a position of final contact with the polygon P_1, passing through the vertex X (and possibly a whole edge $\underline{XY}$) of the inner polygon P_1. Let this final contact line parallel to m be m'.

Since P_1 is convex and $k \geqslant 3$, we know that X is different from B and from D. As before, we may assume that m' is an edge of the outer polygon P_2. Cut both P_2 and P_1 along the line CX to obtain two smaller configurations, each of which consists of an inner convex polygon inside an outer convex polygon, but in which

- each of the inner polygons has at most k edges, and
- in each of the smaller configurations, the inner and outer polygons both share the edge $\underline{CX}$.

Then (by induction on the number of edges of the inner polygon) the perimeter of each inner polygon is no larger than the perimeter of the corresponding outer polygon; so for each inner polygon, the partial perimeter running from C to X (omitting the edge $\underline{CX}$) is no larger than the partial perimeter of the corresponding outer polygon running from C to X (omitting the edge $\underline{CX}$). So when we put the two parts back together again, we see that the perimeter of P_1 is no larger than the perimeter of P_2.

Hence the result holds when P_1 is a triangle; and if the result holds whenever the inner polygon has $k \geq 3$ edges, it also holds whenever the inner polygon has $(k+1)$ edges.

It follows that the result holds whatever the number of edges of the inner polygon may be. QED

213.

(a) Join PQ. Then the lines $y = b$ and $x = d$ meet at R to form the right angled triangle PQR.

Pythagoras' Theorem then implies that $(d-a)^2 + (e-b)^2 = PQ^2$.

(b) Join PQ. The points $P = (a, b, c)$ and $R = (d, e, c)$ lie in the plane $z = c$. If we work exclusively in this plane, then part (a) shows that
$$PR^2 = (d-a)^2 + (e-b)^2.$$
$QR = |f - c|$, and $\triangle PRQ$ has a right angle at R. Hence
$$PQ^2 = PR^2 + RQ^2 = (d-a)^2 + (e-b)^2 + (f-c)^2.$$

214.

(a) area($\triangle ABC$) = $\frac{bc}{2}$, area($\triangle ACD$) = $\frac{cd}{2}$, area($\triangle ABD$) = $\frac{bd}{2}$.

(b) [This can be a long algebraic slog. And the answer can take very different looking forms depending on how one proceeds. Moreover, most of the resulting expressions are not very pretty, and are likely to incorporate errors.

One way to avoid this slog is to appeal to the fact that the modulus of the vector product $\boldsymbol{DB} \times \boldsymbol{DC}$ is equal to the area of the parallelogram spanned by $\boldsymbol{DB}$ and $\boldsymbol{DC}$ – and so is twice the area of $\triangle BCD$.]

(c) However part (b) is approached, it is in fact true that
$$\text{area}(\triangle BCD)^2 = \left(\frac{bc}{2}\right)^2 + \left(\frac{cd}{2}\right)^2 + \left(\frac{bd}{2}\right)^2,$$
so that
$$\text{area}(\triangle ABC)^2 + \text{area}(\triangle ACD)^2 + \text{area}(\triangle ABD)^2 = \text{area}(\triangle BCD)^2.$$

215.

(a) b and c are indeed lengths of arcs of great circles on the unit sphere: that is, arcs of circles of radius 1 (centred at the centre of the sphere). However, back in Chapter 1 we used the 'length' of such circular arcs to define the **angle** (in radians) subtended by the arc at the centre. So b and c are also *angles* (in radians).

(b) The only standard functions of angles are the familiar trig functions (sin, cos).

(c) If $c = 0$, then the output should specify that $a = b$, so c should have no effect on the output. This suggests that we might expect a formula that involves "adding $\sin c$" or "multiplying by $\cos c$".

Similarly when $b = 0$, the output should give $a = c$, so we might expect a formula that involved "adding $\sin b$" or "multiplying by $\cos b$".

In general, we should expect a formula in which b and c appear interchangeably (since the input pair (b, c) could equally well be replaced by the input pair (c, b) and should give the same output value of a).

(d)(i) If $\underline{BC}$ runs along the equator and $\angle B = \angle C = \frac{\pi}{2}$, then $\underline{BA}$ and $\underline{CA}$ run along circles of longitude, so A must be at the North pole. Since A is a right angle, it follows that $a = b = c = \frac{\pi}{2}$. (This tends to rule out the idea that the formula might involve "adding $\sin b$ and adding $\sin c$".)

(ii) Suppose that $\angle B = \frac{\pi}{2}$. Since we can imagine $\underline{AB}$ along the equator, and since there is a right angle at A, it follows that $\underline{AC}$ and $\underline{BC}$ both lie along circles of longitude, and so meet at the North pole. Hence C will be at the North pole, so $a = b = \frac{\pi}{2}$.

The inputs to any spherical version of Pythagoras' Theorem are then $b = \frac{\pi}{2}$, and c. And c is not constrained, so every possible input value of c must lead to the same output $a = \frac{\pi}{2}$. This tends to suggest that the formula involves some multiple of a product combining "$\cos b$" with some function of c. And since the inputs "b" and "c" must appear symmetrically, we might reasonably expect some multiple of "$\cos b \cdot \cos c$".

216.

(a) $\triangle OAB'$ has a right angle at A with $\angle AOB' = \angle AOB = c$. Hence $\underline{AB'} = \tan c$. Similarly $\underline{AC'} = \tan b$.
Hence $\underline{B'C'}^2 = \tan^2 b + \tan^2 c$.

(b) In $\triangle OAB'$ we see that $\underline{OB'} = \sec c$. Similarly $\underline{OC'} = \sec b$. We can now apply the Cosine Rule to $\triangle OB'C'$ to obtain:

$$\begin{aligned} \tan^2 b + \tan^2 c &= \sec^2 b + \sec^2 c - 2\sec b \cdot \sec c \cdot \cos(\angle B'OC') \\ &= \sec^2 b + \sec^2 c - 2\sec b \cdot \sec c \cdot \cos a. \end{aligned}$$

Hence $\cos a = \cos b \cdot \cos c$. QED

217. Construct the plane tangent to the sphere at A. Extend $\underline{OB}$ to meet this plane at B', and extend $\underline{OC}$ to meet the plane at C'.

$\triangle OAB'$ has a right angle at A with $\angle AOB' = \angle AOB = c$. Hence $\underline{AB'} = \tan c$. Similarly $\underline{AC'} = \tan b$.

Hence $\underline{B'C'}^2 = \tan^2 b + \tan^2 c - 2 \cdot \tan b \cdot \tan c \cdot \cos A$.

In $\triangle OAB'$ we see that $\underline{OB'} = \sec c$. Similarly $\underline{OC'} = \sec b$.

We can now apply the Cosine Rule to $\triangle OB'C'$ to obtain:

$$\tan^2 b + \tan^2 c - 2 \cdot \tan b \cdot \tan c \cdot \cos A$$
$$= \sec^2 b + \sec^2 c - 2 \sec b \cdot \sec c \cdot \cos(\angle B'OC')$$
$$= \sec^2 b + \sec^2 c - 2 \sec b \cdot \sec c \cdot \cos a.$$

Hence $\cos a = \cos b \cdot \cos c + \sin b \cdot \sin c \cdot \cos A$. QED

218. We show that $\frac{\sin b}{\sin a} = \sin B$ (where B denotes the angle $\angle ABC$ at the vertex B).

Construct the plane $\boldsymbol{T}$ which is tangent to the sphere at B. Let O be the centre of the sphere; let $\underline{OA}$ produced meet the plane $\boldsymbol{T}$ at A'', and let $\underline{OC}$ produced meet the plane $\boldsymbol{T}$ at C'''.

Imagine $\underline{BA}$ positioned along the equator; then $\underline{BA''}$ is horizontal; $\underline{AC}$ lies on a circle of longitude, so $A''C'''$ is vertical. Hence $\angle C'''BA'' = \angle B$, and $\angle BA''C'''$ is a right angle; so $\sin B = \frac{A''C'''}{BC'''}$.

$\triangle OA''C'''$ is right angled at A''; and $\angle A''OC''' = b$; so $\sin b = \frac{A''C'''}{OC'''}$.

$\triangle OBC'''$ is right angled at B; and $\angle BOC''' = a$; so $\sin a = \frac{BC'''}{OC'''}$.

Hence $\frac{\sin b}{\sin a} = \sin B$ depends only on the angle at B, so $\frac{\sin b}{\sin a} = \frac{\sin b'}{\sin a'} = \sin B$.

219. We show that
$$\frac{\sin a}{\sin A} = \frac{\sin b}{\sin B}.$$

Position the triangle (or rather "rotate the sphere") so that $\underline{AB}$ runs along the equator, with $\underline{AC}$ leading into the northern hemisphere.

(i) If $\angle A$ is a right angle, then
$$\frac{\sin b}{\sin a} = \sin B = \frac{\sin B}{\sin A}$$
by Problem **218**. Hence
$$\frac{\sin a}{\sin A} = \frac{\sin b}{\sin B}.$$
The same is true if $\angle B$ is a right angle.

(ii) If $\angle A$ and $\angle B$ are both less than a right angle, one can draw the circle of longitude from C to some point X on $\underline{AB}$. One can then apply Problem **218** to the two triangles $\triangle CXA$ and $\triangle CXB$ (each with a right angle at X). Let x denote the length of the $\underline{CX}$. Then $\frac{\sin x}{\sin b} = \sin A$, and $\frac{\sin x}{\sin a} = \sin B$.

Hence $\sin b \cdot \sin A = \sin x = \sin a \cdot \sin B$, so
$$\frac{\sin a}{\sin A} = \frac{\sin b}{\sin B}.$$

(iii) If $\angle A$ (say) is greater than a right angle, the circle of longitude from C meets the line BA extended beyond A at a point X (say). If we let $CX = x$, then one can argue similarly using the triangles $\triangle CXA$ and $\triangle CXB$ to get $\frac{\sin x}{\sin a} = \sin B$, and $\frac{\sin x}{\sin b} = \sin A$, whence
$$\frac{\sin a}{\sin A} = \frac{\sin b}{\sin B}.$$

220. Let $X = (x, y)$ be an arbitrary point of the locus.

(i) Then the distance from X to m is equal to y, so setting this equal to XF gives the equation:
$$y^2 = x^2 + (y - 2a)^2,$$
or
$$x^2 = 4a(y - a).$$

(ii) If we change coordinates and choose the line $y = a$ as a new x-axis, then the equation becomes $x^2 = 4aY$. The curve is then tangent to the new x-axis at the (new) origin, and is symmetrical about the y-axis.

221.

(a) Choose the line AB as x-axis, and the perpendicular bisector of AB as the y-axis. Then $A = (-3, 0)$ and $B = (3, 0)$. The point $X = (x, y)$ is a point of the unknown locus precisely when
$$(x+3)^2 + y^2 = \underline{XA}^2 = (2 \cdot \underline{XB})^2 = 2^2((x-3)^2 + y^2)$$
that is, when
$$(x-5)^2 + y^2 = 4^2.$$
This is the equation of a circle with centre $(5, 0)$ and radius $r = 4$.

(b) Choose the line AB as x-axis, and the perpendicular bisector of AB as the y-axis.
If $f = 1$, the locus is the perpendicular bisector of $\underline{AB}$.
We may assume that $f > 1$ (since if $f < 1$, then $\underline{BX} : \underline{AX} = f^{-1} : 1$ and $f^{-1} > 1$, so we may simply swap the labelling of A and B).
Now $A = (-b, 0)$ and $B = (b, 0)$, and the point $X = (x, y)$ is a point of the unknown locus precisely when
$$(x+b)^2 + y^2 = \underline{XA}^2 = (f \cdot \underline{XB})^2 = f^2\left[(x-b)^2 + y^2\right]$$
that is, when
$$x^2(f^2 - 1) - 2bx(f^2 + 1) + y^2(f^2 - 1) + b^2(f^2 - 1) = 0$$

$$\left(x - \frac{b(f^2+1)}{f^2-1}\right)^2 + y^2 = \left(\frac{2fb}{f^2-1}\right)^2.$$

This is the equation of a circle with centre $\left(\frac{b(f^2+1)}{f^2-1}, 0\right)$ and radius $r = \frac{2fb}{f^2-1}$.

222.

(a) Choose the line AB as x-axis, and the perpendicular bisector of AB as the y-axis.
Then $A = (-c, 0)$ and $B = (c, 0)$. The point $X = (x, y)$ is a point of the unknown locus precisely when

$$2a = AX + BX = \sqrt{(x+c)^2 + y^2} + \sqrt{(x-c)^2 + y^2},$$

that is, when

$$2a - \sqrt{(x+c)^2 + y^2} = \sqrt{(x-c)^2 + y^2}.$$
$$\therefore\ 4a^2 - 4a\sqrt{(x+c)^2 + y^2} + [(x+c)^2 + y^2] = [(x-c)^2 + y^2]$$
$$\therefore\ a^2 + cx = a\sqrt{(x+c)^2 + y^2}$$
$$\therefore\ (a^2 - c^2)x^2 + a^2 y^2 = a^2(a^2 - c^2)$$

Setting $\frac{c}{a} = e$ then yields the equation for the locus in the form:

$$\frac{x^2}{a^2} + \frac{y^2}{a^2(1-e^2)} = 1.$$

Note: In the derivation of the equation we squared both sides (twice). This may introduce spurious solutions. So we should check that every solution (x, y) of the final equation satisfies the original condition.

(b) The real number $e < 1$ is given, so we may set the distance from F to m be $\frac{a}{e}(1 - e^2)$. Choose the line through F and perpendicular to m as x-axis. To start with, we choose the line m as y-axis and adjust later if necessary.
Hence F has coordinates $\left(\frac{a}{e}(1-e^2), 0\right)$, and the point $X = (x, y)$ is a point of the unknown locus precisely when

$$\left(x - \frac{a}{e}(1 - e^2)\right)^2 + y^2 = (ex)^2,$$

which can be rearranged as

$$(1 - e^2)x^2 - 2\frac{a}{e}(1 - e^2)x + \left(\frac{a}{e}\right)^2(1 - e^2)^2 + y^2 = 0,$$

and further as

$$(1-e^2)\left[x^2 - 2\frac{a}{e}x + \left(\frac{a}{e}\right)^2\right] + y^2 = \left(\frac{a}{e}\right)^2(1-e^2) - \left(\frac{a}{e}\right)^2(1-e^2)^2$$
$$= \left(\frac{a}{e}\right)^2(e^2 - e^4)$$
$$= a^2(1-e^2)$$
$$\therefore \left(x - \frac{a}{e}\right)^2 + \frac{y^2}{1-e^2} = a^2.$$

If we now move the y-axis to the line $x = \frac{a}{e}$ the equation takes the simpler form:

$$\frac{x^2}{a^2} + \frac{y^2}{a^2(1-e^2)} = 1.$$

(c) This was done in the derivations in the solutions to parts (a) and (b).

223.

(a) The triangle inequality shows that, if $AX > BX$, then $AB + BX \geq AX$; hence the locus is non-empty only when $a \leq c$. If $a = c$, then X must lie on the line AB, but not between A and B, so the locus consists of the two half-lines on AB outside AB. Hence we may assume that $a < c$.

Choose the line AB as x-axis, and the perpendicular bisector of AB as y-axis. Then $A = (-c, 0)$ and $B = (c, 0)$. The point $X = (x, y)$ is a point of the unknown locus precisely when

$$2a = |AX - BX| = \left|\sqrt{(x+c)^2 + y^2} - \sqrt{(x-c)^2 + y^2}\right|.$$

If $AX > BX$, we can drop the modulus signs and calculate as in Problem **222**.

$$2a + \sqrt{(x-c)^2 + y^2} = \sqrt{(x+c)^2 + y^2}.$$
$$\therefore 4a^2 + 4a\sqrt{(x-c)^2 + y^2} + (x-c)^2 + y^2 = (x+c)^2 + y^2$$
$$\therefore a^2 - cx = -a\sqrt{(x-c)^2 + y^2}$$
$$\therefore (c^2 - a^2)x^2 - a^2 y^2 = a^2(c^2 - a^2)$$

Setting $\frac{c}{a} = e \ (> 1)$, then yields the equation for the locus in the form:

$$\frac{x^2}{a^2} - \frac{y^2}{a^2(e^2 - 1)} = 1.$$

Note: In the derivation of the equation we squared both sides (twice). This may introduce spurious solutions. So we should check that every solution (x, y) of the final equation satisfies the original condition. In fact, the squaring process introduces additional solutions in the form of a second branch of the locus, corresponding precisely to points X where $AX < BX$.

(b) The real number $e > 1$ is given, so we may set the distance from F to m be $\frac{a}{e}(e^2 - 1)$. Choose the line through F and perpendicular to m as x-axis. To start with, we choose the line m as y-axis and adjust later if necessary.
Hence F has coordinates $(\frac{a}{e}(e^2 - 1), 0)$, and the point $X = (x, y)$ is a point of the unknown locus precisely when

$$\left(x - \frac{a}{e}(e^2 - 1)\right)^2 + y^2 = (ex)^2$$

$$\therefore \quad (e^2 - 1)x^2 + \frac{2a}{e}(e^2 - 1)x - y^2 = \left(\frac{a}{e}\right)^2 (e^2 - 1)^2$$

$$\therefore \quad (e^2 - 1)\left[x^2 + \frac{2a}{e}x + \left(\frac{a}{e}\right)^2\right] - y^2 = \left(\frac{a}{e}\right)^2 (e^2 - 1)^2 + \left(\frac{a}{e}\right)^2 (e^2 - 1)$$

$$= \left(\frac{a}{e}\right)^2 (e^4 - e^2)$$

$$\therefore \quad (e^2 - 1)\left[x^2 + \frac{2a}{e}x + \left(\frac{a}{e}\right)^2\right] - y^2 = a^2(e^2 - 1)$$

$$\therefore \quad \left(x + \frac{a}{e}\right)^2 - \frac{y^2}{e^2 - 1} = a^2$$

If we now move the y-axis to the line $x = -\frac{a}{e}$ the equation takes the simpler form:

$$\frac{x^2}{a^2} - \frac{y^2}{a^2(e^2 - 1)} = 1.$$

(c) This was done in derivations in the solutions to parts (a) and (b).

224.

(a) When $z = k$ is a constant, the equation reduces to that of a circle

$$x^2 + y^2 = (rk)^2$$

in the plane $z = k$. When the cutting plane is the xy-plane "$z = 0$", the circle has radius 0, so is a single point.

(b)(i) A vertical plane through the apex cuts the cone in a pair of generators crossing at the apex.

(ii) If the cutting plane through the apex is less steep than a generator, then it cuts the cone only at the apex.

If the cutting plane through the apex is parallel to a generator, then it cuts the cone in a generator – a single line (the next paragraph indicates that this line may be better thought of as a pair of "coincident" lines).

What happens when the cutting plane through the apex is steeper than a generator may not be intuitively clear. One way to make sense of this is to treat the cross-section as the set of solutions of two simultaneous equations –

one for the cone, and the other for the plane (say $y = nz$, with $n < r$). This leads to the equation

$$x^2 = (r^2 - n^2)z^2, \quad y = nz,$$

with solution set

$$x = \pm z\sqrt{r^2 - n^2}, \quad xy = nz$$

which specifies a pair of lines crossing at the apex.

A slightly easier way to visualize the cross-section in this case is to let the apex of the double cone be A, and to let X be any other point of the cross-section. Then the line AX is a generator of the cone, so lies on the cone's surface. But A and X also lie in the cutting plane – so the whole line AX must lie in the cutting plane. Hence the cross-section contains the whole line AX.

(c)(i) If the cutting plane passes through the apex and is parallel to a generator of the cone, then we saw in (b) that the cross-section is simply the generator itself.

(ii) Thus we assume that the cutting plane does not pass through the apex, and may assume that it cuts the bottom half of the cone. If V is the point nearest the apex where the cutting plane meets the cone, then the cross-section curve starts at V and becomes wider as we go down the cone. Because the plane is parallel to a generator, the plane never cuts the "other side" of the bottom half of the cone, so the cross section never "closes up" – but continues to open up wider and wider as we go further and further down the bottom half of the cone.

Let S be the sphere, which is inscribed in the cone above the cutting plane, and which is tangent to the cutting plane at F. Let C be the circle of contact between S and the cone. Let A be the apex of the cone, and let the apex angle of the cone be equal to 2θ. Let X be an arbitrary point of the cross-section.

To illustrate the general method, consider first the special case where $X = V$ is the "highest" point of the cross-section. The line segment $\underline{VA}$ is tangent to the sphere S, so crosses the circle C at some point M. Now $\underline{VF}$ lies in the cutting plane, so is also tangent to the sphere S at the point F. Any two tangents to a sphere from the same exterior point are equal, so it follows that $\underline{VM} = \underline{VF}$. Moreover, $\underline{VM}$ is exactly equal to the distance from V to the line m (since

* firstly the line m lies in the horizontal plane through C and so is on the same horizontal level as M, and

* secondly the shortest line VM^* from V to m, runs straight up the cutting plane, which is parallel to a generator – so the angle $\angle MVM^* = 2\theta$, whence $\underline{VM^*} = \underline{VM} = \underline{VF}$).

Now let X be an arbitrary point of the cross-sectional curve, and use a similar argument. First the line XA is always a generator of the cone, so is tangent to the sphere S, and crosses the circle C at some point Y. Moreover, XF is also tangent to the sphere. Hence $\underline{XY} = \underline{XF}$. It remains to see that $\underline{XY}$ is equal to the distance $\underline{XY^*}$ from X to the closest point Y^* on the line m – since

* firstly the two points Y and Y^* both lie on the same horizontal level (namely the horizontal plane through the circle C), and
* secondly both make the same angle θ with the vertical.

Hence the cross-sectional curve is a parabola with focus F and directrix m.

(d)(i) If the cutting plane is less steep than a generator, the cross-section is a closed curve. If V and W are the highest and lowest points of intersection of the plane with the cone, then the cross-section is clearly symmetrical under reflection in the line VW. Intuitively it is tempting to think that the lower end near W should be 'fatter' than the upper part of the curve (giving an egg-shaped cross-section). This turns out to be false, and the correct version was known to the ancient Greeks, though the error was repeated in many careful drawings from the 14th and 15th centuries (e.g. Albrecht Dürer (1471–1528)).

(ii) The derivation is very similar to that in part (c), and we leave the reader to reconstruct it.

An alternative approach is to insert a second sphere S' below the cutting plane, and inflate it until it makes contact with the cone along a circle C' while at the same time touching the cutting plane at a point F'. If X is an arbitrary point of the cross-sectional curve, then XA is tangent to both spheres, so meets the circle C at some point Y and meets the circle C' at some point Y'. Then Y, X, Y' are collinear. Moreover, $\underline{XY} = \underline{XF}$ (since both are tangents to the sphere S from the point X), and $\underline{XY'} = \underline{XF'}$ (since both are tangents to the sphere S' from the point X), so

$$\underline{XF} + \underline{XF'} = \underline{XY} + \underline{XY'} = \underline{YY'}$$

But $\underline{YY'}$ is equal to the slant height of the cone between the two fixed circles C and C', and so is equal to a constant k. Hence, the focus-focus specification in Problem **222** shows that the cross-section is an ellipse.

(e)(i) If the cutting plane is steeper than a generator, the cross-section cuts both the bottom half and the top half of the cone to give two separate parts of the cross-section. Neither part "closes up", so each part opens up more and more widely.

If V is the highest point of the cross-section on the lower half of the cone, and W is the lowest point of the cross-section on the upper half of the cone, then it seems clear that the cross-section is symmetrical under reflection in the line VW. But it is quite unclear that the two halves of the cross-section are exactly congruent (though again it was known to the ancient Greeks).

(ii) The formal derivation is very similar to that in part (c), and we leave the reader to reconstruct it.

An alternative approach is to copy the alternative in (d), and to insert a second sphere S' in the upper part of the cone, on the same side of the cutting plane as the apex, inflate it until it makes contact with the cone along a circle C' while at the same time touching the cutting plane at a point F'. If X is an arbitrary point of the cross-sectional curve, then XA is tangent to both spheres, so meets the circle C at some point Y and meets the circle

C' at some point Y'. Then Y, X, Y' are collinear. Moreover, X, F, and F' all lie on the cutting plane. Now $\underline{XY} = \underline{XF}$ (since both are tangents to the sphere S from the point X), and $\underline{XY'} = \underline{XF'}$ (since both are tangents to the sphere S' from the point X). If X is on the upper half of the cone, then

$$XF - XF' = XY - XY' = YY'.$$

But $\underline{YY'}$ is equal to the slant height of the cone between the two fixed circles C and C', and so is equal to a constant k. Hence, the focus-focus specification in Problem **223** shows that the cross-section is a hyperbola.

225.

(a) (i) 2^1; (ii) $1 = 2^0$
(b) (i) 2^2; (ii) $1 = 2^0$; (iii) $4 = 2 \times 2^0 + 2^1$
(c) (i) 2^3; (ii) $1 = 2^0$; (iii) $12 = 2 \times 4 + 2^2$; (iv) $6 = 2 \times 2^0 + 4$
(d) (i) 2^4; (ii) $1 = 2^0$; (iii) $32 = 2 \times 12 + 2^3$; (iv) $24 = 2 \times 6 + 12$; (v) $8 = 2 \times 2^0 + 6$

226.

(c) (i) If you look carefully at the diagram shown here you should be able to see not only the upper and lower 3D-cubes, but also the four other 3D-cubes formed by joining each 2D-cube in the upper 3D-cube to the corresponding 2D-cube in the lower 3D-cube.

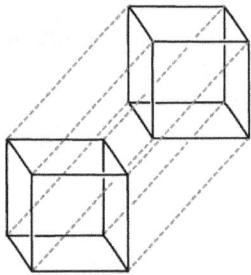 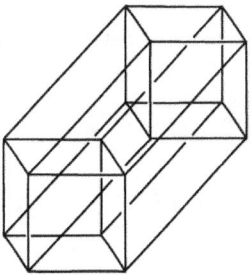

Note: Once we have the 3D-cube expressed in coordinates, we can specify precisely which planes produce which cross-sections in Problem **38**. The plane $x + y + z = 1$ passes through the three neighbours of $(0,0,0)$ and creates an equilateral triangular cross-section. Any plane of the form $z = c$ (where c is a constant between 0 and 1) produces a square cross-section. And the plane $x + y + z = \frac{3}{2}$ is the perpendicular bisector of the line joining $(0,0,0)$ to $(1,1,1)$, and creates a regular hexagon as cross-section.

227.

(a) View the coordinates as (x, y, z). Start at the origin $(0,0,0)$ and travel round 3 edges of the lower 2D-cube "$z = 0$" to the point $(0, 1, 0)$. Copy this path of length 3 on the upper 2D-cube "$z = 1$" (from $(0,0,1)$ to $(0,1,1)$). Then join $(0,0,0)$ to $(0,0,1)$ and join $(0,1,0)$ to $(0,1,1)$. The result

$(0,0,0)$, $(1,0,0)$, $(1,1,0)$, $(0,1,0)$, $(0,1,1)$, $(1,1,1)$, $(1,0,1)$, $(0,0,1)$ (and back to $(0,0,0)$)

has the property that exactly one coordinate changes when we move from each vertex to the next. This is an example of a *Gray code of length* 3.

Note: How many such paths/circuits are there in the 3D-cube? We can certainly count those of the kind described here. Each such circuit has a "direction": the 12 edges of the 3D-cube lie in one of 3 "directions", and each such circuit contains all four edges in one of these 3 directions. Moreover this set of four edges can be completed to a circuit in exactly 2 ways. So there are 3×2 such circuits. In 3D this accounts for all such circuits. But in higher dimensions the numbers begin to explode (in the 4D-cube there are 1344 such circuits).

(b) View the coordinates as (w, x, y, z). Start at the origin $(0,0,0,0)$ and travel round the 8 vertices of the lower 3D-cube "$z = 0$" to the point $(0,0,1,0)$. Then copy this path on the upper 3D-cube "$z = 1$" from $(0,0,0,1)$ to $(0,0,1,1)$. Finally join $(0,0,0,0)$ to $(0,0,0,1)$ and join $(0,0,1,0)$ to $(0,0,1,1)$. The result

$(0,0,0.0)$, $(1,0,0,0)$, $(1,1,0,0)$, $(0,1,0,0)$, $(0,1,1,0)$, $(1,1,1,0)$, $(1,0,1,0)$, $(0,0,1,0)$ $(0,0,1,1)$, $(1,0,1,1)$, $(1,1,1,1)$, $(0,1,1,1)$, $(0,1,0,1)$, $(1,1,0,1)$, $(1,0,0,1)$, $(0,0,0,1)$ (and back to $(0,0,0,0)$)

has the property that exactly one coordinate changes when we move from each vertex to the next. This is an example of a *Gray code of length* 4.

Note: The general construction in dimension $n + 1$ depends on the previous construction in dimension n, so makes use of *mathematical induction* (see Problem **262** in Chapter 6).

VI. Infinity: recursion, induction, infinite descent

> *Mathematical induction*
> *– i.e. proof by recurrence –*
> *is ... imposed on us,*
> *because it is ... the affirmation*
> *of a property of the mind itself.*
> Henri Poincaré (1854–1912)

> *Allez en avant, et la foi vous viendra.*
> *(Press on ahead, and understanding will follow.)*
> Jean le Rond d'Alembert (1717–1783)

Mathematics has been called "the science of infinity". However, infinity is a slippery notion, and many of the techniques which are characteristic of modern mathematics were developed precisely to tame this slipperiness. This chapter introduces some of the relevant ideas and techniques.

There are aspects of the story of infinity in mathematics which we shall not address. For example, astronomers who study the night sky and the movements of the planets and stars soon note its immensity, and its apparently 'fractal' nature – where increasing the detail or magnification reveals more or less the same level of complexity on different scales. And it is hard then to avoid the question of whether the stellar universe is finite or infinite.

In the *mental* universe of mathematics, once counting, and the process of halving, become routinely iterative processes, questions about infinity and infinitesimals are almost inevitable. However, mathematics recognises the conceptual gulf between the finite and the infinite (or infinitesimal), and rejects the lazy use of "infinity" as a metaphor for what is simply "very large". Large finite numbers are still numbers; and long finite sums are conceptually quite different from sums that "go on for ever". Indeed, in the third century BC, Archimedes (c. 287–212 BC) wrote a small booklet called *The Sand Reckoner*, dedicated to King Gelon, in which he introduced the arithmetic of powers (even though the ancient Greeks had no convenient notation for writing such numbers), in order to demonstrate that – contrary to what some people had claimed – the number of grains of sand in the known universe must be finite (he derived an upper bound of approximately 8×10^{63}).

The influence wielded by ideas of infinity on mathematics has been profound, even if we now view some of these ideas as flights of fancy –

- from Zeno of Elea (c. 495 BC – c. 430 BC), who presented his paradoxes to highlight the dangers inherent in reasoning sloppily with infinity,
- through Giordano Bruno (1548–1600), who declared that there were infinitely many inhabited universes, and who was burned at the stake when he refused to retract this and other "heresies",
- to Georg Cantor (1845–1918) whose groundbreaking work in developing a true "mathematics of infinity" was inextricably linked to his religious beliefs.

In contrast, we focus here on the delights of the mathematics, and in particular on how an initial doorway into "ideas of infinity" can be forged from careful reasoning with finite entities. Readers who would like to explore what we pass over in silence could do worse than to start with the essay on "infinity" in the MacTutor History of Mathematics archive:

http://www-history.mcs.st-and.ac.uk/HistTopics/Infinity.html.

The simplest infinite processes begin with *recursion* – a process where we repeat exactly the same operation over and over again (in principle, continuing for ever). For example, we may start with 0, and repeat the operation "add 1", to generate the sequence:

$$0, 1, 2, 3, 4, 5, 6, 7, \ldots.$$

Or we may start with $2^0 = 1$ and repeat the operation "multiply by 2", to generate:

$$1, 2, 4, 8, 16, 32, 64, 128, \ldots.$$

Or we may start with $1.000000\cdots$, and repeat the steps involved in "dividing by 7" to generate the infinite decimal for $\frac{1}{7}$:

$$\frac{1}{7} = 0.1428571428571428571\cdots.$$

We can then vary this idea of "recursion" by allowing each operation to be "essentially" (rather than exactly) the same, as when we define *triangular* numbers by "adding n" at the nth stage to generate the sequence:

$$0, 1, 3, 6, 10, 15, 21, 28, \ldots.$$

In other words, the sequence of *triangular numbers* is defined by a *recurrence relation*:

$T_0 = 0$; and
when $n \geqslant 1$, $T_n = T_{n-1} + n$.

We can vary this idea further by allowing more complicated recurrence relations – such as that which defines the *Fibonacci numbers*:

$F_0 = 0$, $F_1 = 1$; and
when $n \geqslant 1$, $F_{n+1} = F_n + F_{n-1}$.

All of these "images of infinity" hark back to the familiar counting numbers.

- We know how the counting numbers begin (with 0, or with 1); and
- we know that we can "add 1" over and over again to get ever larger counting numbers.

The intuition that this process is, in principle, *never-ending* (so is never actually completed), yet somehow manages to count all positive integers, is what Poincaré called a "property of the mind itself": that is, the idea that we can define an **infinite** sequence, or process, or chain of deductions (involving digits, or numbers, or objects, or statements, or truths) by

- specifying how it *begins*, and by then
- specifying in a uniform way "how to construct *the next term*", or "how to perform *the next step*".

This idea is what lies behind "**proof** by mathematical induction", where we prove that some assertion $\mathbf{P}(n)$ holds **for all** $\boldsymbol{n \geqslant 1}$ – so demonstrating infinitely many separate statements at a single blow. The validity of this method of proof depends on a fundamental property of the *positive integers*, or of the counting sequence

$$\text{"1, 2, 3, 4, 5, ..."},$$

namely:

The Principle of Mathematical Induction: If a subset S of the positive integers

- contains the integer "1",

and has the property that

- whenever an integer k is in the set S, then the next integer $k+1$ is always in S too,

then S contains **all** positive integers.

6.1. Proof by mathematical induction I

When students first meet "proof by induction", it is often explained in a way that leaves them feeling distinctly uneasy, because it appears to break the fundamental taboo:

> never assume what you are trying to prove.

This tends to leave the beginner in the position described by d'Alembert's quote at the start of the chapter: they may "press on" in the hope that "understanding will follow", but a doubt often remains. So we encourage readers who have already met proof by induction to take a step back, and to try to understand afresh how it really works. This may require you to study the solutions (Section 6.10), and to be prepared to learn to write out proofs more carefully than, and rather differently from, what you are used to.

When we wish to prove a general result which involves a parameter n, where n can be **any positive integer**, we are really trying to prove *infinitely many results all at once*. If we tried to prove such a collection of results in turn, "one at a time", not only would we never finish, we would *scarcely get started* (since completing the first million, or billion, cases leaves just as much undone as at the start). So our only hope is:

- to think of the sequence of assertions in a uniform way, as consisting of infinitely many different, but similar-looking, statements $\mathbf{P}(n)$, one for each n separately (with each statement depending on a particular n); and

- to recognise that the **overall** result to be proved is not just a single statement $\mathbf{P}(n)$, but the *compound statement* that "$\mathbf{P}(n)$ is true, **for all** $n \geqslant 1$".

Once the result to be proved has been formulated in this way, we can

- use bare hands to check that the very first statement is true (usually $\mathbf{P}(1)$); and

- try to find some way of demonstrating that,
  - as soon as we know that "$\mathbf{P}(k)$ is true, for some (particular, but unspecified) $k \geqslant 1$",
  - we can prove in a uniform way that the next result $\mathbf{P}(k+1)$ is then automatically true.

Having implemented the first of the two induction steps, we know that $\mathbf{P}(1)$ is true.

The second bullet point above then comes into play and assures us that (since we know that $\mathbf{P}(1)$ is true), $\mathbf{P}(2)$ must be true.

And once we know that **P**(2) is true, the second bullet point assures us that **P**(3) is also true.

And once we know that **P**(3) is true, the second bullet point assures us that **P**(4) is also true.

And so on *for ever*.

We can then conclude that the *whole sequence* of infinitely many statements are all true – namely that:

"**every** statement **P**(n) is true",

or that

"**P**(n) is true, **for all** $n \geqslant 1$."

In other words, if we define S to be the set of positive integers n for which the statement **P**(n) is true, then S contains the element "1", and whenever k is in S, so is $k + 1$; hence, by the *Principle of Mathematical Induction* we know that S contains all positive integers.

At this stage we should acknowledge an important didactical (rather than mathematical) ploy in our recommended layout here. It is important to underline the distinction between

(i) the *individual* statements **P**(n) which are the separate ingredients in the overall statement to be proved, namely:

"**P(n)** is true, **for all** $n \geqslant 1$",

where infinitely many individual statements have been compressed into a single compound statement, and

(ii) the *induction step*, where we

– assume some *particular* **P**(k) is known to be true, and

– show that the next statement **P**(k + 1) is then automatically true.

To underline this distinction we consistently use a different "dummy variable" (namely "k") in the latter case. This distinction is a psychological ploy rather than a logical necessity. However, we recommend that readers should imitate this distinction (at least initially).

6.2. 'Mathematical induction' and 'scientific induction'

The idea of a "list that goes on for ever" arose in the sequence of powers of 4 back in Problem **16**, where we asked

Do the two sequences arising from successive powers of 4:

- the *leading* digits:

$$4, 2, 6, 2, 1, 4, 2, 6, 2, 1, 4, \ldots,$$

and

- the *units* digits:
$$4, 6, 4, 6, 4, 6, 4, 6, \ldots,$$

really "repeat for ever" as they seem to?

This example illustrates the most basic misconception that sometimes arises concerning *mathematical induction* – namely to confuse it with the kind of pattern spotting that is often called 'scientific induction'.

In science (as in everyday life), we routinely infer that something that is observed to occur repeatedly, apparently without exception (such as the sun rising every morning; or the Pole star seeming to be fixed in the night sky) may be taken as a "fact". This kind of "scientific induction" makes perfect sense when trying to understand the world around us – even though the inference is not warranted in a strictly logical sense.

Proof by mathematical induction is quite different. Admittedly, it often requires intelligent guesswork at a preliminary stage to make a guess that allows us to formulate precisely what it is that we should be trying to prove. But this initial guess is separate from the proof, which remains a strictly deductive construction. For example,

> the fact that "1", "1 + 3", "1 + 3 + 5", "1 + 3 + 5 + 7", etc. all appear to be successive squares gives us an idea that perhaps the identity
>
> **P**(n): $1 + 3 + 5 + \cdots + (2n - 1) = n^2$
>
> is true, **for all $n \geqslant 1$**.

This guess is needed before we can start the proof by mathematical induction. **But the process of guessing is not part of the proof.** And until we construct the "proof by induction" (Problem **231**), we cannot be sure that our guess is correct.

The danger of confusing 'mathematical induction' and 'scientific induction' may be highlighted to some extent if we consider the *proof* in Problem **76** above that "we can always construct ever larger prime numbers", and contrast it with an observation (see Problem **228** below) that is often used in its place – even by authors who should know better.

In Problem **76** we gave a strict construction by *mathematical induction*:

- we first showed how to begin (with $p_1 = 2$ say);
- then we showed how, given any finite list of distinct prime numbers $p_1, p_2, p_3, \ldots, p_k$, it is always possible to construct a **new** prime p_{k+1} (as the *smallest* prime number dividing $N_k = p_1 \times p_2 \times p_3 \times \cdots \times p_k + 1$).

This construction was very carefully worded, so as to be logically correct.

In contrast, one often finds lessons, books and websites that present the essential idea in the above proof, but "simplify" it into a form that encourages anti-mathematical "pattern-spotting" which is all-too-easily misconstrued. For example, some books present the sequence

$(2;) \ 2+1 = \mathbf{3}; \ 2 \times 3 + 1 = \mathbf{7}; \ 2 \times 3 \times 5 + 1 = \mathbf{31}; \ 2 \times 3 \times 5 \times 7 + 1 = \mathbf{211}; \ldots$

as a way of generating more and more primes.

Problem 228

(a) Are 3, 7, 31, 211 all prime?

(b) Is $2 \times 3 \times 5 \times 7 \times 11 + 1$ prime?

(c) Is $2 \times 3 \times 5 \times 7 \times 11 \times 13 + 1$ prime? △

We have already met two excellent historical examples of the dangers of plausible pattern-spotting in connection with Problem **118**. There you proved that:

"if $2^n - 1$ is prime, then n must be prime."

You then showed that $2^2 - 1$, $2^3 - 1$, $2^5 - 1$, $2^7 - 1$ are all prime, but that $2^{11} - 1 = 2047 = 23 \times 89$ is **not**. This underlines the need to avoid jumping to (possibly false) conclusions, and never to confuse a statement with its converse.

In the same problem you showed that:

"if $a^b + 1$ is to be prime and $a \neq 1$, then a must be even, and b must be a power of 2."

You then looked at the simplest family of such candidate primes namely the sequence of Fermat numbers f_n:

$f_0 = 2^1 + 1 = 3, f_1 = 2^2 + 1 = 5, f_2 = 2^4 + 1 = 17, f_3 = 2^8 + 1 = 257, f_4 = 2^{16} + 1.$

It turned out that, although f_0, f_1, f_2, f_3, f_4 are all prime, and although Fermat (1601–1665) claimed (in a letter to Marin Mersenne (1588–1648)) that **all** Fermat numbers f_n are prime, we have yet to discover a sixth Fermat prime!

There are times when a mathematician may appear to guess a general result on the basis of what looks like very modest evidence (such as noticing that it appears to be true in a few small cases). Such "informed guesses" are almost always rooted in other experience, or in some unnoticed feature of the particular situation, or in some striking analogy: that is, an apparent pattern strikes a chord for reasons that go way beyond the mere numbers. However those with less experience need to realise that apparent patterns or trends are often no more than numerical accidents.

Pell's equation (John Pell (1611–1685)) provides some dramatic examples.

- If we evaluate the expression "$n^2 + 1$" for $n = 1, 2, 3, \ldots$, we may notice that the outputs $2, 5, 10, 17, 26, \ldots$ never give a perfect square. And this is to be expected, since the next square after n^2 is
$$(n+1)^2 = n^2 + 2n + 1,$$
and this is always greater than $n^2 + 1$.

- However, if we evaluate "$991n^2 + 1$" for $n = 1, 2, 3, \ldots$, we may observe that the outputs never seem to include a perfect square. But this time there is no obvious reason why this should be so – so we may anticipate that this is simply an accident of "small" numbers. And we should hesitate to change our view, even though this accident goes on happening for a very, very, very long time: the smallest value of n for which $991n^2 + 1$ gives rise to a perfect square is apparently
$$n = 12\,055\,735\,790\,331\,359\,447\,442\,538\,767.$$

6.3. Proof by mathematical induction II

Even where there is no confusion between *mathematical induction* and 'scientific induction', students often fail to appreciate the essence of "proof by induction". Before illustrating this, we repeat the basic structure of such a proof.

A mathematical result, or formula, often involves a parameter n, where n can be any positive integer. In such cases, what is presented as a single result, or formula, is a short way of writing an *infinite family of results*. The proof that such a result is correct therefore requires us to prove *infinitely many results at once*. We repeat that our only hope of achieving such a mind-boggling feat is

- to formulate the stated result for each value of n separately: that is, as a statement $\mathbf{P}(n)$ which depends on n;

- then to use bare hands to check the "beginning" - namely that the simplest case (usually **P**(1)) is true;
- finally to find some way of demonstrating that,
  - as soon as we know that **P**(k) is true, for some (unknown) $k \geqslant 1$,
  - we can prove that the next result **P**($k+1$) is then automatically true.

We can then conclude that

"**every** statement **P**(n) is true",

or that

"**P**(n) is true, **for all** $n \geqslant 1$".

Problem 229 Prove the statement:

"$5^{2n+2} - 24n - 25$ is divisible by 576, for all $n \geqslant 1$". △

When trying to construct proofs in private, one is free to write anything that helps as 'rough work'. But the intended thrust of Problem **229** is two-fold:

- to introduce the habit of distinguishing clearly between
  (i) the statement **P**(n) for a particular n, and
  (ii) the statement to be proved – namely "**P**(n) is true, for all $n \geqslant 1$"; and
- to draw attention to the "induction step" (i.e. the third bullet point above), where
  (i) we assume that some unspecified **P**(k) is known to be true, and
  (ii) seek to prove that the next statement **P**($k+1$) must then be true.

The central lesson in completing the "induction step" is to recognize that:

> to prove that **P**($k+1$) is true,
> **one has to start by looking at what P($k+1$) says.**

In Problem **229** **P**($k+1$) says that

"$5^{2(k+1)+2} - 24(k+1) - 25$ is divisible by 576".

Hence one has to start the induction step with the relevant expression

$$5^{2(k+1)+2} - 24(k+1) - 25,$$

and look for some way to rearrange this into a form where one can use $\mathbf{P}(k)$ (which we assume is already known to be true, and so are free to use).

It is in general a false strategy to work the other way round – by "starting with $\mathbf{P}(k)$, and then fiddling with it to try to get $\mathbf{P}(k+1)$". (This strategy can be made to work in the simplest cases; but it does not work in general, and so is a bad habit to get into.) So the induction step should **always** start with the hypothesis of $\mathbf{P}(k+1)$.

The next problem invites you to prove the formula for the sum of the angles in any polygon. The result is well-known; yet we are fairly sure that the reader will never have seen a correct proof. So the intention here is for you to recognise the basic character of the result, to identify the flaws in what you may until now have accepted as a proof, and to try to find some way of producing a general proof.

Problem 230 Prove by induction the statement:

"for each $n \geqslant 3$, the angles of any n-gon in the plane have sum equal to $(n-2)\pi$ radians." △

The formulation certainly involves a parameter $n \geqslant 3$; so you clearly need to begin by formulating the statement $\mathbf{P}(n)$. For the proof to have a chance of working, finding the right formulation involves a modest twist! So if you get stuck, it may be worth checking the first couple of lines of the solution.

No matter how $\mathbf{P}(n)$ is formulated, you should certainly know how to prove the statement $\mathbf{P}(3)$ (essentially the formula for the sum of the angles in a triangle). But it is far from obvious how to prove the "induction step":

"if we know that $\mathbf{P}(k)$ is true for some particular $k \geqslant 1$, then $\mathbf{P}(k+1)$ must also be true".

When tackling the induction step, we certainly cannot start with $\mathbf{P}(k)$! The statement $\mathbf{P}(k+1)$ says something about polygons with $k+1$ sides: and there is no way to obtain a typical $(k+1)$-gon by fiddling with some statement about polygons with k sides. (If you start with a k-gon, you can of course draw a triangle on one side to get a $(k+1)$-gon; but this is a very special construction, and there is no easy way of knowing whether all $(k+1)$-gons can be obtained in this way from some k-gon.) The whole thrust of mathematical induction is that we must always start the induction step by thinking about the *hypothesis* of $\mathbf{P}(k+1)$ – that is in this case, by considering an arbitrary

$(k+1)$-gon and then finding some guaranteed way of "reducing" it in order to make use of **P**(k).

The next two problems invite you to prove some classical algebraic identities. Most of these may be familiar. The challenge here is to think carefully about the way you lay out your induction proof, to learn from the examples above, and (later) to learn from the detailed solutions provided.

Problem 231 Prove by induction the statement:

"$1 + 3 + 5 + \cdots + (2n-1) = n^2$ holds, for all $n \geqslant 1$". △

The summation in Problem **231** was known to the ancient Greeks. The mystical *Pythagorean* tradition (which flourished in the centuries after Pythagoras) explored the character of integers through the 'spatial figures' which they formed. For example, if we arrange each successive integer as a new line of dots in the plane, then the sum "$1 + 2 + 3 + \cdots + n$" can be seen to represent a *triangular* number. Similarly, if we arrange each odd number $2k-1$ in the sum "$1 + 3 + 5 + \cdots + (2n-1)$" as a "$k$-by-$k$ reverse L-shape", or *gnomon* (a word which we still use to refer to the L-shaped piece that casts the shadow on a sundial), then the accumulated L-shapes build up an n by n square of dots – the "1" being the dot in the top left hand corner, the "3" being the reverse L-shape of 3 dots which make this initial "1" into a 2 by 2 square, the "5" being the reverse L-shape of 5 dots which makes this 2 by 2 square into a 3 by 3 square, and so on. Hence the sum "$1 + 3 + 5 + \cdots + (2n-1)$" can be seen to represent a *square* number.

There is much to be said for such geometrical illustrations; but there is no escape from the fact that they hide behind an *ellipsis* (the three dots which we inserted in the sum between "5" and "$2n-1$", which were then summarised when arranging the reverse L-shapes by ending with the words "and so on"). Proof by mathematical induction, and its application in Problem **231**, constitute a formal way of avoiding both the appeal to pictures, and the hidden *ellipsis*.

Problem 232 The sequence

$$2, 5, 13, 35, \ldots$$

is defined by its first two terms $u_0 = 2$, $u_1 = 5$, and by the recurrence relation:

$$u_{n+2} = 5u_{n+1} - 6u_n.$$

(a) Guess a closed formula for the nth term u_n.

(b) Prove by induction that your guess in (a) is correct for all $n \geqslant 0$. △

Problem 233 The sequence of Fibonacci numbers

$$0, 1, 1, 2, 3, 5, 8, 13, \ldots$$

is defined by its first two terms $F_0 = 0$, $F_1 = 1$, and by the recurrence relation:
$$F_{n+2} = F_{n+1} + F_n \text{ when } n \geqslant 0.$$
Prove by induction that, for all $n \geqslant 0$,

$$F_n = \frac{\alpha^n - \beta^n}{\sqrt{5}}, \text{ where } \alpha = \frac{1+\sqrt{5}}{2} \text{ and } \beta = \frac{1-\sqrt{5}}{2}. \qquad \triangle$$

Problem 234 Prove by induction that

$$5^{2n+1} \cdot 2^{n+2} + 3^{n+2} \cdot 2^{2n+1}$$

is divisible by 19, for all $n \geqslant 0$. $\qquad \triangle$

Problem 235 Use mathematical induction to prove that each of these identities holds, for all $n \geqslant 1$:

(a) $1 + 2 + 3 + \cdots + n = \frac{n(n+1)}{2}$

(b) $\frac{1}{1 \cdot 2} + \frac{1}{2 \cdot 3} + \frac{1}{3 \cdot 4} + \cdots + \frac{1}{n(n+1)} = 1 - \frac{1}{n+1}$

(c) $1 + q + q^2 + q^3 + \cdots + q^{n-1} = \frac{1}{1-q} - \frac{q^n}{1-q}$

(d) $0 \cdot 0! + 1 \cdot 1! + 2 \cdot 2! + \cdots + (n-1) \cdot (n-1)! = n! - 1$

(e) $(\cos\theta + i\sin\theta)^n = \cos n\theta + i\sin n\theta.$ $\qquad \triangle$

Problem 236 Prove by induction the statement:

"$(1 + 2 + 3 + \cdots + n)^2 = 1^3 + 2^3 + 3^3 + \cdots + n^3$, for all $n \geqslant 1$". $\qquad \triangle$

We now know that, for all $n \geqslant 1$:

$$1 + 1 + 1 + \cdots + 1 \ (n \text{ terms}) = n.$$

And if we sum these "outputs" (that is, the first n natural numbers), we get the nth triangular number:

$$1 + 2 + 3 + \cdots + n = \frac{n(n+1)}{2} = T_n.$$

6.3. Proof by mathematical induction II

The next problem invites you to find the sum of these "outputs": that is, to find the sum of the first n triangular numbers.

Problem 237

(a) Experiment and guess a formula for the sum of the first n triangular numbers:
$$T_1 + T_2 + T_3 + \cdots + T_n = 1 + 3 + 6 + \cdots + \frac{n(n+1)}{2}.$$

(b) Prove by induction that your guessed formula is correct for all $n \geqslant 1$. △

We now know closed formulae for

"$1 + 2 + 3 + \cdots + n$"

and for

"$1 \cdot 2 + 2 \cdot 3 + 3 \cdot 4 + \cdots + (n-1)n$".

The next problem hints firstly that these identities are part of something more general, and secondly that these results allow us to find identities for the sum of the first n squares:

$$1^2 + 2^2 + 3^2 + \cdots + n^2$$

for the first n cubes:
$$1^3 + 2^3 + 3^3 + \cdots + n^3$$

and so on.

Problem 238

(a) Note that
$$1 \cdot 2 + 2 \cdot 3 + 3 \cdot 4 + \cdots + n(n+1) = 1 \cdot (1+1) + 2 \cdot (2+1) + 3 \cdot (3+1) + \cdots + n \cdot (n+1).$$

Use this and the result of Problem **237** to derive a formula for the sum:
$$1^2 + 2^2 + 3^2 + \cdots + n^2.$$

(b) Guess and prove a formula for the sum
$$1 \cdot 2 \cdot 3 + 2 \cdot 3 \cdot 4 + 3 \cdot 4 \cdot 5 + \cdots + (n-2)(n-1)n.$$

Use this to derive a closed formula for the sum:
$$1^3 + 2^3 + 3^3 + \cdots + n^3.$$ △

It may take a bit of effort to digest the statement in the next problem. It extends the idea behind the "method of undetermined coefficients" that is discussed in **Note 2** to the solution of Problem **237**(a).

Problem 239

(a) Given $n + 1$ distinct real numbers

$$a_0, a_1, a_2, \ldots, a_n,$$

find all possible polynomials of degree n which satisfy

$$f(a_0) = f(a_1) = f(a_2) = \cdots = f(a_{n-1}) = 0, \ f(a_n) = b$$

for some specified number b.

(b) For each $n \geq 1$, prove the following statement:

Given two labelled sets of $n + 1$ real numbers

$$a_0, a_1, a_2, \ldots, a_n,$$

and

$$b_0, b_1, b_2, \ldots, b_n,$$

where the a_i are all distinct (but the b_i need not be), there exists a polynomial f_n of degree n, such that

$$f_n(a_0) = b_0, \ f_n(a_1) = b_1, \ f_n(a_2) = b_2, \ \ldots, \ f_n(a_n) = b_n. \qquad \triangle$$

We end this subsection with a mixed bag of three rather different induction problems. In the first problem the induction step involves a simple construction of a kind we will meet later.

Problem 240 A country has only 3 cent and 4 cent coins.

(a) Experiment to decide what seems to be the largest amount, N cents, that **cannot** be paid directly (without receiving change).

(b) Prove by induction that "n cents can be paid directly for each $n > N$".
$\qquad \triangle$

Problem 241

(a) Solve the equation $z + \frac{1}{z} = 1$. Calculate z^2, and check that $z^2 + \frac{1}{z^2}$ is also an integer.

(b) Solve the equation $z + \frac{1}{z} = 2$. Calculate z^2, and check that $z^2 + \frac{1}{z^2}$ is also an integer.

(c) Solve the equation $z + \frac{1}{z} = 3$. Calculate z^2, and check that $z^2 + \frac{1}{z^2}$ is also an integer.

(d) Solve the equation $z + \frac{1}{z} = k$, where k is an integer. Calculate z^2, and check that $z^2 + \frac{1}{z^2}$ is also an integer.

(e) Prove that if a number z has the property that $z + \frac{1}{z}$ is an integer, then $z^n + \frac{1}{z^n}$ is also an integer for each $n \geqslant 1$. △

Problem 242 Let p be any prime number. Use induction to prove:

"$n^p - n$ is divisible by p for all $n \geqslant 1$". △

6.4. Infinite geometric series

Elementary mathematics is predominantly about equations and identities. But it is often impossible to capture important mathematical relations in the form of exact equations. This is one reason why inequalities become more central as we progress; another reason is because inequalities allow us to make precise statements about certain infinite processes.

One of the simplest infinite process arises in the formula for the "sum" of an infinite *geometric series*:

$$1 + r + r^2 + r^3 + \cdots + r^n + \cdots \quad \text{(for ever)}.$$

Despite the use of the familiar-looking "+" signs, this can be no ordinary addition. Ordinary addition is defined for **two** summands; and by repeating the process, we can add **three** summands (thanks in part to the associative law of addition). We can then add four, or **any finite number** of summands. But this does not allow us to "add" infinitely many terms as in the above sum. To get round this we combine ordinary addition (of finitely many terms) and simple inequalities to find a new way of giving a meaning to the above "endless sum". In Problem **116** you used the factorisation

$$r^{n+1} - 1 = (r-1)(1 + r + r^2 + r^3 + \cdots + r^n)$$

to derive the closed formula:

$$1 + r + r^2 + r^3 + \cdots + r^n = \frac{1 - r^{n+1}}{1 - r}.$$

This formula for the sum of a finite geometric series can be rewritten in the form
$$1 + r + r^2 + r^3 + \cdots + r^n = \frac{1}{1-r} - \frac{r^{n+1}}{1-r}.$$
At first sight, this may not look like a clever move! However, it separates the part that is independent of n from the part on the RHS that depends on n; and it allows us to see how the second part behaves as n gets large:

when $|r| < 1$, successive powers of r get smaller and smaller and converge rapidly towards 0,

so the above form of the identity may be interpreted as having the form:
$$1 + r + r^2 + r^3 + \cdots + r^n = \frac{1}{1-r} - (\text{an ``error term''}).$$
Moreover if $|r| < 1$, then the "error term" converges towards 0 as $n \to \infty$. In particular, if $1 > r > 0$, the error term is always positive, so we have proved, for all $n \geqslant 1$:
$$1 + r + r^2 + r^3 + \cdots + r^n < \frac{1}{1-r}$$
and

the difference between the two sides tends rapidly to 0 as $n \to \infty$.

We then make the natural (but bold) step to interpret this, when $|r| < 1$, as offering a new **definition** which explains precisely what is meant by the **endless** sum
$$1 + r + r^2 + r^3 + \cdots \text{ (\textbf{for ever})},$$
declaring that, when $|r| < 1$,
$$1 + r + r^2 + r^3 + \cdots \text{ (\textbf{for ever})} = \frac{1}{1-r}.$$
More generally, if we multiply every term by a, we see that
$$a + ar + ar^2 + ar^3 + \cdots \text{ (\textbf{for ever})} = \frac{a}{1-r}.$$

Problem 243 Interpret the recurring decimal $0.037037037\cdots$ (for ever) as an infinite geometric series, and hence find its value as a fraction. △

6.5. Some classical inequalities

Problem 244 Interpret the following endless processes as infinite geometric series.

(a) A square cake is cut into four quarters, with two perpendicular cuts through the centre, parallel to the sides. Three people receive one quarter each – leaving a smaller square piece of cake. This smaller piece is then cut in the same way into four quarters, and each person receives one (even smaller) piece – leaving an even smaller residual square piece, which is then cut in the same way. And so on for ever. What fraction of the original cake does each person receive as a result of this endless process?

(b) I give you a whole cake. Half a minute later, you give me half the cake back. One quarter of a minute later, I return one quarter of the cake to you. One eighth of a minute later you return one eighth of the cake to me. And so on. Adding the successive time intervals, we see that

$$\frac{1}{2} + \frac{1}{4} + \frac{1}{8} + \cdots \text{ (for ever)} = 1,$$

so the whole process is completed in exactly 1 minute. How much of the cake do I have at the end, and how much do you have? △

Problem 245 When John von Neumann (1903–1957) was seriously ill in hospital, a visitor tried (rather insensitively) to distract him with the following elementary mathematics problem.

> Have you heard the one about the two trains and the fly? Two trains are on a collision course on the same track, each travelling at 30 km/h. A super-fly starts on Train A when the trains are 120 km apart, and flies at a constant speed of 50 km/h – from Train A to Train B, then back to Train A, and so on. Eventually the two trains collide and the fly is squashed. How far did the fly travel before this sad outcome? △

6.5. Some classical inequalities

The fact that our formula for the sum of a geometric series gives us an *exact* sum is very unusual – and hence very precious. For almost all other infinite series – no matter how natural, or beautiful, they may seem – you can be fairly sure that there is no obvious exact formula for the value of the sum. Hence in those cases where we happen to know the exact value, you may infer that it took the best efforts of some of the finest mathematical minds to discover what we know.

One way in which we can make a little progress in estimating the value of an infinite series is to obtain an *inequality* by comparing the given sum with a geometric series.

Problem 246

(a)(i) Explain why
$$\frac{1}{3^2} < \frac{1}{2^2},$$
so
$$\frac{1}{2^2} + \frac{1}{3^2} < \frac{2}{2^2} = \frac{1}{2}.$$

(ii) Explain why $\frac{1}{5^2}, \frac{1}{6^2}, \frac{1}{7^2}$ are all $< \frac{1}{4^2}$, so
$$\frac{1}{4^2} + \frac{1}{5^2} + \frac{1}{6^2} + \frac{1}{7^2} < \frac{4}{4^2} = \frac{1}{4}.$$

(b) Use part (a) to prove that
$$\frac{1}{1^2} + \frac{1}{2^2} + \frac{1}{3^2} + \cdots + \frac{1}{n^2} < 2, \quad \text{for all } n \geqslant 1.$$

(c) Conclude that the endless sum
$$\frac{1}{1^2} + \frac{1}{2^2} + \frac{1}{3^2} + \cdots + \frac{1}{n^2} + \cdots \quad \text{(for ever)}$$
has a definite value, and that this value lies somewhere between $\frac{17}{12}$ and 2.
△

The next problem presents a rather different way of deriving a similar equality. Once the relevant inequality has been guessed, or given (see Problem **247**(a) and (b)), the proof by mathematical induction is often relatively straightforward. And after a little thought about Problem **246**, it should be clear that much of the inaccuracy in the general inequality arises from the rather poor approximations made for the first few terms (when $n = 1$, when $n = 2$, when $n = 3$, etc.); hence by keeping the first few terms as they are, and only approximating for $n \geqslant 2$, or $n \geqslant 3$, or $n \geqslant 4$, we can often prove a sharper result.

Problem 247

(a) Prove by induction that
$$\frac{1}{1^2} + \frac{1}{2^2} + \frac{1}{3^2} + \cdots + \frac{1}{n^2} \leqslant 2 - \frac{1}{n}, \quad \text{for all } n \geqslant 1.$$

(b) Prove by induction that
$$\frac{1}{1^2}+\frac{1}{2^2}+\frac{1}{3^2}+\cdots+\frac{1}{n^2} < 1.68 - \frac{1}{n}, \text{ for all } n \geqslant 4.\qquad\triangle$$

The infinite sum
$$\frac{1}{1^2}+\frac{1}{2^2}+\frac{1}{3^2}+\cdots+\frac{1}{n^2}+\cdots \text{ (for ever)}$$
is a historical classic, and has many instructive stories to tell. Recall that, in Problems **54, 62, 63, 236, 237, 238** you found closed formulae for the sums
$$1+2+3+\cdots+n$$
$$1^2+2^2+3^2+\cdots+n^2$$
$$1^3+2^3+3^3+\cdots+n^3$$
and for the sums
$$1\times 2+2\times 3+3\times 4+\cdots+(n-1)n$$
$$1\times 2\times 3+2\times 3\times 4+3\times 4\times 5+\cdots+(n-2)(n-1)n.$$
Each of these expressions has a "natural" feel to it, and invites us to believe that there must be an equally natural compact answer representing the sum. In Problem **235** you took this idea one step further by finding a beautiful closed expression for the sum
$$\frac{1}{1\cdot 2}+\frac{1}{2\cdot 3}+\frac{1}{3\cdot 4}+\cdots+\frac{1}{n(n+1)} = 1-\frac{1}{n+1}.$$
When we began to consider *infinite* series, we found the elegant closed formula
$$1+r+r^2+r^3+\cdots+r^n = \frac{1}{1-r}-\frac{r^{n+1}}{1-r}.$$
We then observed that the final term on the RHS could be viewed as an "error term", indicating the amount by which the LHS **differs from** $\frac{1}{1-r}$, and noticed that, for any given value of r between -1 and $+1$, this error term "tends towards 0 as the power n increases". We interpreted this as indicating that one could assign a value to the endless sum
$$1+r+r^2+r^3+\cdots \text{ (for ever)} = \frac{1}{1-r}.$$
In the same way, in the elegant closed formula
$$\frac{1}{1\cdot 2}+\frac{1}{2\cdot 3}+\frac{1}{3\cdot 4}+\cdots+\frac{1}{n(n+1)} = 1-\frac{1}{n+1}$$

the final term on the RHS indicates the amount by which the finite sum on the **LHS differs from 1**; and since this "error term" tends towards 0 as n increases, we may assign a value to the endless sum

$$\frac{1}{1\cdot 2} + \frac{1}{2\cdot 3} + \frac{1}{3\cdot 4} + \cdots \text{ (for ever)} = 1.$$

It is therefore natural to ask whether other infinite series, such as

$$\frac{1}{1^2} + \frac{1}{2^2} + \frac{1}{3^2} + \cdots + \frac{1}{n^2} + \cdots \text{ (for ever)}$$

may also be assigned some natural finite value. And since the series is purely numerical (without any variable parameters, such as the "r" in the geometric series formula), this answer should be a strictly *numerical* answer. And it should be **exact** – though all we have managed to prove so far (in Problems **246** and **247**) is that this numerical answer lies somewhere between $\frac{17}{12}$ and 1.68.

This question arose naturally in the middle of the seventeenth century, when mathematicians were beginning to explore all sorts of infinite series (or "sums that go on for ever"). With a little more work in the spirit of Problems **246** and **247** one could find a much more accurate approximate value. But what is wanted is an *exact* expression, not an unenlightening decimal approximation. This aspiration has a serious mathematical basis, and is not just some purist preference for elegance. The actual decimal value is very close to

$$1.649934\cdots.$$

But this conveys no structural information. One is left with no hint as to **why** the sum has this value. In contrast, the eventual form of the exact expression suggests connections whose significance remains of interest to this day.

The greatest minds of the seventeenth and early eighteenth century tried to find an exact value for the infinite sum – and failed. The problem became known as the *Basel problem* (after Jakob Bernoulli (1654–1705) who popularised the problem in 1689 – one of several members of the Bernoulli family who were all associated with the University in Basel). The problem was finally solved in 1735 – in truly breathtaking style – by the young Leonhard Euler (1707–1783) (who was at the time also in Basel). The answer

$$\frac{\pi^2}{6}$$

illustrates the final sentence of the preceding paragraph in unexpected ways, which we are still trying to understand.

6.5. Some classical inequalities

In the next problem you are invited to apply similar ideas to an even more important series. Part (a) provides a relatively crude first analysis. Part (b) attacks the same question; but it does so using algebra and induction (rather than the formula for the sum of a geometric series) in a way that is then further refined in part (c).

Problem 248

(a)(i) Choose a suitable r and prove that
$$\frac{1}{1!} + \frac{1}{2!} + \cdots + \frac{1}{n!} < 1 + r + r^2 + \cdots + r^{n-1} < 2.$$

(ii) Conclude that
$$\frac{1}{0!} + \frac{1}{1!} + \frac{1}{2!} + \cdots + \frac{1}{n!} < 3, \text{ for every } n \geqslant 0,$$
and hence that the endless sum
$$\frac{1}{0!} + \frac{1}{1!} + \frac{1}{2!} + \cdots + \frac{1}{n!} + \cdots \text{ (for ever)}$$
can be assigned a value "e" satisfying $2 < e \leqslant 3$.

(b)(i) Prove by induction that
$$\frac{1}{0!} + \frac{1}{1!} + \frac{1}{2!} + \cdots + \frac{1}{n!} \leqslant 3 - \frac{1}{n.n!}, \text{ for all } n \geqslant 1.$$

(ii) Use part (i) to conclude that the endless sum
$$\frac{1}{0!} + \frac{1}{1!} + \frac{1}{2!} + \cdots + \frac{1}{n!} + \cdots \text{ (for ever)}$$
can be assigned a definite value "e", and that this value lies somewhere between 2.5 and 3.

(c) (It may help to read the **Note** at the start of the solution to part (c) before attempting parts (c), (d).)

(i) Prove by induction that
$$\frac{1}{0!} + \frac{1}{1!} + \frac{1}{2!} + \cdots + \frac{1}{n!} \leqslant 2.75 - \frac{1}{n.n!}, \text{ for all } n \geqslant 2.$$

(ii) Use part (i) to conclude that the endless sum
$$\frac{1}{0!} + \frac{1}{1!} + \frac{1}{2!} + \cdots + \frac{1}{n!} + \cdots \text{ (for ever)}$$

can be assigned a definite value "e", and that this value lies somewhere between 2.6 and 2.75.

(d)(i) Prove by induction that

$$\frac{1}{0!} + \frac{1}{1!} + \frac{1}{2!} + \cdots + \frac{1}{n!} \leqslant 2.722\cdots \text{ (for ever)} - \frac{1}{n.n!}, \text{ for all } n \geqslant 3.$$

(ii) Use part (i) to conclude that the endless sum

$$\frac{1}{0!} + \frac{1}{1!} + \frac{1}{2!} + \cdots + \frac{1}{n!} + \cdots \text{ (for ever)}$$

can be assigned a definite value "e", and that this value lies somewhere between 2.708 and 2.7222$\cdots$ (for ever). △

We end this section with one more inequality in the spirit of this section, and two rather different inequalities whose significance will become clear later.

Problem 249 Prove by induction that

$$\tfrac{1}{\sqrt{1}} + \tfrac{1}{\sqrt{2}} + \tfrac{1}{\sqrt{3}} + \cdots + \tfrac{1}{\sqrt{n}} \geqslant \sqrt{n}, \text{ for all } n \geqslant 1.$$ △

Problem 250 Let a, b be real numbers such that $a \neq b$, and $a + b > 0$. Prove by induction that

$$2^{n-1}(a^n + b^n) \geqslant (a + b)^n, \text{ for all } n \geqslant 1.$$ △

Problem 251 Let x be any real number $\geqslant -1$. Prove by induction that

$$(1 + x)^n \geqslant 1 + nx, \text{ for all } n \geqslant 1.$$ △

6.6. The harmonic series

> *The great foundation of mathematics is*
> **the principle of contradiction**, or **of identity**,
> *that is to say that a statement*
> *cannot be true and false at the same time,*
> *and that thus A **is** A, **and cannot be not** A.*
> *And this single principle is enough to prove*
> *the whole of arithmetic and the whole of geometry,*
> *that is to say all mathematical principles.*
> Gottfried W. Leibniz (1646–1716)

6.6. The harmonic series

We have seen how some infinite series, or sums that go on for ever, can be assigned a finite value for their sum:

$$1 + r + r^2 + r^3 + \cdots \text{ (for ever)} = \frac{1}{1-r}$$

$$\frac{1}{1\cdot 2} + \frac{1}{2\cdot 3} + \frac{1}{3\cdot 4} + \cdots + \frac{1}{n(n+1)} + \cdots \text{ (for ever)} = 1$$

$$\frac{1}{1^2} + \frac{1}{2^2} + \frac{1}{3^2} + \cdots + \frac{1}{n^2} + \cdots \text{ (for ever)} = \frac{\pi^2}{6}$$

$$\frac{1}{0!} + \frac{1}{1!} + \frac{1}{2!} + \cdots + \frac{1}{n!} + \cdots \text{ (for ever)} = e.$$

We say that these series *converge* (meaning that they can be assigned a finite value).

This section is concerned with another very natural series, the so-called *harmonic series*

$$\frac{1}{1} + \frac{1}{2} + \frac{1}{3} + \cdots + \frac{1}{n} + \cdots \text{ (for ever)}.$$

It is not entirely clear why this is called the harmonic series. The natural overtones that arise in connection with plucking a stretched string (as with a guitar or a harp) have wavelengths that are $\frac{1}{2}$ the basic wavelength, or $\frac{1}{3}$ of the basic wavelength, and so on. It is also true that, just as each term of an arithmetic series is the *arithmetic mean* of its two neighbours, and each term of a geometric series is the *geometric mean* of its two neighbours, so each term of the harmonic series after the first is equal to the *harmonic mean* (see Problems **85**, **89**) of its two neighbours:

$$\frac{1}{k} = \frac{2}{\left(\frac{1}{k-1}\right)^{-1} + \left(\frac{1}{k+1}\right)^{-1}}.$$

Unlike the first two series above, there is no obvious closed formula for the finite sum

$$s_n = \frac{1}{1} + \frac{1}{2} + \frac{1}{3} + \cdots + \frac{1}{n}.$$

Certainly the sequence of successive sums

$$s_1 = 1,\ s_2 = \frac{3}{2},\ s_3 = \frac{11}{6},\ s_4 = \frac{25}{12},\ s_5 = \frac{137}{60}, \ldots$$

does not suggest any general pattern.

Problem 252 Suppose we denote by S the "value" of the endless sum

$$\frac{1}{1} + \frac{1}{2} + \frac{1}{3} + \cdots + \frac{1}{n} + \cdots \quad \text{(for ever)}$$

(i) Write out the endless sum corresponding to "$\frac{1}{2}S$".

(ii) Remove the terms of this endless sum from the endless sum S, to obtain another endless sum corresponding to "$S - \frac{1}{2}S$" $= \frac{1}{2}S$.

(iii) Compare the first term of the series in (i) (namely $\frac{1}{2}$) with the first term in the series in (ii) (namely 1); compare the second term in the series in (i) with the second term in the series in (ii); and so on. What do you notice?
△

The Leibniz quotation above emphasizes that the reliability of mathematics stems from a single principle – namely the refusal to tolerate a contradiction. We have already made explicit use of this principle from time to time (see, for example, the solution to Problem **125**). The message is simple: whenever we hit a contradiction, we know that we have "gone wrong" – either by making an error in calculation or logic, or by beginning with a false assumption. In Problem **252** the observations you were expected to make are paradoxical: you obtained two different series, which both correspond to "$\frac{1}{2}S$", but every term in one series is larger than the corresponding term in the other! What one can conclude may not be entirely clear. But it is certainly clear that something is wrong: we have somehow created a contradiction. The three steps ((i), (ii), (iii)) appear to be relatively sensible. But the final observation "$\frac{1}{2}S < \frac{1}{2}S$" (since $\frac{1}{2} < 1$, $\frac{1}{4} < \frac{1}{3}$, etc.) makes no sense. And the only obvious assumption we have made is to assume that the endless sum

$$\frac{1}{1} + \frac{1}{2} + \frac{1}{3} + \cdots + \frac{1}{n} + \cdots \quad \text{(for ever)}$$

can be assigned a value "S", which can then be manipulated *as though it were a number*.

The conclusion would seem to be that, whether or not the endless sum has a meaning, it cannot be assigned a value in this way. We say that the series *diverges*. Each finite sum

$$\frac{1}{1} + \frac{1}{2} + \frac{1}{3} + \cdots + \frac{1}{n}$$

has a value, and these values "grow more and more slowly" as n increases:

- the first term immediately makes the sum $= 1$
- it takes 4 terms to obtain a sum > 2;
- it takes 11 terms to obtain a sum > 3; and

- it takes 12 367 terms before the series reaches a sum > 10.

However, this slow growth is not enough to guarantee that the corresponding endless sum corresponds to a finite numerical value.

The danger signals should already have been apparent in Problem **249**, where you proved that

$$\frac{1}{\sqrt{1}} + \frac{1}{\sqrt{2}} + \frac{1}{\sqrt{3}} + \cdots + \frac{1}{\sqrt{n}} \geq \sqrt{n}$$

The nth term $\frac{1}{\sqrt{n}}$ tends to 0 as n increases; so the finite sums grow ever more slowly as n increases. However, the LHS can be made larger than any integer K simply by taking K^2 terms. Hence there is no way to assign a finite value to the endless sum

$$\frac{1}{\sqrt{1}} + \frac{1}{\sqrt{2}} + \frac{1}{\sqrt{3}} + \cdots + \frac{1}{\sqrt{n}} + \cdots \quad \text{(for ever)}.$$

Problem 253

(a)(i) Explain why
$$\frac{1}{2} + \frac{1}{3} < 1.$$

(ii) Explain why
$$\frac{1}{4} + \frac{1}{5} + \frac{1}{6} + \frac{1}{7} < 1.$$

(iii) Extend parts (i) and (ii) to prove that
$$\frac{1}{1} + \frac{1}{2} + \frac{1}{3} + \cdots + \frac{1}{2^n - 1} < n, \quad \text{for all } n \geq 2.$$

(iv) Finally use the fact that, when $n \geq 3$,
$$\frac{1}{2^n} < \frac{1}{2} - \frac{1}{3}$$
to modify the proof in (iii) slightly, and hence show that
$$\frac{1}{1} + \frac{1}{2} + \frac{1}{3} + \cdots + \frac{1}{2^n} < n, \quad \text{for all } n \geq 3.$$

(b)(i) Explain why
$$\frac{1}{3} + \frac{1}{4} > \frac{1}{2}.$$

308 Infinity: recursion, induction, infinite descent

(ii) Explain why
$$\frac{1}{5} + \frac{1}{6} + \frac{1}{7} + \frac{1}{8} > \frac{1}{2}.$$

(iii) Extend parts (i) and (ii) to prove that
$$\frac{1}{1} + \frac{1}{2} + \frac{1}{3} + \cdots + \frac{1}{2^n} > 1 + \frac{n}{2}, \text{ for all } n \geqslant 2.$$

(c) Combine parts (a) and (b) to show that, for all $n \geqslant 2$, we have the two inequalities
$$1 + \frac{n}{2} < \frac{1}{1} + \frac{1}{2} + \frac{1}{3} + \cdots + \frac{1}{2^n} < n.$$

Conclude that the endless sum
$$\frac{1}{1} + \frac{1}{2} + \frac{1}{3} + \cdots + \frac{1}{n} + \cdots \quad \text{(for ever)}$$

cannot be assigned a finite value. △

The result in Problem **253**(c) has an unexpected consequence.

Problem 254 Imagine that you have an unlimited supply of identical rectangular strips of length 2. (Identical empty plastic CD cases can serve as a useful illustration, provided one focuses on their rectangular side profile, rather than the almost square frontal cross-section.) The goal is to construct a 'stack' in such a way as to stick out as far as possible beyond a table edge. One strip balances exactly at its midpoint, so can protrude a total distance of 1 without tipping over.

(a) Arrange a stack of n strips of length 2, one on top of the other, with the bottom strip protruding distance $\frac{1}{n}$ beyond the edge of the table, the second strip from the bottom protruding $\frac{1}{n-1}$ beyond the leading edge of the bottom strip, the third strip from the bottom protruding $\frac{1}{n-2}$ beyond the leading edge of the strip below it, and so on until the $(n-1)^{\text{th}}$ strip from the bottom protrudes distance $\frac{1}{2}$ beyond the leading edge of the strip below it, and the top strip protrudes distance 1 beyond the leading edge of the strip below it (see Figure 10). Prove that a stack of n identical strips arranged in this way will just avoid tipping over the table edge.

(b) Conclude that we can choose n so that an arrangement of n strips can (in theory) protrude as far beyond the edge of the table as we wish – without tipping. △

The next problem illustrates, in the context of the harmonic series, what is in fact a completely general phenomenon: an endless sum of *steadily decreasing*

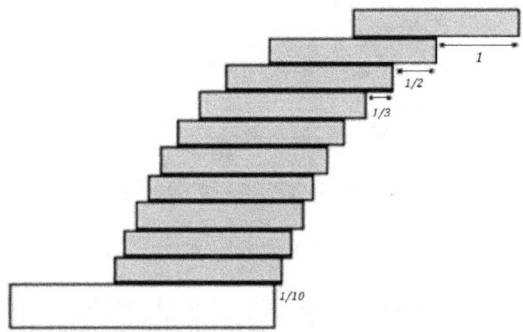

Figure 10: Overhanging strips, $n = 10$.

positive terms may converge or diverge; but **provided** the terms themselves converge to 0, then the the corresponding "alternating sum" – where the same terms are combined but with alternately positive and negative signs – **always** converges.

Problem 255

(a) Let
$$s_n = \frac{1}{1} - \frac{1}{2} + \frac{1}{3} - \frac{1}{4} + \frac{1}{5} - \cdots \pm \frac{1}{n}$$
(where the final operation is "+" if n is odd and "−" if n is even).

(i) Prove that
$$s_{2n-2} < s_{2n} < s_{2m+1} < s_{2m-1},$$
for all $m, n \geqslant 1$.

(ii) Conclude that the endless alternating sum
$$\frac{1}{1} - \frac{1}{2} + \frac{1}{3} - \frac{1}{4} + \frac{1}{5} - \cdots \quad \text{(for ever)}$$
can be assigned a value s that lies somewhere between $s_6 = \frac{37}{60}$ and $s_5 = \frac{47}{60}$.

(b) Let
$$a_1, a_2, a_3, \ldots$$
be an endless, decreasing sequence of positive terms (that is, $a_{n+1} < a_n$ for all $n \geqslant 1$). Suppose that the sequence of terms a_n converges to 0 as $n \to \infty$.

(i) Let
$$s_n = a_1 - a_2 + a_3 - a_4 + a_5 - \cdots \pm a_n$$
(where the final operation is "+" if n is odd and "−" if n is even). Prove that
$$s_{2n-2} < s_{2n} < s_{2m+1} < s_{2m-1}, \quad \text{for all } m, n \geq 1.$$

(ii) Conclude that the endless alternating sum
$$a_1 - a_2 + a_3 - a_4 + a_5 - \cdots \quad \text{(for ever)}$$
can be assigned a value s that lies somewhere between $s_2 = a_1 - a_2$ and $s_3 = a_1 - a_2 + a_3$. △

Just as with the series
$$\frac{1}{1^2} + \frac{1}{2^2} + \frac{1}{3^2} + \cdots + \frac{1}{n^2} + \cdots \quad \text{(for ever)}$$
$$\frac{1}{0!} + \frac{1}{1!} + \frac{1}{2!} + \cdots + \frac{1}{n!} + \cdots \quad \text{(for ever)},$$
we can show relatively easily that
$$\frac{1}{1} - \frac{1}{2} + \frac{1}{3} - \frac{1}{4} + \frac{1}{5} - \cdots \quad \text{(for ever)}$$
can be assigned a value s. It is far less clear whether this value has a familiar name! (It is in fact equal to the natural logarithm of 2: "$\log_e 2$".) A similarly intriguing series is the alternating series of odd terms from the harmonic series:
$$\frac{1}{1} - \frac{1}{3} + \frac{1}{5} - \frac{1}{7} + \frac{1}{9} - \cdots \quad \text{(for ever)}$$
You should be able to show that this endless series can be assigned a value somewhere between $s_2 = \frac{2}{3}$ and $s_3 = \frac{13}{15}$; but you are most unlikely to guess that its value is equal to $\frac{\pi}{4}$. This was first discovered in 1674 by Leibniz (1646–1716). One way to obtain the result is using the integral of $(1+x^2)^{-1}$ from 0 to 1: on the one hand the integral is equal to $\arctan x$ evaluated when $x = 1$ (that is, $\frac{\pi}{4}$); on the other hand, we can expand the integrand as a power series $1 - x^2 + x^4 - x^6 + \cdots$, integrate term by term, and prove that the resulting series converges when $x = 1$. (It does indeed converge, though it does so very, very slowly.)

The fact that the alternating harmonic series has the value $\log_e 2$ seems to have been first shown by Euler (1707–1783), using the power series expansion for $\log(1 + x)$.

6.7. Induction in geometry, combinatorics and number theory

We turn next to a mixed collection of problems designed to highlight a range of applications.

Problem 256 Let $f_1 = 2$, $f_{k+1} = f_k(f_k + 1)$. Prove by induction that f_k has at least k distinct prime factors. △

Problem 257

(a) Prove by induction that n points on a straight line divide the line into $n + 1$ parts.

(b)(i) By experimenting with small values of n, guess a formula R_n for the maximum number of regions which can be created in the plane by an array of n straight lines.

 (ii) Prove by induction that n straight lines in the plane divide the plane into at most R_n regions.

(c)(i) By experimenting with small values of n, guess a formula S_n for the maximum number of regions which can be created in 3-dimensions by an array of n planes.

 (ii) Prove by induction that n planes in 3-dimensions divide space into at most S_n regions. △

Problem 258 Given a square, prove that, for each $n \geqslant 6$, the initial square can be cut into n squares (of possibly different sizes). △

Problem 259 A tree is a connected graph, or network, consisting of vertices and edges, but with no cycles (or circuits). Prove that a tree with n vertices has exactly $n - 1$ edges. △

The next problem concerns *spherical polyhedra*. A spherical polyhedron is a polyhedral surface in 3-dimensions, which can be inflated to form a sphere (where we assume that the faces and edges can stretch as required). For example, a cube is a spherical polyhedron; but the surface of a picture frame is not. A spherical polyhedron has

- *faces* (flat 2-dimensional polygons, which can be stretched to take the form of a disc),

- *edges* (1-dimensional line segments, where exactly two faces meet), and
- *vertices* (0-dimensional points, where several faces and edges meet, in such a way that they form a single cycle around the vertex).

Each face must clearly have at least 3 edges; and there must be at least three edges and three faces meeting at each vertex.

If a spherical polyhedron has V vertices, E edges, and F faces, then the numbers V, E, F satisfy Euler's formula

$$V - E + F = 2.$$

For example, a cube has $V = 8$ vertices, $E = 12$ edges, and $F = 6$ faces, and $8 - 12 + 6 = 2$.

Problem 260

(a)(i) Describe a spherical polyhedron with exactly 6 edges.

(ii) Describe a spherical polyhedron with exactly 8 edges.

(b) Prove that no spherical polyhedron can have exactly 7 edges.

(c) Prove that for every $n \geqslant 8$, there exists a spherical polyhedron with n edges. △

Problem 261 A map is a (finite) collection of regions in the plane, each with a boundary, or border, that is 'polygonal' in the sense that it consists of a single sequence of distinct vertices and – possibly curved – edges, that separates the plane into two parts, one of which is the polygonal region itself. A map can be properly coloured if each region can be assigned a colour so that each pair of neighbouring regions (sharing an edge) always receive different colours. Prove that the regions of such a map can be properly coloured with just two colours if and only if an even number of edges meet at each vertex. △

Problem 262 (Gray codes) There are 2^n sequences of length n consisting of 0s and 1s. Prove that, for each $n \geqslant 2$, these sequences can be arranged in a cyclic list such that any two neighbouring sequences (including the last and the first) differ in exactly one coordinate position. △

Problem 263 (Calkin-Wilf tree) The binary tree in the plane has a distinguished vertex as 'root', and is constructed inductively. The root is joined to two new vertices; and each new vertex is then joined to two further new vertices – with the construction process continuing for ever (Figure 11).

Label the vertices of the binary tree with positive fractions as follows:

- the root is given the label $\frac{1}{1}$

- whenever we know the label $\frac{i}{j}$ of a 'parent' vertex, we label its 'left descendant' as $\frac{i}{i+j}$, and its 'right descendant' $\frac{i+j}{j}$.

(a) Prove that every positive rational $\frac{r}{s}$ occurs once and only once as a label, and that it occurs in its lowest terms.

(b) Prove that the labels are left-right symmetric in the sense that labels in corresponding left and right positions are reciprocals of each other. △

Problem 264 A collection of n intervals on the x-axis is such that every pair of intervals have a point in common. Prove that all n intervals must then have at least one point in common. △

6.8. Two problems

Problem 265 Several identical tanks of water sit on a horizontal base. Each pair of tanks is connected with a pipe at ground level controlled by a valve, or tap. When a valve is opened, the water level in the two connected tanks becomes equal (to the average, or mean, of the initial levels). Suppose we start with tank T which contains the least amount of water. The aim is to open and close valves in a sequence that will lead to the final water level in tank T being as high as possible. In what order should we make these connections? △

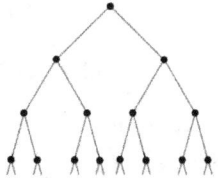

Figure 11: A (rooted) binary tree.

Problem 266 I have two flasks. One is 'empty', but still contains a residue of a dangerous chemical; the other contains a fixed amount of solvent that can be used to wash away the remaining traces of the dangerous chemical. What is the best way to use the fixed quantity of solvent? Should I use it all at once to wash out the first flask? Or should I first wash out the flask using just half of the solvent, and then repeat with the other half? Or is there a better way of using the available solvent to remove as much as possible of the dangerous chemical? △

6.9. Infinite descent

In this final section we touch upon an important variation on mathematical induction. This variation is well-illustrated by the next (probably familiar) problem.

Problem 267 Write out for yourself the following standard proof that $\sqrt{2}$ is irrational.

(i) Suppose to the contrary that $\sqrt{2}$ is rational. Then $\sqrt{2} = \frac{m}{n}$ for some positive integers m, n. Prove first that m must be even.

(ii) Write $m = 2m'$, where m' is also an integer. Show that n must also be even.

(iii) How does this lead to a contradiction? △

Problem **267** has the classic form of a proof which reaches a contradiction by *infinite descent*.

1. We start with a claim which we wish to prove is true. Often when we do not know how to begin, it makes sense to ask what would happen if the claim were *false*. This then guarantees that there must be some *counterexample*, which satisfies the given hypothesis, but which fails to satisfy the asserted conclusion.

2. *Infinite descent* becomes an **option** whenever each such counterexample gives rise to some *positive integer* parameter n (such as the denominator in Problem **267**(i)).

3. *Infinite descent* becomes a **reality**, if one can prove that the existence of the initial counterexample leads to a construction that produces a counterexample *with a smaller value n' of the parameter n*, since repeating this step then gives rise to an endlessly decreasing sequence

$$n > n' > n'' > \cdots > 0$$

of positive integers, which is impossible (since such a chain can have length at most n).

4. Hence the initial assumption that the claim was false must itself be false – so the claim must be true (as required).

Proof by "infinite descent" is an invaluable tool. But it is important to realise that the method is essentially a variation on proof by mathematical induction. As a first step in this direction it is worth reinterpreting Problem **267** as an induction proof.

Problem 268 Let $\mathbf{P}(n)$ be the statement:

"$\sqrt{2}$ cannot be written as a fraction with positive denominator $\leqslant n$".

(i) Explain why $\mathbf{P}(1)$ is true.

(ii) Suppose that $\mathbf{P}(k)$ is true for some $k \geqslant 1$. Use the proof in Problem **267** to show that $\mathbf{P}(k+1)$ must then be true as well.

(iii) Conclude that $\mathbf{P}(n)$ is true for all $n \geqslant 1$, whence $\sqrt{2}$ must be irrational. △

Problem **268** shows that, in the particular case of Problem **267** one can translate the standard proof that "$\sqrt{2}$ is irrational" into a proof by induction. But much more is true. The contradiction arising in step 3. above is an application of an important principle, namely

The Least Element Principle: Every non-empty set S of positive integers has a smallest element.

The Least Element Principle is equivalent to *The Principle of Mathematical Induction* which we stated at the beginning of the chapter:

The Principle of Mathematical Induction: If a subset S of the positive integers

- contains the integer "1", and has the property that
- whenever an integer k is in the set S, then the next integer $k+1$ is always in S too,

then S contains **all** positive integers.

Problem 269

(a) Assume the *Least Element Principle*. Suppose a subset T of the positive integers contains the integer "1", and that whenever k is in the set T, then $k+1$ is also in the set T. Let S be the set of all positive integers which are **not** in the set T. Conclude that S must be empty, and hence that T contains all positive integers.

(b) Assume the *Principle of Mathematical Induction*. Let T be a non-empty set of positive integers, and suppose that T does **not** have a smallest element. Let S be the set of all positive integers which do **not** belong to the set T. Prove that "1" must belong to S, and that whenever the positive integer k belongs to S, then so does $k+1$. Derive the contradiction that T must be empty, contrary to assumption. Conclude that T must in fact have a smallest element. △

To round off this final chapter you are invited to devise a rather different proof of the irrationality of $\sqrt{2}$.

Problem 270 This sequence of constructions presumes that we know – for example, by Pythagoras' Theorem – that, in any square $OPQR$, the ratio

$$\text{"diagonal : side"} = OQ : OP = \sqrt{2} : 1.$$

Let $OPQR$ be a square. Let the circle with centre Q and passing through P meet OQ at P'. Construct the perpendicular to OQ at P', and let this meet OR at Q'.

(i) Explain why $OP' = P'Q'$. Construct the point R' to complete the square $OP'Q'R'$.

(ii) Join QQ'. Explain why $P'Q' = RQ'$.

(iii) Suppose $OQ : OP = m : n$. Conclude that $OQ' : OP' = 2n - m : m - n$.

(iv) Prove that $n < m < 2n$, and hence that $0 < m - n < n$, $0 < 2n - m$.

(v) Explain how, if m, n can be chosen to be positive integers, the above sequence of steps sets up an "infinite descent" – which is impossible. △

6.10. Chapter 6: Comments and solutions

Note: It is important to separate the underlying idea of "induction" from the formal way we have chosen to present proofs. As ever in mathematics, the ideas are what matter most. But the process of struggling with (and slowly coming to understand why we need) the formal structure behind the written proofs is part of the way the ideas are tamed and organised.

Readers should not be intimidated by the physical extent of the solutions to this chapter. As explained in the main text it is important for all readers to review the way they approach induction proofs: so we have erred in favour of completeness – knowing that as each reader becomes more confident, s/he will increasingly compress, or abbreviate, some of the steps.

228.

(a) Yes.

(b) Yes.
$$2 \times 3 \times 5 \times 7 \times 11 + 1 = 2311,$$
and $\sqrt{2311} = 48.07\ldots$, so we only need to check prime factors up to 47.

(c) No.
$$2 \times 3 \times 5 \times 7 \times 11 \times 13 + 1 = 30\,031,$$
and $\sqrt{30\,031} = 173.29\ldots$ so we might have to check all 40 possible prime factors up to 173. However, we only have to start at 17 [Why?], and checking with a calculator is very quick. In fact 30 031 factorises rather prettily as 59×509.

229. Note: The statement in the problem includes the *quantifier* "**for all** $n \geqslant 1$".

What is to be proved is the compound statement

"$\mathbf{P}(n)$ is true **for all** $n \geqslant 1$".

In contrast, each individual statement $\mathbf{P}(n)$ refers to a single value of n.

It is essential to be clear when you are dealing with the *compound* statement, and when you are referring to some *particular* statement $\mathbf{P}(1)$, or $\mathbf{P}(n)$, or $\mathbf{P}(k)$.

Let $\mathbf{P}(n)$ be the statement:

"$5^{2n+2} - 24n - 25$ is divisible by 576".

- $\mathbf{P}(1)$ is the statement:

"$5^4 - 24 \times 1 - 25$ is divisible by 576".

That is:

"$625 - 49 = 576$ is divisible by 576",

which is evidently true.

- Now suppose that we know $\mathbf{P}(k)$ is true for some $k \geqslant 1$. We must show that $\mathbf{P}(k+1)$ is then also true.

  To prove that $\mathbf{P}(k+1)$ is true, we have to consider the statement $\mathbf{P}(k+1)$. It is no use starting with $\mathbf{P}(k)$. However, since we know that $\mathbf{P}(k)$ is true, we are free to use it at any stage if it turns out to be useful.

  To prove that $\mathbf{P}(k+1)$ is true, we have to show that

  "$5^{2(k+1)+2} - 24(k+1) - 25$ is divisible by 576".

  So we must start with $5^{2(k+1)+2} - 24(k+1) - 25$ and try to "simplify" it (knowing that "simplify" in this case means "rewrite it in a way that involves $5^{2k+2} - 24k - 25$").

  $$\begin{aligned} 5^{2(k+1)+2} &- 24(k+1) - 25 \\ &= [5^{2k+4}] - 24k - 24 - 25 \\ &= [5^2(5^{2k+2} - 24k - 25) + 5^2 \cdot (24k) + 5^2 \cdot 25] \\ &\quad - 24k - (24 + 25) \\ &= 5^2(5^{2k+2} - 24k - 25) + [(5^2 - 1) \times (24k)] \\ &\quad + [5^2 \times 25 - 24 - 25] \\ &= 5^2(5^{2k+2} - 24k - 25) + 24^2 k + [5^2 \times 25 - 24 - 25] \\ &= 5^2(5^{2k+2} - 24k - 25) + 24^2 k + [5^4 - 24 - 25]. \end{aligned}$$

  The first term on the RHS is a multiple of $(5^{2k+2} - 24k - 25)$, so is divisible by 576 (since we know that $\mathbf{P}(k)$ is true); the second term on the RHS is a multiple of $24^2 = 576$; and the third term on the RHS is the expression arising in $\mathbf{P}(1)$, which we saw is equal to 576.
  ∴ the whole RHS is divisible by 576
  ∴ the LHS is divisible by 576, so $\mathbf{P}(k+1)$ is true.

Hence

- $\mathbf{P}(1)$ is true; and
- whenever $\mathbf{P}(k)$ is true for some $k \geqslant 1$, we have proved that $\mathbf{P}(k+1)$ must be true.

∴ $\mathbf{P}(n)$ is true for all $n \geqslant 1$. QED

230. Let $\mathbf{P}(n)$ be the statement:

  "the angles of any p-gon, for any value of p with $3 \leqslant p \leqslant n$, have sum exactly $(p-2)\pi$ radians".

1. $\mathbf{P}(3)$ is the statement:

   "the angles of any triangle have sum π radians".

   This is a known fact: given triangle $\triangle ABC$, draw the line XAY through A parallel to BC, with X on the same side of AC as B and Y on the same side of AB as C. Then $\angle XAB = \angle CBA$ and $\angle YAC = \angle BCA$ (alternate angles), so

$$\angle B + \angle A + \angle C = \angle XAB + \angle A + \angle YAC = \angle XAY = \pi.$$

2. Now we suppose that $\mathbf{P}(k)$ is known to be true for some $k \geqslant 3$. We must show that $\mathbf{P}(k+1)$ is then necessarily true.

To prove that $\mathbf{P}(k+1)$ is true, we have to consider the statement $\mathbf{P}(k+1)$: that is,

"the angles of a p-gon, for any value of p with $3 \leqslant p \leqslant k+1$, have sum exactly $(p-2)\pi$ radians".

This can be reworded by splitting it into two parts:

"the angles of any p-gon for $3 \leqslant p \leqslant k$ have sum exactly $(p-2)\pi$ radians;"

and

"the angles of any $(k+1)$-gon have sum exactly $((k+1)-2)\pi$ radians".

The first part of this revised version is precisely the statement $\mathbf{P}(k)$, which we suppose is known to be true. So the crux of the matter is to prove the second part – namely that the angles of any $(k+1)$-gon have sum $(k-1)\pi$.

Let $A_0 A_1 A_2 \cdots A_k$ be any $(k+1)$-gon.

[**Note:** The usual move at this point is to say "draw the chord $A_k A_1$ to cut the polygon into the triangle $A_k A_1 A_0$ (with angle sum π (by $\mathbf{P}(3)$), and the k-gon $A_1 A_2 \cdots A_k$ (with angle sum $(k-2)\pi$ (by $\mathbf{P}(k)$)), whence we can add to see that $A_0 A_1 A_2 \cdots A_k$ has angle sum $((k+1)-2)\pi$. However, this only works

- if the triangle $A_k A_1 A_0$ "sticks out" rather than in, and
- if no other vertices or edges of the $(k+1)$-gon encroach into the part that is cut off – which can only be guaranteed if the polygon is "convex".

So what is usually presented as a "proof" does not work *in general*.

If we want to prove the general result – for polygons of all shapes – we have to get round this unwarranted assumption. Experiment may persuade you that "there is always some vertex that sticks out and which can be safely "cut off"; but it is not at all clear how to prove this fact (we know of no simple proof). So we have to find another way.]

Consider the vertex A_1, and its two neighbours A_0 and A_2.

Imagine each half-line, which starts at A_1, and which sets off into the *interior* of the polygon. Because the polygon is finite, each such half-line defines a line segment $\underline{A_1 X}$, where X is the next point of the polygon which the half line hits (that is, X is one of the vertices A_m, or a point on one of the sides $\underline{A_m A_{m+1}}$).

Consider the locus of all such points X as the half line swings from $\underline{A_1 A_0}$ (produced) to $\underline{A_1 A_2}$ (produced). There are two possibilities: either

(a) these points X all belong to a single side of the polygon; or
(b) they don't.

(a) In the first case none of the vertices or sides of the polygon $A_0A_1A_2\cdots A_k$ intrude into the interior of the triangle $A_0A_1A_2$, so the chord $\underline{A_0A_2}$ separates the $(k+1)$-gon $A_0A_1A_2\cdots A_k$ into a triangle $A_0A_1A_2$ and a k-gon $A_0A_2A_3\cdots A_k$. The angle sum of $A_0A_1A_2\cdots A_k$ is then equal to the sum of (i) the angle sum of the triangle $A_0A_1A_2$ and (ii) the angle sum of $A_0A_2A_3\cdots A_k$ – which are equal to π and $(k-2)\pi$ respectively (by $\mathbf{P}(k)$). So the angle sum of the $(k+1)$-gon $A_0A_2A_3\cdots A_k$ is equal to $((k+1)-2)\pi$ as required.

(b) In the second case, as the half-line A_1X rotates from A_1A_0 to A_1A_2, the point X must at some instant switch from lying on one side of the polygon to lying on another side; at the very instant where X switches sides, $X = A_m$ must be a vertex of the polygon.

Because of the way the point X was chosen, the chord $A_1X = A_1A_m$ does not meet any other point of the $(k+1)$-gon $A_0A_1A_2\cdots A_k$, and so splits the $(k+1)$-gon into an m-gon $A_1A_2A_3\cdots A_m$ (with angle sum $(m-2)\pi$ by $\mathbf{P}(k)$) and a $(k-m+3)$-gon $A_mA_{m+1}A_{m+2}\cdots A_kA_0A_1$ (with angle sum $(k-m+1)\pi$ by $\mathbf{P}(k)$). So the $(k+1)$-gon $A_0A_1A_2\cdots A_k$ has angle sum $((k+1)-2)\pi$ as required.

Hence $\mathbf{P}(k+1)$ is true.
∴ $\mathbf{P}(n)$ is true for all $n \geqslant 3$. QED

231. Let $\mathbf{P}(n)$ be the statement
$$1 + 3 + 5 + \cdots + (2n-1) = n^2.$$

- LHS of $\mathbf{P}(1) = 1$; RHS of $\mathbf{P}(1) = 1^2$. Since these two are equal, $\mathbf{P}(1)$ is true.
- Suppose that $\mathbf{P}(k)$ is true for some particular (unspecified) $k \geqslant 1$; that is, we know that, for this particular k,
$$1 + 3 + 5 + \cdots + (2k-1) = k^2.$$

We wish to prove that $\mathbf{P}(k+1)$ must then be true.

Now $\mathbf{P}(k+1)$ is an equation, so we start with the LHS of $\mathbf{P}(k+1)$ and try to simplify it in an appropriate way to get the RHS of $\mathbf{P}(k+1)$:

$$\begin{aligned}\text{LHS of } \mathbf{P}(k+1) &= 1 + 3 + 5 + \cdots + (2(k+1) - 1) \\ &= (1 + 3 + 5 + \cdots + (2k-1)) + (2k+1).\end{aligned}$$

If we now use $\mathbf{P}(k)$, which we are supposing to be true, then the first bracket is equal to k^2, so this sum is equal to:

$$\begin{aligned}&= k^2 + (2k+1) \\ &= (k+1)^2 \\ &= \text{RHS of } \mathbf{P}(k+1).\end{aligned}$$

Hence $\mathbf{P}(k+1)$ holds.

Combining these two bullet points then shows that "$\mathbf{P}(n)$ holds, for all $n \geqslant 1$".

QED

232.

(a) The only way to learn is by trying and failing; then trying again and failing slightly better! So don't give up too quickly. It is natural to try to relate each term to the one before. You may then notice that each term is slightly less than 3 times the previous term.

(b) **Note:** The recurrence relation for u_n involves the two previous terms. So when we assume that $\mathbf{P}(k)$ is known to be true and try to prove $\mathbf{P}(k+1)$, the recurrence relation for u_{k+1} will involve u_k and u_{k-1}, so $\mathbf{P}(n)$ needs to be formulated to ensure that we can use closed expressions for both these terms. For the same reason, the induction proof has to start by showing that both $\mathbf{P}(0)$ and $\mathbf{P}(1)$ are true.

Let $\mathbf{P}(n)$ be the statement:

"$u_m = 2^m + 3^m$ for all m, $0 \leqslant m \leqslant n$".

- LHS of $\mathbf{P}(0) = u_0 = 2$; RHS of $\mathbf{P}(0) = 2^0 + 3^0 = 1 + 1$. Since these two are equal, $\mathbf{P}(0)$ is true.

  $\mathbf{P}(1)$ combines $\mathbf{P}(0)$, and the equality of $u_1 = 5$ and $2^1 + 3^1$; since these two are equal, $\mathbf{P}(1)$ is true.

- Suppose that $\mathbf{P}(k)$ is true for some particular (unspecified) $k \geqslant 1$; that is, we know that, for this particular k,

  "$u_m = 2^m + 3^m$ for all m, $0 \leqslant m \leqslant k$."

We wish to prove that $\mathbf{P}(k+1)$ must then be true.

Now $\mathbf{P}(k+1)$ requires us to prove that

"$u_m = 2^m + 3^m$ for all m, $0 \leqslant m \leqslant k+1$."

Most of this is guaranteed by $\mathbf{P}(k)$, which we assume to be true. It remains for us to check that the equality holds for u_{k+1}. We know that

$$u_{k+1} = 5u_k - 6u_{k-1}.$$

And we may use $\mathbf{P}(k)$, which we are supposing to be true, to conclude that:

$$\begin{aligned} u_{k+1} &= 5\left(2^k + 3^k\right) - 6\left(2^{k-1} + 3^{k-1}\right) \\ &= (10-6)2^{k-1} + (15-6)3^{k-1} \\ &= 2^{k+1} + 3^{k+1}. \end{aligned}$$

Hence $\mathbf{P}(k+1)$ holds.

Combining these two bullet points then shows that "$\mathbf{P}(n)$ holds, for all $n \geqslant 0$".

QED

233. Let $\mathbf{P}(n)$ be the statement:

$$\text{``}F_m = \frac{\alpha^m - \beta^m}{\sqrt{5}} \text{ for all } m, 0 \leq m \leq n\text{''},$$

where $\alpha = \frac{1+\sqrt{5}}{2}$ and $\beta = \frac{1-\sqrt{5}}{2}$.

- LHS of $\mathbf{P}(0) = F_0 = 0$; RHS of $\mathbf{P}(0) = \frac{1-1}{\sqrt{5}} = 0$. Since these two are equal, $\mathbf{P}(0)$ is true.
  LHS of $\mathbf{P}(1) = F_1 = 1$; RHS of $\mathbf{P}(1) = \frac{\alpha - \beta}{\sqrt{5}} = 1$. Since these two are equal, $\mathbf{P}(1)$ is true.

- Suppose that $\mathbf{P}(k)$ is true for some particular (unspecified) $k \geq 1$; that is, we know that, for this particular k,

$$\text{``}F_m = \frac{\alpha^m - \beta^m}{\sqrt{5}} \text{ for all } m, 0 \leq m \leq k.\text{''}$$

We wish to prove that $\mathbf{P}(k+1)$ must then be true.

Now $\mathbf{P}(k+1)$ requires us to prove that

$$\text{``}F_m = \frac{\alpha^m - \beta^m}{\sqrt{5}} \text{ for all } m, 0 \leq m \leq k+1.\text{''}$$

Most of this is guaranteed by $\mathbf{P}(k)$, which we assume to be true. It remains to check this for F_{k+1}. We know that

$$F_{k+1} = F_k + F_{k-1}.$$

And we may use $\mathbf{P}(k)$, which we are supposing to be true to conclude that:

$$\begin{aligned} F_{k+1} &= \frac{\alpha^k - \beta^k}{\sqrt{5}} + \frac{\alpha^{k-1} - \beta^{k-1}}{\sqrt{5}} \\ &= \frac{\alpha^k + \alpha^{k-1}}{\sqrt{5}} - \frac{\beta^k + \beta^{k-1}}{\sqrt{5}} \\ &= \frac{\alpha^{k+1} - \beta^{k+1}}{\sqrt{5}} \end{aligned}$$

(since α and β are roots of the equation $x^2 - x - 1 = 0$)
Hence $\mathbf{P}(k+1)$ holds.

Combining these two bullet points then shows that "$\mathbf{P}(n)$ holds, for all $n \geq 1$".
QED

Note: You may understand the above solution and yet wonder how such a formula could be discovered. The answer is fairly simple. There is a general theory about linear recurrence relations which guarantees that the set of all solutions of a second order recurrence (that is, a recurrence in which each term depends on the two previous terms) is "two dimensional" (that is, it is just like the familiar 2D plane, where every vector (p, q) is a combination of the two "base vectors" $(1, 0)$ and $(0, 1)$:

$$(p, q) = p(1, 0) + q(0, 1)).$$

Once we know this, it remains:

- to find two *special* solutions (like the vectors $(1,0)$ and $(0,1)$ in the plane), which we do here by looking for sequences of the form "$1, x, x^2, x^3, \ldots$" that satisfy the recurrence, which implies that $1 + x = x^2$, so $x = \alpha$, or $x = \beta$;
- then to choose a linear combination $F_m = p\alpha^m + q\beta^m$ of these two power solutions to give the correct first two terms: $0 = F_0 = p + q$, $1 = F_1 = p\alpha + q\beta$, so $p = \frac{1}{\sqrt{5}}$, $q = -\frac{1}{\sqrt{5}}$.

234. Let $\mathbf{P}(n)$ be the statement:

$$\text{``}5^{2n+1} \cdot 2^{n+2} + 3^{n+2} \cdot 2^{2n+1} \text{ is divisible by 19''}.$$

- $\mathbf{P}(0)$ is the statement: "$5 \times 4 + 9 \times 2$ is divisible by 19", which is true.
- Now suppose that we know that $\mathbf{P}(k)$ is true for some $k \geqslant 0$. We must show that $\mathbf{P}(k+1)$ is then also true.

To prove that $\mathbf{P}(k+1)$ is true, we have to show that

$$\text{``}5^{2k+3} \cdot 2^{k+3} + 3^{k+3} \cdot 2^{2k+3} \text{ is divisible by 19''}.$$

$$\begin{aligned}
5^{2k+3} \cdot 2^{k+3} + 3^{k+3} \cdot 2^{2k+3} &= 5^2 \cdot 2 \left(5^{2k+1} \cdot 2^{k+2} + 3^{k+2} \cdot 2^{2k+1} \right) \\
&\quad - 5^2 \cdot 2 \cdot 3^{k+2} \cdot 2^{2k+1} + 3^{k+3} \cdot 2^{2k+3} \\
&= 5^2 \cdot 2 \cdot \left(5^{2k+1} \cdot 2^{k+2} + 3^{k+2} \cdot 2^{2k+1} \right) \\
&\quad - \left(5^2 - 3 \cdot 2 \right) 3^{k+2} \cdot 2^{2k+2}.
\end{aligned}$$

The bracket in the first term on the RHS is divisible by 19 (by $\mathbf{P}(k)$), and the bracket in the second term is equal to 19. Hence both terms on the RHS are divisible by 19, so the RHS is divisible by 19. Therefore the LHS is also divisible by 19, so $\mathbf{P}(k+1)$ is true.

$\therefore \mathbf{P}(n)$ is true for all $n \geqslant 0$.   QED

235.

Note: The proofs of identities such as those in Problem **235**, which are given in many introductory texts, ignore the lessons of the previous two problems. In particular,

- they often fail to distinguish between
  - the single statement $\mathbf{P}(n)$ for a particular n, and
  - the "quantified" result to be proved ("for all $n \geqslant 1$"),

and

- they proceed in the 'wrong' direction (e.g. starting with the identity $\mathbf{P}(n)$ and "adding the same to both sides").

This latter strategy is psychologically and didactically misleading – even though it can be justified logically when proving very simple identities. In these very simple cases, each statement $\mathbf{P}(n)$ to be proved is unusual in that it refers to **exactly one configuration, or equation, for each n**. And since there is exactly one configuration for each n, the configuration or identity for $k+1$ can often be obtained by fiddling with the configuration for k. In contrast, in Problem **230**, for each value of n, there is a bewildering variety of possible polygons with n vertices, ranging from *regular* polygons to the most convoluted, re-entrant shapes: the statement $\mathbf{P}(n)$ makes an assertion about **all** such configurations, and there is no way of knowing whether we can obtain all such configurations for $k+1$ in a uniform way by fiddling with some configuration for k.

Readers should try to write each proof in the intended spirit, and to learn from the solutions – since this style has been chosen to highlight what mathematical induction is really about, and it is this approach that is needed in serious applications.

(a) Let $\mathbf{P}(n)$ be the statement:
$$\text{``}1 + 2 + 3 + \cdots + n = \tfrac{n(n+1)}{2}\text{''}.$$

- LHS of $\mathbf{P}(1) = 1$; RHS of $\mathbf{P}(1) = \tfrac{1 \cdot (1+1)}{2} = 1$. Since these two are equal, $\mathbf{P}(1)$ is true.

- Suppose that $\mathbf{P}(k)$ is true for some particular (unspecified) $k \geqslant 1$; that is, we know that, for this particular k,
$$\text{``}1 + 2 + 3 + \cdots + k = \tfrac{k(k+1)}{2}\text{''}.$$
We wish to prove that $\mathbf{P}(k+1)$ must then be true.

Now $\mathbf{P}(k+1)$ is an equation, so we start with the LHS of $\mathbf{P}(k+1)$ and try to simplify it in an appropriate way to get the RHS of $\mathbf{P}(k+1)$:

$$\begin{aligned}\text{LHS of } \mathbf{P}(k+1) &= 1 + 2 + 3 + \cdots + k + (k+1) \\ &= (1 + 2 + 3 + \cdots + k) + (k+1).\end{aligned}$$

If we now use $\mathbf{P}(k)$, which we are supposing to be true, then the first bracket is equal to $\tfrac{k(k+1)}{2}$, so the complete sum is equal to:

$$\begin{aligned} &= \tfrac{k(k+1)}{2} + (k+1) \\ &= \tfrac{(k+1)(k+2)}{2} \\ &= \text{RHS of } \mathbf{P}(k+1).\end{aligned}$$

Hence $\mathbf{P}(k+1)$ holds.

If we combine these two bullet points, then we have proved that "$\mathbf{P}(n)$ holds for all $n \geqslant 1$". QED

(b) Let $\mathbf{P}(n)$ be the statement:
$$\text{``}\tfrac{1}{1 \cdot 2} + \tfrac{1}{2 \cdot 3} + \tfrac{1}{3 \cdot 4} + \cdots + \tfrac{1}{n \cdot (n+1)} = 1 - \tfrac{1}{n+1}\text{''}.$$

6.10. Chapter 6: Comments and solutions   325

- LHS of $\mathbf{P}(1) = \frac{1}{1\cdot 2} = \frac{1}{2}$; RHS of $\mathbf{P}(1) = 1 - \frac{1}{2} = \frac{1}{2}$. Since these two are equal, $\mathbf{P}(1)$ is true.
- Suppose that $\mathbf{P}(k)$ is true for some particular (unspecified) $k \geqslant 1$; that is, we know that, for this particular k,

  "$\frac{1}{1\cdot 2} + \frac{1}{2\cdot 3} + \frac{1}{3\cdot 4} + \cdots + \frac{1}{k\cdot (k+1)} = 1 - \frac{1}{k+1}$".

  We wish to prove that $\mathbf{P}(k+1)$ must then be true.

  Now $\mathbf{P}(k+1)$ is an equation, so we start with the LHS of $\mathbf{P}(k+1)$ and try to simplify it in an appropriate way to get the RHS of $\mathbf{P}(k+1)$:

  $$\begin{aligned}
  \text{LHS of } \mathbf{P}(k+1) &= \frac{1}{1\cdot 2} + \frac{1}{2\cdot 3} + \frac{1}{3\cdot 4} + \cdots + \frac{1}{(k+1)(k+2)} \\
  &= \left[\frac{1}{1\cdot 2} + \frac{1}{2\cdot 3} + \frac{1}{3\cdot 4} + \frac{1}{k(k+1)}\right] \\
  &\quad + \frac{1}{(k+1)(k+2)}.
  \end{aligned}$$

  If we now use $\mathbf{P}(k)$, which we assume is true, then the first bracket is equal to $1 - \frac{1}{k+1}$, so the complete sum is equal to:

  $$\begin{aligned}
  &= \left[1 - \frac{1}{k+1}\right] + \frac{1}{(k+1)(k+2)} \\
  &= 1 - \left[\frac{1}{k+1} - \frac{1}{(k+1)(k+2)}\right] \\
  &= 1 - \frac{1}{k+2} \\
  &= \text{RHS of } \mathbf{P}(k+1).
  \end{aligned}$$

  Hence $\mathbf{P}(k+1)$ holds.

  If we combine these two bullet points, we have proved that "$\mathbf{P}(n)$ holds for all $n \geqslant 1$". QED

(c) **Note:** If $q = 1$, then the LHS is equal to n, but the RHS makes no sense. So we assume $q \neq 1$.

Let $\mathbf{P}(n)$ be the statement:

"$1 + q + q^2 + q^3 + \cdots + q^{n-1} = \frac{1}{1-q} - \frac{q^n}{1-q}$".

- LHS of $\mathbf{P}(1) = 1$; RHS of $\mathbf{P}(1) = \frac{1}{1-q} - \frac{q}{1-q} = 1$. Since these two are equal, $\mathbf{P}(1)$ is true.
- Suppose that $\mathbf{P}(k)$ is true for some particular (unspecified) $k \geqslant 1$; that is, we know that, for this particular k,

  "$1 + q + q^2 + q^3 + \cdots + q^{k-1} = \frac{1}{1-q} - \frac{q^k}{1-q}$".

  We wish to prove that $\mathbf{P}(k+1)$ must then be true.

Now $\mathbf{P}(k+1)$ is an equation, so we start with the LHS of $\mathbf{P}(k+1)$ and try to simplify it in an appropriate way to get the RHS of $\mathbf{P}(k+1)$:

$$\begin{aligned}\text{LHS of } \mathbf{P}(k+1) &= 1 + q + q^2 + q^3 + \cdots + q^k \\ &= \left(1 + q + q^2 + q^3 + \cdots + q^{k-1}\right) + q^k.\end{aligned}$$

If we now use $\mathbf{P}(k)$, which we assume is true, then the first bracket is equal to

$$\frac{1}{1-q} - \frac{q^k}{1-q}$$

so the complete sum is equal to:

$$\begin{aligned} &= \frac{1}{1-q} - \left[\frac{q^k}{1-q} - q^k\right] \\ &= \frac{1}{1-q} - \frac{q^{k+1}}{1-q} \\ &= \text{RHS of } \mathbf{P}(k+1).\end{aligned}$$

Hence $\mathbf{P}(k+1)$ holds.

If we combine these two bullet points, we have proved that "$\mathbf{P}(n)$ holds for all $n \geqslant 1$". **QED**

(d) **Note:** The statement to be proved starts with a term involving "0!". The definition

$$n! = 1 \times 2 \times 3 \times \cdots \times n$$

does not immediately tell us how to interpret "0!". The correct interpretation emerges from the fact that several different thoughts all point in the same direction.

(i) When $n > 0$, then to get from $n!$ to $(n+1)!$ we multiply by $(n+1)$. If we extend this to $n = 0$, then "to get from 0! to 1!, we have to multiply by 1" – which suggests that $0! = 1$.

(ii) When $n > 0$, $n!$ counts the number of *permutations* of n symbols, or the number of different linear orders of n objects (i.e. how many different ways they can be arranged in a line). If we extend this to $n = 0$, we see that there is just one way to arrange 0 objects (namely, sit tight and do nothing).

(iii) The definition of $n!$ as a *product* suggests that 0! involves a "product with no terms" at all. Now when we "add no terms together" it makes sense to interpret the result as "$= 0$" (perhaps because if this "sum of no terms" were added to some other sum, it would make no difference). In the same spirit, the *product* of no terms should be taken to be "$= 1$" (since if this empty product were included at the end of some other product, it would make no difference to the result).

Let $\mathbf{P}(n)$ be the statement:

"$0 \cdot 0! + 1 \cdot 1! + 2 \cdot 2! + \cdots + (n-1) \cdot (n-1)! = n! - 1$".

- LHS of $\mathbf{P}(1) = 0 \cdot 0! = 0$; RHS of $\mathbf{P}(1) = 1! - 1 = 0$. Since these two are equal, $\mathbf{P}(1)$ is true.
- Suppose that $\mathbf{P}(k)$ is true for some particular (unspecified) $k \geqslant 1$; that is, we know that, for this particular k,

  "$0 \cdot 0! + 1 \cdot 1! + 2 \cdot 2! + \cdots + (k-1) \cdot (k-1)! = k! - 1$".

We wish to prove that $\mathbf{P}(k+1)$ must then be true.

Now $\mathbf{P}(k+1)$ is an equation, so we start with the LHS of $\mathbf{P}(k+1)$ and try to simplify it in an appropriate way to get the RHS of $\mathbf{P}(k+1)$:

$$\begin{aligned}
\text{LHS of } \mathbf{P}(k+1) &= 0 \cdot 0! + 1 \cdot 1! + 2 \cdot 2! + \cdots + k \cdot k! \\
&= [0 \cdot 0! + 1 \cdot 1! + 2 \cdot 2! + \cdots + (k-1) \cdot (k-1)!] + k.k!.
\end{aligned}$$

If we now use $\mathbf{P}(k)$, which we assume is true, then the first bracket is equal to $k! - 1$, so the complete sum is equal to:

$$\begin{aligned}
&= (k! - 1) + k \cdot k! \\
&= (k+1) \cdot k! - 1 \\
&= (k+1)! - 1 = \text{RHS of } \mathbf{P}(k+1).
\end{aligned}$$

Hence $\mathbf{P}(k+1)$ holds.

If we combine these two bullet points, we have proved that "$\mathbf{P}(n)$ holds for all $n \geqslant 1$". QED

(e) Let $\mathbf{P}(n)$ be the statement:

"$(\cos\theta + i\sin\theta)^n = \cos n\theta + i\sin n\theta$"

- LHS of $\mathbf{P}(1) = (\cos\theta + i\sin\theta)^1$; RHS of $\mathbf{P}(1) = \cos\theta + i\sin\theta$. Since these two are equal, $\mathbf{P}(1)$ is true.
- Suppose that $\mathbf{P}(k)$ is true for some particular (unspecified) $k \geqslant 1$; that is, we know that, for this particular k,

  "$(\cos\theta + i\sin\theta)^k = \cos k\theta + i\sin k\theta$".

We wish to prove that $\mathbf{P}(k+1)$ must then be true.

Now $\mathbf{P}(k+1)$ is an equation, so we start with the LHS of $\mathbf{P}(k+1)$ and try to simplify it in an appropriate way to get the RHS of $\mathbf{P}(k+1)$:

$$\begin{aligned}
\text{LHS of } \mathbf{P}(k+1) &= (\cos\theta + i\sin\theta)^{k+1} \\
&= (\cos\theta + i\sin\theta)^k (\cos\theta + i\sin\theta).
\end{aligned}$$

If we now use $\mathbf{P}(k)$, which we assume is true, then the first bracket is equal to $(\cos k\theta + i\sin k\theta)$, so the complete expression is equal to:

$$\begin{aligned}
&= (\cos k\theta + i\sin k\theta)(\cos\theta + i\sin\theta) \\
&= [\cos k\theta \cdot \cos\theta - \sin k\theta \cdot \sin\theta] + i[\cos k\theta \cdot \sin\theta + \sin k\theta \cdot \cos\theta] \\
&= \cos(k+1)\theta + i\sin(k+1)\theta = \text{RHS of } \mathbf{P}(k+1).
\end{aligned}$$

Hence **P**(k + 1) holds.

If we combine these two bullet points, we have proved that "**P**(n) holds for all $n \geq 1$".
QED

236. Let **P**(n) be the statement:

"$(1 + 2 + 3 + \cdots + n)^2 = 1^3 + 2^3 + 3^3 + \cdots + n^3$".

- LHS of **P**(1) = 1^2; RHS of **P**(1) = 1^3. Since these two are equal, **P**(1) is true.
- Suppose that **P**(k) is true for some particular (unspecified) $k \geq 1$; that is, we know that, for this particular k,

  "$(1 + 2 + 3 + \cdots + k)^2 = 1^3 + 2^3 + 3^3 + \cdots + k^3$".

We wish to prove that **P**(k + 1) must then be true.

Now **P**(k+1) is an equation, so we start with one side of **P**(k+1) and try to simplify it in an appropriate way to get the other side of **P**(k + 1). In this instance, the RHS of **P**(k + 1) is the most promising starting point (because we know a formula for the kth triangular number, and so can see how to simplify it):

$$\begin{aligned} \text{RHS of } \mathbf{P}(k+1) &= 1^3 + 2^3 + 3^3 + \cdots + k^3 + (k+1)^3 \\ &= \left[1^3 + 2^3 + 3^3 + \cdots + k^3\right] + (k+1)^3. \end{aligned}$$

If we now use **P**(k), which we assume is true, then the first bracket is equal to

$$(1 + 2 + 3 + \cdots + k)^2,$$

so the complete RHS is:

$$\begin{aligned} &= (1 + 2 + 3 + \cdots + k)^2 + (k+1)^3 \\ &= \left[\frac{k(k+1)}{2}\right]^2 + (k+1)^3 \\ &= \frac{1}{4}(k+1)^2 \left[k^2 + 4k + 4\right] \\ &= \left[\frac{(k+1)(k+2)}{2}\right]^2 \\ &= (1 + 2 + 3 + \cdots + (k+1))^2 \\ &= \text{LHS of } \mathbf{P}(k+1). \end{aligned}$$

Hence **P**(k + 1) holds.

If we combine these two bullet points, we have proved that "**P**(n) holds for all $n \geq 1$".
QED

Note: A slightly different way of organizing the proof can sometimes be useful. Denote the two sides of the equation in the statement **P**(n) by $f(n)$ and $g(n)$ respectively. Then

- $f(1) = 1^2 = 1^3 = g(1)$; and

- simple algebra allows one to check that, for each $k \geqslant 1$,
$$f(k+1) - f(k) = (k+1)^3 = g(k+1) - g(k).$$
It then follows (by induction) that $f(n) = g(n)$ for all $n \geqslant 1$.

237.

(a) $T_1 = 1$, $T_1 + T_2 = 1 + 3 = 4$, $T_1 + T_2 + T_3 = 1 + 3 + 6 = 10$. These may not be very suggestive. But
$$T_1 + T_2 + T_3 + T_4 = 20 = 5 \times 4,$$
$$T_1 + T_2 + T_3 + T_4 + T_5 = 35 = 5 \times 7,$$
and
$$T_1 + T_2 + T_3 + T_4 + T_5 + T_6 = 56 = 7 \times 8$$
may eventually lead one to guess that
$$T_1 + T_2 + T_3 + \cdots + T_n = \frac{n(n+1)(n+2)}{6}.$$

Note 1: This will certainly be easier to guess if you remember what you found in Problem **17** and Problem **63**.

Note 2: There is another way to help in such guessing. Suppose you notice that

– adding values for $k = 1$ up to $k = n$ of a polynomial of degree 0 (such as $p(x) = 1$) gives an answer that is a "polynomial of degree 1",
$$1 + 1 + \cdots + 1 = n,$$
and

– adding values for $k = 1$ up to $k = n$ of a polynomial of degree 1 (such as $p(x) = x$) gives an answer that is a "polynomial of degree 2",
$$1 + 2 + 3 + \cdots + n = \frac{n(n+1)}{2}.$$
Then you might guess that the sum
$$T_1 + T_2 + T_3 + \cdots + T_n$$
will give an answer that is a polynomial of degree 3 in n. Suppose that
$$T_1 + T_2 + T_3 + \cdots + T_n = An^3 + Bn^2 + Cn + D.$$
We can then use small values of n to set up equations which must be satisfied by A, B, C, D and solve them to find A, B, C, D:

– when $n = 0$, we get $D = 0$;

- when $n = 1$, we get $A + B + C = 1$;
- when $n = 2$, we get $8A + 4B + 2C = 4$;
- when $n = 3$, we get $27A + 9B + 3C = 10$.

This method assumes that we know that the answer is a polynomial and that we know its degree: it is called "the method of undetermined coefficients".

There are various ways of improving the basic method (such as writing the polynomial $An^3 + Bn^2 + Cn + D$ in the form

$$Pn(n-1)(n-2) + Qn(n-1) + Rn + S,$$

which tailors it to the values $n = 0, 1, 2, 3$ that one intends to substitute).

(b) Let $\mathbf{P}(n)$ be the statement:

"$T_1 + T_2 + T_3 + \cdots + T_n = \frac{n(n+1)(n+2)}{6}$".

- LHS of $\mathbf{P}(1) = T_1 = 1$; RHS of $\mathbf{P}(1) = \frac{1 \times 2 \times 3}{6} = 1$. Since these two are equal, $\mathbf{P}(1)$ is true.

- Suppose that $\mathbf{P}(k)$ is true for some particular (unspecified) $k \geqslant 1$; that is, we know that, for this particular k,

$$T_1 + T_2 + T_3 + \cdots + T_k = \frac{k(k+1)(k+2)}{6}.$$

We wish to prove that $\mathbf{P}(k+1)$ must then be true.

Now $\mathbf{P}(k+1)$ is an equation, so we start with the LHS of $\mathbf{P}(k+1)$ and try to simplify it in an appropriate way to get the RHS of $\mathbf{P}(k+1)$:

$$\begin{aligned} \text{LHS of } \mathbf{P}(k+1) &= T_1 + T_2 + T_3 + \cdots + T_k + T_{k+1} \\ &= [T_1 + T_2 + T_3 + \cdots + T_k] + T_{k+1}. \end{aligned}$$

If we now use $\mathbf{P}(k)$, which we assume is true, then the first bracket is equal to

$$\frac{k(k+1)(k+2)}{6}.$$

so the complete sum is equal to:

$$\begin{aligned} &= \frac{k(k+1)(k+2)}{6} + \frac{(k+1)(k+2)}{2} \\ &= \frac{(k+1)(k+2)(k+3)}{6} \\ &= \text{RHS of } \mathbf{P}(k+1). \end{aligned}$$

Hence $\mathbf{P}(k+1)$ holds.

If we combine these two bullet points, we have proved that "$\mathbf{P}(n)$ holds for all $n \geqslant 1$".

QED

6.10. Chapter 6: Comments and solutions

Note: The triangular numbers $T_1, T_2, T_3, \ldots, T_k, \ldots T_n$ are also equal to the binomial coefficients $\binom{k+1}{2}$. And the sum of these binomial coefficients is another binomial coefficient $\binom{n+2}{3}$, so the result in Problem **237** can be written as:

$$\binom{2}{2} + \binom{3}{2} + \binom{4}{2} + \cdots + \binom{n+1}{2} = \binom{n+2}{3}.$$

You might like to interpret Problem **237** in the language of binomial coefficients, and prove it by repeated use of the basic Pascal triangle relation (Pascal (1623–1662)):

$$\binom{k}{r} + \binom{k}{r+1} = \binom{k+1}{r+1}.$$

Start by rewriting

$$\binom{n+2}{3} = \binom{n+1}{2} + \binom{n+1}{3}.$$

238.

(a) We know from Problem **237**(b) that

$$1 \cdot 2 + 2 \cdot 3 + 3 \cdot 4 + \cdots + n(n+1) = \frac{n(n+1)(n+2)}{3}.$$

Also

$$\begin{aligned}
1 \cdot 2 + 2 \cdot 3 + 3 \cdot 4 + \cdots + n(n+1) &= 1 \cdot (1+1) + 2 \cdot (2+1) + 3 \cdot (3+1) \\
&\quad + \cdots + n \cdot (n+1) \\
&= (1^2 + 1) + (2^2 + 2) + (3^2 + 3) \\
&\quad + \cdots + (n^2 + n) \\
&= (1^2 + 2^2 + 3^2 + \cdots + n^2) \\
&\quad + (1 + 2 + 3 + \cdots + n).
\end{aligned}$$

Therefore

$$\begin{aligned}
1^2 + 2^2 + 3^2 + \cdots + n^2 &= \frac{n(n+1)(n+2)}{3} - \frac{n(n+1)}{2} \\
&= \frac{n(n+1)(2n+1)}{6}.
\end{aligned}$$

(b) **Guess:**

$$1 \cdot 2 \cdot 3 + 2 \cdot 3 \cdot 4 + 3 \cdot 4 \cdot 5 + \cdots + n(n+1)(n+2) = \frac{n(n+1)(n+2)(n+3)}{4}.$$

The proof by induction is entirely routine, and is left for the reader.

$$
\begin{aligned}
1\cdot 2\cdot 3 + 2\cdot 3\cdot 4 + \cdots + n(n+1)(n+2) &= 1\cdot(1+1)(1+2) + 2\cdot(2+1)(2+2) \\
&\quad + \cdots + n\cdot(n+1)(n+2) \\
&= (1^3 + 3\cdot 1^2 + 2\cdot 1) \\
&\quad + (2^3 + 3\cdot 2^2 + 2\cdot 2) \\
&\quad + \cdots + (n^3 + 3n^2 + 2n) \\
&= (1^3 + 2^3 + \cdots + n^3) \\
&\quad + 3(1^2 + 2^2 + \cdots + n^2) \\
&\quad + 2(1 + 2 + \cdots + n).
\end{aligned}
$$

Therefore

$$
\begin{aligned}
1^3 + 2^3 + 3^3 + \cdots + n^3 &= \frac{n(n+1)(n+2)(n+3)}{4} \\
&\quad - 3\left[\frac{n(n+1)(2n+1)}{6}\right] \\
&\quad - n(n+1) \\
&= \left[\frac{n(n+1)}{2}\right]^2.
\end{aligned}
$$

239.

(a) Let $f(x)$ be any such polynomial. If $f(a_k) = 0$, then we know (by the *Remainder Theorem*) that $f(x)$ has $(x - a_k)$ as a factor. Since the a_k are all distinct, and $f(a_k) = 0$ for each k, $0 \leq k \leq n-1$, we have

$$f(x) = (x - a_0)(x - a_1)(x - a_2) \cdots (x - a_{n-1}) \cdot g(x)$$

for some polynomial $g(x)$. And since we are told that $f(x)$ has degree n, $g(x)$ has degree 0, so is a constant. Hence every such polynomial of degree n has the form

$$C \cdot (x - a_0)(x - a_1)(x - a_2) \cdots (x - a_{n-1}).$$

Since $f(a_n) = b$, we can substitute to find C:

$$C = \frac{b}{(a_n - a_0)(a_n - a_1)(a_n - a_2) \cdots (a_n - a_{n-1})}.$$

(b) Let **P**(n) be the statement:

"Given any two labelled sets of $n+1$ real numbers $a_0, a_1, a_2, \ldots, a_n$, and $b_0, b_1, b_2, \ldots, b_n$, where the a_i are all distinct (but the b_i need not be), there exists a polynomial f_n of degree n, such that $f_n(a_0) = b_0$, $f_n(a_1) = b_1$, $f_n(a_2) = b_2$, ..., $f_n(a_n) = b_n$."

- When $n = 0$, we may choose $f_0(x) = b_0$ to be the constant polynomial. Hence **P**(0) is true.

- Suppose that $\mathbf{P}(k)$ is true for some particular (unspecified) $k \geqslant 0$; that is, we know that, for this particular k:

  "Given any two labelled sets of $k+1$ real numbers $a_0, a_1, a_2, \ldots, a_k$, and $b_0, b_1, b_2, \ldots, b_k$, where the a_i are all distinct (but the b_i need not be), there exists a polynomial f_k of degree k, such that $f_k(a_0) = b_0$, $f_k(a_1) = b_1$, $f_k(a_2) = b_2, \ldots, f_k(a_k) = b_k$."

  We wish to prove that $\mathbf{P}(k+1)$ must then be true.

  Now $\mathbf{P}(k+1)$ is the statement:

  "Given any two labelled sets of $(k+1)+1$ real numbers $a_0, a_1, \ldots, a_{k+1}$, and $b_0, b_1, b_2, \ldots, b_{k+1}$, where the a_i are all distinct (but the b_i need not be), there exists a polynomial f_{k+1} of degree $k+1$, such that
  $$f_{k+1}(a_0) = b_0, \ f_{k+1}(a_1) = b_1, \ f_{k+1}(a_2) = b_2, \ \ldots, \ f_{k+1}(a_{k+1}) = b_{k+1}."$$

  So to prove that $\mathbf{P}(k+1)$ holds, we must start by considering

  any two labelled sets of $(k+1)+1$ real numbers
  $$a_0, a_1, a_2, \ldots, a_{k+1}, \text{ and } b_0, b_1, b_2, \ldots, b_{k+1},$$
  where the a_i are all distinct (but the b_i need not be).

  We must then somehow construct a polynomial function f_{k+1} of degree $k+1$ with the required property.

  Because we are supposing that $\mathbf{P}(k)$ is known to be true, we can focus on the first $k+1$ numbers in each of the two lists – $a_0, a_1, a_2, \ldots, a_k$, and $b_0, b_1, b_2, \ldots, b_k$ – and we can then be sure that there is a polynomial f_k of degree k such that
  $$f_k(a_0) = b_0, f_k(a_1) = b_1, f_k(a_2) = b_2, \ldots, f_k(a_k) = b_k.$$

  The next step is slightly indirect: we make use of the polynomial f_{k+1} *which we are still trying to construct*, and focus on the polynomial
  $$f(x) = f_{k+1}(x) - f_k(x)$$
  satisfying
  $$f(a_0) = f(a_1) = \cdots = f(a_k) = 0, f(a_{k+1}) = b_{k+1} - f_k(a_{k+1}) = b \quad \text{(say)}.$$

  Part (a) guarantees the existence of such a polynomial $f(x)$ of degree $k+1$ and tells us exactly what this polynomial function $f(x)$ is equal to. Hence we can construct the required polynomial $f_{k+1}(x)$ by setting it equal to $f(x) + f_k(x)$, which proves that $\mathbf{P}(k+1)$ is true.

  If we combine these two bullet points, we have proved that "$\mathbf{P}(n)$ holds for all $n \geqslant 1$". QED

240.

(a) 5 cents cannot be made; $6 = 3+3$; $7 = 3+4$; $8 = 4+4$; $9 = 3+3+3$; etc.

Guess: Every amount $> N = 5$ can be paid directly (without receiving change).

(b) Let **P**(n) be the statement:

> "n cents can be paid directly (without change) using 3 cent and 4 cent coins".
>
> – **P**(6) is the statement: "6 cents can be paid directly". And $6 = 3 + 3$, so **P**(6) is true.
> – Now suppose that we know **P**(k) is true for some $k \geqslant 6$. We must show that **P**($k+1$) must then be true.
>
> To prove **P**($k+1$) we consider the statement **P**($k+1$):
>
> "$k+1$ cents can be paid directly".
>
> We know that **P**(k) is true, so we know that "k cents can be paid directly".
>
> – If a direct method of paying k cents involves at least one 3 cent coin, then we can replace one 3 cent coin by a 4 cent coin to produce a way of paying $k+1$ cents.
>
> Hence we only need to worry about a situation in which the only way to pay k cents directly involves no 3 cent coins at all – that is, paying k cents uses only 4 cent coins. But then there must be at least two 4 cent coins (since $k \geqslant 6$), and we can replace two 4 cent coins by three 3 cent coins to produce a way of paying $k+1$ cents directly.
>
> Hence
>
> - **P**(6) is true; and
> - whenever **P**(k) is true for some $k \geqslant 6$, we know that **P**($k+1$) is also true.
>
> ∴ **P**(n) is true for all $n \geqslant 6$. QED

241.

(a) $z^2 - z + 1 = 0$, so $z = \frac{1 \pm \sqrt{-3}}{2}$ (these are the two primitive sixth roots of unity). ∴ $z^2 = \frac{-1 \pm \sqrt{-3}}{2}$ (the two primitive cube toots of unity), and
$$z^2 + \frac{1}{z^2} = -1.$$

(b) $z^2 - 2z + 1 = 0$, so $z = 1$ (repeated root). ∴ $z^2 = 1$ and $z^2 + \frac{1}{z^2} = 2$.

(c), (d) $z^2 - 3z + 1 = 0$, so $z = \frac{3 \pm \sqrt{5}}{2}$.

As soon as one starts calculating z^2 and $\frac{1}{z^2}$, it becomes clear that it is time to p-a-u-s-e and think.
$$\left(z + \frac{1}{z}\right)^2 = \left(z^2 + \frac{1}{z^2}\right) + 2,$$
so whenever $z + \frac{1}{z} = k$ is an integer,
$$z^2 + \left(\frac{1}{z}\right)^2 = k^2 - 2$$

6.10. Chapter 6: Comments and solutions   335

is also an integer.

(e) Let $\mathbf{P}(n)$ be the statement:

"if z has the property that $z + \frac{1}{z}$ is an integer, then $z^m + \frac{1}{z^m}$ is also an integer for all m, $0 \leq m \leq n$".

- $\mathbf{P}(0)$ and $\mathbf{P}(1)$ are clearly both true; and $\mathbf{P}(2)$ was proved in part (d) above.
- Suppose that $\mathbf{P}(k)$ is true for some particular (unspecified) $k \geq 2$; that is, we know that, for this particular k:

    "if z has the property that $z + \frac{1}{z}$ is an integer, then $z^m + \frac{1}{z^m}$ is also an integer for all m, $0 \leq m \leq k$".

We wish to prove that $\mathbf{P}(k+1)$ must then be true.

If $z + \frac{1}{z}$ is an integer, then, by $\mathbf{P}(k)$,

"$z^m + \frac{1}{z^m}$ is also an integer for all m, $0 \leq m \leq k$".

So to prove that $\mathbf{P}(k+1)$ holds, we only need to show that

"$z^{k+1} + \frac{1}{z^{k+1}}$ is an integer".

By the Binomial Theorem:

$$\left(z + \frac{1}{z}\right)^{k+1} = \left(z^{k+1} + \frac{1}{z^{k+1}}\right) + \binom{k+1}{1}\left(z^{k-1} + \frac{1}{z^{k-1}}\right)$$
$$+ \binom{k+1}{2}\left(z^{k-3} + \frac{1}{z^{k-3}}\right) + \cdots$$

The LHS is an integer (since $z + \frac{1}{z}$ is an integer), and (by $\mathbf{P}(k)$) every term on the RHS is an integer except possibly the first. Hence the first term is the difference of two integers, so must also be an integer.

Hence $\mathbf{P}(k+1)$ is true.

If we combine these two bullet points, we have proved that "$\mathbf{P}(n)$ holds for all $n \geq 1$". QED

Note: If $k + 1 = 2m$ is even, the expansion of $\left(z + \frac{1}{z}\right)^{k+1}$ has an odd number of terms, so the RHS of the above re-grouped expansion ends with the term $\binom{2m}{m} \cdot z^m \cdot \left(\frac{1}{z}\right)^m$, which is also an integer.

242.

Note: In the solution to Problem **241** we included the condition on z as part of the statement $\mathbf{P}(n)$.

In Problem **242** the result to be proved has a similar background hypothesis – "Let p be a prime number". It may make the induction clearer if, as in the statement of the Problem, this hypothesis is stated *before* starting the induction proof.

Let p be any prime number. We let $\mathbf{P}(n)$ be the statement:

"$n^p - n$ is divisible by p ".

- **P**(1) is true (since $1^p - 1 = 0 = 0 \times p$, which is divisible by p).
- Suppose that **P**(k) is true for some particular (unspecified) $k \geqslant 1$; that is, we know that, for this particular k:

  "$k^p - k$ is divisible by p".

We wish to prove **P**($k+1$) – that is,

"$(k+1)^p - (k+1)$ is divisible by p"

must then be true. Using the Binomial Theorem again we see that

$$(k+1)^p - (k+1) = \left[k^p + \binom{p}{p-1}k^{p-1} + \binom{p}{p-2}k^{p-2} + \cdots + \binom{p}{1}k + 1\right]$$
$$-(k+1)$$
$$= (k^p - k) + \left[\binom{p}{p-1}k^{p-1} + \binom{p}{p-2}k^{p-2} + \cdots + \binom{p}{1}k\right].$$

By **P**(k), the first bracket on the RHS is divisible by p; and in each of the other terms each of the binomial coefficients $\binom{p}{r}$, $0 < r < p$,

- is an integer, and
- has a factor "p" in the numerator and no such factor in the denominator.

Hence each term in the second bracket is a multiple of p. So the RHS (and hence the LHS) is divisible by p.

Hence **P**($k+1$) is true.

If we combine these two bullet points, we have proved that "**P**(n) holds for all $n \geqslant 1$". QED

243.

$$0.037037037\ldots \text{ (for ever)} = \frac{37}{1\,000} + \frac{37}{1\,000\,000} + \frac{37}{1\,000\,000\,000} + \cdots \text{ (for ever)}.$$

This is a geometric series with first term $a = \frac{37}{1000}$ and common ratio $r = \frac{1}{1000}$, and so has sum

$$\frac{a}{1-r} = \frac{37}{999} = \frac{1}{27}.$$

244.

(a) Each person receives in total:

$$\frac{1}{4} + \left(\frac{1}{4}\right)^2 + \left(\frac{1}{4}\right)^3 + \left(\frac{1}{4}\right)^4 + \cdots \text{ (for ever)} = \frac{1}{3}$$

(here $a = \frac{1}{4} = r$).

(b) You have
$$1 - \frac{1}{2} + \frac{1}{4} - \frac{1}{8} + \cdots \text{ (for ever)} = \frac{2}{3}$$
(here $a = 1$, $r = -\frac{1}{2}$); I have
$$\frac{1}{2} - \frac{1}{4} + \frac{1}{8} - \cdots \text{ (for ever)} = \frac{1}{3}$$
(here $a = \frac{1}{2}$, $r = -\frac{1}{2}$).

245. The trains are 120 km apart, and the fly travels at 50 km/h towards Train B, which is initially 120 km away and travelling at 30 km/h.

The relative speed of the fly and Train B is 80 km/h, so it takes $\frac{3}{2}$ hours before they meet. In this time Train A and Train B have each travelled 45 km, so they are now 30 km apart. The fly then turns right round and flies back to Train A.

The relative speed of the fly and Train A is then also 80 km/h, so it takes just $\frac{3}{8}$ hours (or 22.5 minutes) for the fly to return to Train A. Train A and Train B have each travelled $\frac{45}{4}$ km in this time, so they are now $\frac{30}{4}$ km apart. The fly then turns round and flies straight back to Train B.

Train B is $\frac{30}{4}$ km away and the relative speed of the fly and Train B is again 80 km/h, so the journey takes $\frac{3}{32}$ hours (or 5.625 minutes).

Continuing in this way, we see that the fly takes
$$\frac{3}{2} + \frac{3}{8} + \frac{3}{32} + \frac{3}{128} + \cdots \text{ (for ever)} = 2 \text{ hours.}$$
Hence the fly travels 100 km before being squashed.

Note: The two trains are approaching each other at 60 km/h, so they crash in exactly 2 hours – during which time the fly travels 100 km.

246.

(a)(i) $3^2 > 2^2$; therefore
$$\frac{1}{3^2} < \frac{1}{2^2},$$
so
$$\frac{1}{2^2} + \frac{1}{3^2} < \frac{2}{2^2} = \frac{1}{2}.$$

(ii) $7^2 > 6^2 > 5^2 > 4^2$; therefore
$$\frac{1}{7^2} < \frac{1}{6^2} < \frac{1}{5^2} < \frac{1}{4^2},$$
so
$$\frac{1}{4^2} + \frac{1}{5^2} + \frac{1}{6^2} + \frac{1}{7^2} < \frac{4}{4^2} = \frac{1}{4}.$$

(b) The argument in part (a) gives an upper bound for each bracketed expression in the sum
$$\left(\frac{1}{1^2}\right) + \left(\frac{1}{2^2} + \frac{1}{3^2}\right) + \left(\frac{1}{4^2} + \frac{1}{5^2} + \frac{1}{6^2} + \frac{1}{7^2}\right) + \left(\frac{1}{8^2} + \cdots + \frac{1}{15^2}\right) + \cdots$$
Replacing each bracket by its upper bound, we see that the sum is
$$< \frac{1}{1^2} + \frac{2}{2^2} + \frac{4}{4^2} + \frac{8}{8^2} + \cdots$$
$$= 1 + \frac{1}{2} + \frac{1}{4} + \frac{1}{8} + \cdots \text{ (for ever)}$$
$$= 2.$$

(c) The finite partial sums
$$S_n = \frac{1}{1^2} + \frac{1}{2^2} + \frac{1}{3^2} + \cdots + \frac{1}{n^2}$$

- **increase steadily** as we take more and more terms, and
- every partial sum S_n is **less than 2**.

It is clear that these partial sums form a sequence
$$1 = S_1 < S_2 < S_3 < \cdots < S_n < S_{n+1} < \cdots < 2.$$
It follows that there is some (unknown) number $S \leqslant 2$ to which the partial sums converge as $n \to \infty$, and we take this (unknown) exact value S to be the exact value of the endless sum
$$\frac{1}{1^2} + \frac{1}{2^2} + \frac{1}{3^2} + \cdots + \frac{1}{n^2} + \cdots \text{ (for ever)}$$
To see, for example, that $S > \frac{17}{12}$, notice that
$$S > S_4$$
$$= \frac{1}{1^2} + \frac{1}{2^2} + \frac{1}{3^2} + \frac{1}{4^2}$$
$$= 1 + \frac{1}{4} + \frac{1}{9} + \frac{1}{16}$$
$$> \frac{17}{12}.$$

Note 1: The claim that
"an increasing sequence of partial sums S_n, all less than 2, must converge to some number $S \leqslant 2$"
is a fundamental property of the real numbers – called *completeness*.

Note 2: Just as one can obtain better and better lower bounds for S – like "$\frac{17}{12} < S$", so one can improve the upper bound "$S < 2$". For example, if in part (b) we avoid replacing the third term $\frac{1}{9}$ by $\frac{1}{4}$, we get a better upper bound "$S < \frac{67}{36}$", which is $\frac{5}{36}$ less than 2.

247.

(a) Let **P**(n) be the statement:

"$\frac{1}{1^2} + \frac{1}{2^2} + \frac{1}{3^2} + \cdots + \frac{1}{n^2} \leqslant 2 - \frac{1}{n}$".

- Then LHS of **P**(1) = $\frac{1}{1^2}$ = 1, and RHS of **P**(1) = $2 - 1 = 1$. Hence **P**(1) is true.
- Suppose we know that **P**(k) is true for some $k \geqslant 1$. We want to prove that **P**(k + 1) holds.

$$\begin{aligned}
\text{LHS of } \mathbf{P}(k+1) &= \frac{1}{1^2} + \frac{1}{2^2} + \frac{1}{3^2} + \cdots + \frac{1}{k^2} + \frac{1}{(k+1)^2} \\
&= \left[\frac{1}{1^2} + \frac{1}{2^2} + \frac{1}{3^2} + \cdots + \frac{1}{k^2}\right] + \frac{1}{(k+1)^2} \\
&\leqslant \left[2 - \frac{1}{k}\right] + \frac{1}{(k+1)^2} \\
&= 2 - \left[\frac{1}{k} - \frac{1}{(k+1)^2}\right] \\
&< 2 - \frac{1}{k+1}.
\end{aligned}$$

Hence **P**(k + 1) holds.
∴ **P**(1) holds; and whenever **P**(k) is known to be true, **P**(k + 1) is also true.
∴ **P**(n) is true, for all $n \geqslant 1$. QED

(b) Let **P**(n) be the statement:

"$\frac{1}{1^2} + \frac{1}{2^2} + \frac{1}{3^2} + \cdots + \frac{1}{n^2} < 1.68 - \frac{1}{n}$".

- Then
$$\text{LHS of } \mathbf{P}(4) = \frac{1}{1^2} + \frac{1}{2^2} + \frac{1}{3^2} + \frac{1}{4^2} = 1.423611111\cdots,$$
and RHS of **P**(4) = 1.43. Hence **P**(4) is true.

- Suppose we know that **P**(k) is true for some $k \geqslant 4$. We want to prove that **P**(k + 1) holds.

$$\begin{aligned}
\text{LHS of } \mathbf{P}(k+1) &= \frac{1}{1^2} + \frac{1}{2^2} + \frac{1}{3^2} + \cdots + \frac{1}{k^2} + \frac{1}{(k+1)^2} \\
&= \left(\frac{1}{1^2} + \frac{1}{2^2} + \frac{1}{3^2} + \cdots + \frac{1}{k^2}\right) + \frac{1}{(k+1)^2} \\
&< \left[1.68 - \frac{1}{k}\right] + \frac{1}{(k+1)^2} \\
&= 1.68 - \left[\frac{1}{k} - \frac{1}{(k+1)^2}\right] \\
&< 1.68 - \frac{1}{k+1}.
\end{aligned}$$

Hence **P**(k + 1) holds.
∴ **P**(1) holds; and whenever **P**(k) is known to be true, **P**(k + 1) is also true.

∴ **P**(n) is true, for all $n \geq 1$. QED

248.

(a)(i) $n! = n \times (n-1) \times (n-2) \times \cdots \times 3 \times 2 \times 1 \geq 2 \times 2 \times 2 \times \cdots \times 2 \times 1 = 2^{n-1}$ whenever $n \geq 1$.

∴ $\frac{1}{n!} \leq \left(\frac{1}{2}\right)^{n-1}$ for all $n \geq 1$.

∴ $\frac{1}{0!} + \frac{1}{1!} + \frac{1}{2!} + \cdots + \frac{1}{n!} \leq 1 + \left[1 + \frac{1}{2} + \left(\frac{1}{2}\right)^2 + \cdots + \left(\frac{1}{2}\right)^{n-1}\right] < 3$ for all $n \geq 0$.

(ii) As we go on adding more and more terms, each finite sum

$$\frac{1}{0!} + \frac{1}{1!} + \frac{1}{2!} + \cdots + \frac{1}{n!}$$

is bigger than the previous sum. Since every finite sum

$$\frac{1}{0!} + \frac{1}{1!} + \frac{1}{2!} + \cdots + \frac{1}{n!} < 3,$$

the sums increase, but never reach 3, so they accumulate closer and closer to a value "e" ≤ 3. Moreover, this value "e" is certainly larger than the sum of the first two terms $\frac{1}{0!} + \frac{1}{1!} = 2$, so $2 < e \leq 3$.

(b)(i) Let **P**(n) be the statement:

"$\frac{1}{0!} + \frac{1}{1!} + \frac{1}{2!} + \cdots + \frac{1}{n!} \leq 3 - \frac{1}{n \cdot n!}$".

- LHS of **P**(1) = 2 = RHS of **P**(1). Hence **P**(1) is true.
- Suppose we know that **P**(k) is true for some $k \geq 1$. We want to prove that **P**(k + 1) holds.

$$\begin{aligned}
\text{LHS of } \mathbf{P}(k+1) &= \frac{1}{0!} + \frac{1}{1!} + \frac{1}{2!} + \cdots + \frac{1}{(k+1)!} \\
&= \left[\frac{1}{0!} + \frac{1}{1!} + \frac{1}{2!} + \cdots + \frac{1}{k!}\right] + \frac{1}{(k+1)!} \\
&\leq 3 - \frac{1}{k \cdot k!} + \frac{1}{(k+1)!} \\
&= 3 - \frac{1}{k(k+1)!} \\
&< 3 - \frac{1}{(k+1) \cdot (k+1)!}
\end{aligned}$$

Hence **P**(k + 1) holds.

∴ **P**(1) holds; and whenever **P**(k) is known to be true, **P**(k + 1) is also true.
∴ **P**(n) is true, for all $n \geq 1$. QED

(ii) [The reasoning here uses the constant "3" while ignoring the refinement "$3 - \frac{1}{n \cdot n!}$", and so sounds exactly like part (a)(ii).] As we add more terms, each finite sum

$$\frac{1}{0!} + \frac{1}{1!} + \frac{1}{2!} + \cdots + \frac{1}{n!}$$

is bigger than the previous sum. Since every finite sum

$$\frac{1}{0!} + \frac{1}{1!} + \frac{1}{2!} + \cdots + \frac{1}{n!} < 3,$$

the partial sums increase, but never reach 3; so they accumulate closer and closer to a value "e" $\leqslant 3$. Moreover, this value "e" is certainly larger than the sum of the first three terms $\frac{1}{0!} + \frac{1}{1!} + \frac{1}{2!} = 2.5$, so $2.5 < e \leqslant 3$.

(c) **Note:** Examine carefully the role played by the number "3" in the above induction proof in (b)(ii). It is needed precisely to validate the statement $\mathbf{P}(1)$: since $\frac{1}{0!} + \frac{1}{1!} = 3 - \frac{1}{1 \times 1!}$". But the number "3" plays no active part in the second induction step, and could be replaced by any other number we choose.

The exact value "e" of the infinite series is not really affected by what happens when $n = 1$. Suppose we ask: "What number C_2 should replace "3" if we only want to prove that

$$\frac{1}{0!} + \frac{1}{1!} + \frac{1}{2!} + \cdots + \frac{1}{n!} \leqslant C_2 - \frac{1}{n \cdot n!}, \text{ for all } n \geqslant 2?$$

The only part of the induction proof where C_2 then matters is when we try to check that $\mathbf{P}(2)$ holds; so we must choose the smallest possible C_2 to satisfy

$$\frac{1}{0!} + \frac{1}{1!} + \frac{1}{2!} \leqslant C_2 - \frac{1}{2 \cdot 2!}:$$

that is, $C_2 = 2.75$.

(i) Let $\mathbf{P}(n)$ be the statement:
"$\frac{1}{0!} + \frac{1}{1!} + \frac{1}{2!} + \cdots + \frac{1}{n!} \leqslant 2.75 - \frac{1}{n \cdot n!}$".

- LHS of $\mathbf{P}(2) = 2.5$; RHS of $\mathbf{P}(2) = 2.75 - \frac{1}{4}$. Hence $\mathbf{P}(2)$ is true.
- Suppose we know that $\mathbf{P}(k)$ is true for some $k \leqslant 2$. We want to prove that $\mathbf{P}(k+1)$ holds.

$$\begin{aligned}
\text{LHS of } \mathbf{P}(k+1) &= \frac{1}{0!} + \frac{1}{1!} + \frac{1}{2!} + \cdots + \frac{1}{(k+1)!} \\
&= \left[\frac{1}{0!} + \frac{1}{1!} + \frac{1}{2!} + \cdots + \frac{1}{k!} \right] + \frac{1}{(k+1)!} \\
&\leqslant 2.75 - \frac{1}{k \cdot k!} + \frac{1}{(k+1)!} \\
&= 2.75 - \frac{1}{k(k+1)!} \\
&< 2.75 - \frac{1}{(k+1)(k+1)!}
\end{aligned}$$

Hence $\mathbf{P}(k+1)$ holds.

∴ $\mathbf{P}(2)$ holds; and whenever $\mathbf{P}(k)$ is known to be true, $\mathbf{P}(k+1)$ is also true.
∴ $\mathbf{P}(n)$ is true, for all $n \geqslant 2$. QED

(ii) As we add more terms, each finite sum

$$\frac{1}{0!} + \frac{1}{1!} + \frac{1}{2!} + \cdots + \frac{1}{n!}$$

is bigger than the previous sum.
Since every finite sum

$$\frac{1}{0!} + \frac{1}{1!} + \frac{1}{2!} + \cdots + \frac{1}{n!} < 2.75,$$

the finite sums increase, but never reach 2.75, so they accumulate closer and closer to a value "e" $\leqslant 2.75$. Moreover, this value "e" is certainly larger than the sum of the first four terms

$$\frac{1}{0!} + \frac{1}{1!} + \frac{1}{2!} + \frac{1}{3!} > 2.66,$$

so $2.66 < e \leqslant 2.75$.

(d)(i) Let $\mathbf{P}(n)$ be the statement:
"$\frac{1}{0!} + \frac{1}{1!} + \frac{1}{2!} + \cdots + \frac{1}{n!} \leqslant 2.7222\cdots$ (for ever) $- \frac{1}{n.n!}$".

- LHS of $\mathbf{P}(3) = \frac{1}{0!} + \frac{1}{1!} + \frac{1}{2!} + \frac{1}{3!} = 2.666\cdots$ (for ever);
  RHS of $\mathbf{P}(3) = 2.7222\cdots$ (for ever) $- \frac{1}{18} = 2.666\cdots$ (for ever).
  Hence $\mathbf{P}(3)$ is true.

- Suppose we know that $\mathbf{P}(k)$ is true for some $k \geqslant 3$. We want to prove that $\mathbf{P}(k+1)$ holds.

$$\begin{aligned}
\text{LHS of } \mathbf{P}(k+1) &= \frac{1}{0!} + \frac{1}{1!} + \frac{1}{2!} + \cdots + \frac{1}{(k+1)!} \\
&= \left[\frac{1}{0!} + \frac{1}{1!} + \frac{1}{2!} + \cdots + \frac{1}{k!}\right] + \frac{1}{(k+1)!} \\
&\leqslant 2.7222\cdots \text{ (for ever)} - \frac{1}{k \cdot k!} + \frac{1}{(k+1)!} \\
&= 2.7222\cdots \text{ (for ever)} - \frac{1}{k(k+1)!} \\
&< 2.7222\cdots \text{ (for ever)} - \frac{1}{(k+1)(k+1)!}
\end{aligned}$$

Hence $\mathbf{P}(k+1)$ holds.
∴ $\mathbf{P}(3)$ holds; and whenever $\mathbf{P}(k)$ is known to be true, $\mathbf{P}(k+1)$ is also true.
∴ $\mathbf{P}(n)$ is true, for all $n \geqslant 3$. QED

(ii) As we add more terms, each finite sum

$$\frac{1}{0!} + \frac{1}{1!} + \frac{1}{2!} + \cdots + \frac{1}{n!}$$

is bigger than the previous sum.

Since every finite sum
$$\frac{1}{0!} + \frac{1}{1!} + \frac{1}{2!} + \cdots + \frac{1}{n!} < 2.7222\cdots \text{ (for ever)},$$
the finite sums increase, but never reach $2.7222\cdots$ (for ever), so they accumulate closer and closer to a value "e" $\leqslant 2.7222\cdots$ (for ever). Moreover, this value "e" is certainly larger than the sum of the first five terms
$$\frac{1}{0!} + \frac{1}{1!} + \frac{1}{2!} + \frac{1}{3!} + \frac{1}{4!} > 2.708,$$
so $2.708 < e \leqslant 2.7222\cdots$ (for ever).

Note: This process of refinement can continue indefinitely. But we only have to go one further step to pin down the value of "e" with surprising accuracy. The next step uses the same proof to show that
$$\text{``}\frac{1}{0!} + \frac{1}{1!} + \frac{1}{2!} + \cdots + \frac{1}{n!} \leqslant 2.7185 - \frac{1}{n \cdot n!}, \text{ for all } n \geqslant 4\text{''},$$
and to conclude that the endless sum
$$\frac{1}{0!} + \frac{1}{1!} + \frac{1}{2!} + \cdots + \frac{1}{n!} + \cdots \text{ (for ever)}$$
has a definite value "e" that lies somewhere between 2.716 and 2.71875. We could then repeat the same proof to show that
$$\frac{1}{0!} + \frac{1}{1!} + \frac{1}{2!} + \cdots + \frac{1}{n!} \leqslant 2.718333\cdots \text{ (for ever)} - \frac{1}{n \cdot n!}, \text{ for all } n \geqslant 5,$$
and use the lower bound $2.7177\ldots$ from the first seven terms to conclude that the endless sum
$$\frac{1}{0!} + \frac{1}{1!} + \frac{1}{2!} + \cdots + \frac{1}{n!} + \cdots \text{ (for ever)}$$
has a definite value "e" that lies somewhere between 2.7177 and $2.718333\cdots$ (for ever). And so on.

249. Let $\mathbf{P}(n)$ be the statement:
$$\text{``}\frac{1}{\sqrt{1}} + \frac{1}{\sqrt{2}} + \frac{1}{\sqrt{3}} + \cdots + \frac{1}{\sqrt{n}} \geqslant \sqrt{n}\text{''}.$$

- LHS of $\mathbf{P}(1) = 1 =$ RHS of $\mathbf{P}(1)$. Hence $\mathbf{P}(1)$ is true.

- Suppose we know that $\mathbf{P}(k)$ is true for some $k \geq 1$. We want to prove that $\mathbf{P}(k+1)$ holds.

$$\begin{aligned}
\text{LHS of } \mathbf{P}(k+1) &= \frac{1}{\sqrt{1}} + \frac{1}{\sqrt{2}} + \frac{1}{\sqrt{3}} + \cdots + \frac{1}{\sqrt{k+1}} \\
&= \left(\frac{1}{\sqrt{1}} + \frac{1}{\sqrt{2}} + \frac{1}{\sqrt{3}} + \cdots + \frac{1}{\sqrt{k}}\right) + \frac{1}{\sqrt{(k+1)}} \\
&\geq \sqrt{k} + \frac{1}{\sqrt{k+1}} \\
&\geq \sqrt{k+1} \quad \left(\text{since } \frac{1}{\sqrt{k+1}} \geq \frac{1}{\sqrt{k+1}+\sqrt{k}}\right).
\end{aligned}$$

Hence $\mathbf{P}(k+1)$ holds.

∴ $\mathbf{P}(1)$ holds; and whenever $\mathbf{P}(k)$ is known to be true, $\mathbf{P}(k+1)$ is also true.
∴ $\mathbf{P}(n)$ is true, for all $n \geq 1$. QED

250. Let a, b be real numbers such that $a \neq b$, and $a + b > 0$.

One of a, b is then the greater, and we may suppose this is a – so that $a > b$. If $a > b > 0$, then $a^n > b^n > 0$ for all n; if $b < 0$, then $a + b > 0$ implies that $a = |a| > |b|$, so $a^n > b^n$ for all n.

Let $\mathbf{P}(n)$ be the statement:

"$\dfrac{a^n + b^n}{2} \geq \left(\dfrac{a+b}{2}\right)^n$".

- LHS of $\mathbf{P}(1) = \frac{a+b}{2} =$ RHS of $\mathbf{P}(1)$. Hence $\mathbf{P}(1)$ is true.
- Suppose we know that $\mathbf{P}(k)$ is true for some $k \geq 1$. We want to prove that $\mathbf{P}(k+1)$ holds.

$$\begin{aligned}
\text{RHS of } \mathbf{P}(k+1) &= \left(\frac{a+b}{2}\right)^{k+1} \\
&= \frac{a+b}{2} \cdot \left(\frac{a+b}{2}\right)^k \\
&\leq \frac{a+b}{2} \cdot \frac{a^k + b^k}{2} \quad \text{(by } \mathbf{P}(k)) \\
&= \frac{a^{k+1} + b^{k+1}}{4} + \frac{ab^k + ba^k}{4} \\
&< \frac{a^{k+1} + b^{k+1}}{2} \quad (\text{since } (a^k - b^k)(a - b) > 0).
\end{aligned}$$

Hence $\mathbf{P}(k+1)$ holds.
∴ $\mathbf{P}(1)$ holds; and whenever $\mathbf{P}(k)$ is known to be true, $\mathbf{P}(k+1)$ is also true.
∴ $\mathbf{P}(n)$ is true, for all $n \geq 1$. QED

251. Let x be any real number $\geqslant -1$.
If $x = -1$, then $(1+x)^n = 0 \geqslant 1 - n = 1 + nx$, for all $n \geqslant 1$.
Thus we may assume that $x > -1$, so $1 + x > 0$.
Let $\mathbf{P}(n)$ be the statement: "$(1+x)^n \geqslant 1 + nx$".

- LHS of $\mathbf{P}(1) = 1 + x =$ RHS of $\mathbf{P}(1)$. Hence $\mathbf{P}(1)$ is true.
- Suppose we know that $\mathbf{P}(k)$ is true for some $k \geqslant 1$. We want to prove that $\mathbf{P}(k+1)$ holds.

$$\begin{aligned}
\text{LHS of } \mathbf{P}(k+1) &= (1+x)^{k+1} \\
&= (1+x) \cdot (1+x)^k \\
&\geqslant (1+x) \cdot (1+kx) \quad (\text{by } \mathbf{P}(k), \text{ since } 1 + x > 0) \\
&= 1 + (k+1)x + kx^2 \\
&\geqslant 1 + (k+1)x
\end{aligned}$$

Hence $\mathbf{P}(k+1)$ holds.
$\therefore$ $\mathbf{P}(1)$ holds; and whenever $\mathbf{P}(k)$ is known to be true, $\mathbf{P}(k+1)$ is also true.
$\therefore$ $\mathbf{P}(n)$ is true, for all $n \geqslant 1$. QED

252. The problem is discussed after the statement of Problem **252** in the main text.

253.

(a)(i) $3 > 2$, so $\frac{1}{3} < \frac{1}{2}$.
$\therefore \frac{1}{2} + \frac{1}{3} < \frac{1}{2} + \frac{1}{2} = 1$.

(ii) $5, 6, 7 > 4$; hence $\frac{1}{5}, \frac{1}{6}, \frac{1}{7}$ are all $< \frac{1}{4}$.
$\therefore \frac{1}{4} + \frac{1}{5} + \frac{1}{6} + \frac{1}{7} < \frac{1}{4} + \frac{1}{4} + \frac{1}{4} + \frac{1}{4} = 1$.

(iii) Let $\mathbf{P}(n)$ be the statement:
"$\frac{1}{1} + \frac{1}{2} + \frac{1}{3} + \cdots + \frac{1}{2^n - 1} < n$".
Then
- $\mathbf{P}(2)$ is true by (i), since

$$\frac{1}{1} + \frac{1}{2} + \frac{1}{3} < 1 + \left(\frac{1}{2} + \frac{1}{2}\right) = 2.$$

- Suppose that $\mathbf{P}(k)$ is true for some $k \geqslant 2$.

$$\text{LHS of } \mathbf{P}(k+1) = \left[\frac{1}{1} + \frac{1}{2} + \frac{1}{3} + \cdots + \frac{1}{2^k - 1}\right]$$
$$+ \left[\frac{1}{2^k} + \cdots + \frac{1}{2^{k+1} - 1}\right].$$

The first bracket is $< k$ (by $\mathbf{P}(k)$); and each of the 2^k terms in the second bracket is $\leqslant \frac{1}{2^k}$, so the whole bracket is $\leqslant 1$.
Hence the LHS of $\mathbf{P}(k+1) < k+1$, so $\mathbf{P}(k+1)$ is true.
Hence $\mathbf{P}(n)$ is true for all $n \geqslant 2$.

(iv) At the very first stage (part (i)) we replaced $\frac{1}{2} + \frac{1}{3}$ by $\frac{1}{2} + \frac{1}{2} = 1$. Hence when $n \geqslant 2$, we know that the two sides of $\mathbf{P}(n)$ differ by at least $\frac{1}{2} - \frac{1}{3}$. This difference is greater than $\frac{1}{2^n}$ when $n \geqslant 3$, so (iv) follows.

(b)(i) $3 < 4$, so $\frac{1}{3} > \frac{1}{4}$.
$\therefore \frac{1}{3} + \frac{1}{4} > \frac{1}{4} + \frac{1}{4} = \frac{1}{2}$.

(ii) $5, 6, 7 < 8$; hence $\frac{1}{5}, \frac{1}{6}, \frac{1}{7}$ are all $> \frac{1}{8}$.
$\therefore \frac{1}{5} + \frac{1}{6} + \frac{1}{7} + \frac{1}{8} > \frac{1}{8} + \frac{1}{8} + \frac{1}{8} + \frac{1}{8} = \frac{1}{2}$.

(iii) Let $\mathbf{P}(n)$ be the statement:
"$\frac{1}{1} + \frac{1}{2} + \frac{1}{3} + \cdots + \frac{1}{2^n} > 1 + \frac{n}{2}$".
Then

- $\mathbf{P}(2)$ is true by (i), since

$$\frac{1}{1} + \frac{1}{2} + \frac{1}{3} + \frac{1}{4} = 1 + \frac{1}{2} + \left(\frac{1}{3} + \frac{1}{4}\right)$$
$$> 1 + \frac{1}{2} + \left(\frac{1}{4} + \frac{1}{4}\right)$$
$$= 1 + 2 \times \frac{1}{2}.$$

- Suppose that $\mathbf{P}(k)$ is true for some $k \geqslant 2$.
LHS of $\mathbf{P}(k+1) = \left[\frac{1}{1} + \frac{1}{2} + \frac{1}{3} + \cdots + \frac{1}{2^k}\right] + \left[\frac{1}{2^k+1} + \cdots + \frac{1}{2^{k+1}}\right]$.
The first bracket is $> 1 + \frac{k}{2}$ (by $\mathbf{P}(k)$);
and each of the 2^k terms in the second bracket is $\geqslant \frac{1}{2^{k+1}}$, so the whole bracket is $\geqslant \frac{1}{2}$.
Hence the LHS of $\mathbf{P}(k+1) > 1 + \frac{k+1}{2}$, so $\mathbf{P}(k+1)$ is true.
Hence $\mathbf{P}(n)$ is true for all $n \geqslant 2$.

254.

(a) We use induction. Let $\mathbf{P}(n)$ be the statement:

"n identical rectangular strips of length 2 balance *exactly* on the edge of a table if the successive protrusion distances (first beyond the edge of the table, then beyond the leading edge of the strip immediately below, and so on) are the terms

$$\frac{1}{n}, \frac{1}{n-1}, \frac{1}{n-2}, \ldots, \frac{1}{3}, \frac{1}{2}, \frac{1}{1}$$

of the finite harmonic series in reverse order."

- When $n = 1$, a single strip which protrudes distance 1 beyond the edge of the table has its centre of gravity exactly over the edge of the table. Hence $\mathbf{P}(1)$ is true.
- Suppose that we know that $\mathbf{P}(k)$ is true for some $k \geqslant 1$.
  Let $k + 1$ identical strips be arranged as described in the statement $\mathbf{P}(k+1)$. The statement $\mathbf{P}(k)$ guarantees that the top k strips would exactly balance if the leading edge of the bottom strip were in fact the edge of the table; hence the combined centre of gravity of the top k strips is positioned exactly over the leading edge of the bottom strip.
  Let M be the mass of each strip; since the leading edge of the bottom strip is distance $\frac{1}{k+1}$ beyond the edge of the table, the top k strips produce a combined moment about the edge of the table equal to $kM \times \frac{1}{k+1}$.
  The centre of gravity of the bottom strip is distance $1 - \frac{1}{k+1} = \frac{k}{k+1}$ from the edge of the table in the opposite direction; hence it contributes a moment about the edge of the table equal to $M \times \left(-\frac{k}{k+1}\right)$.

  $\therefore$ the total moment of the whole stack about the edge of the table is equal to zero, so the centre of gravity of the combined stack of $k + 1$ strips lies exactly over the edge of the table. Hence $\mathbf{P}(k+1)$ is true.

Hence $\mathbf{P}(n)$ is true for all $n \geqslant 1$.

(b) Problem **253**(b)(iii) now guarantees that a stack of 2^n strips can protrude a distance $> 1 + \frac{n}{2}$ beyond the edge of the table.

Note: Ivars Petersen's *Mathematical Tourist* blog contains an entry in 2009

http://mathtourist.blogspot.com/2009/01/overhang.html

which explores how one can obtain large overhangs with fewer strips if one is allowed to use strips to *counterbalance* those that protrude beyond the edge of the table.

255.

(a)(i) Let $\mathbf{P}(n)$ be the statement:

"$s_{2p-2} < s_{2p} < s_{2q+1} < s_{2q-1}$ for all p, q such that $1 \leqslant p, q \leqslant n$".

- $\mathbf{P}(1)$ is true (since s_0 is the empty sum, so

$$0 = s_0 < s_2 = \frac{1}{2} < s_3 = \frac{5}{6} < s_1 = 1.$$

- Suppose that $\mathbf{P}(k)$ is true for some $k \geqslant 1$. Then most of the inequalities in the statement $\mathbf{P}(k+1)$ are part of the statement $\mathbf{P}(k)$; the only outstanding inequalities which remain to be proved are:

$$s_{2k} < s_{2k+2} < s_{2k+3} < s_{2k+1}.$$

which are true, since
$$s_{2k+3} = s_{2k+2} + \frac{1}{2k+3} = s_{2k+1} - \frac{1}{2k+2} + \frac{1}{2k+3}$$
and
$$s_{2k+2} = s_{2k} + \frac{1}{2k+1} - \frac{1}{2k+2}.$$
Hence **P**(k + 1) is true.

Hence **P**(n) is true for all $n \geqslant 2$.

(ii) The "even sums" $s_0, s_2, s_4, \ldots$ are increasing, but all are less than $s_1 = 1$, so they approach some value $s_{\text{even}} \leqslant 1$.
The "odd sums" $s_1, s_3, s_5, \ldots$ are decreasing, but all are greater than $s_2 = \frac{1}{2}$, so they approach some value $s_{\text{odd}} \geqslant \frac{1}{2}$.
The "even sums" $s_0, s_2, s_4, \ldots$ are increasing, but all are less than $s_5 = \frac{47}{60}$, so they approach some value $s_{\text{even}} \leqslant \frac{47}{60}$.
The "odd sums" $s_1, s_3, s_5, \ldots$ are decreasing, but all are greater than $s_6 = \frac{37}{60}$, so they approach some value $s_{\text{odd}} \geqslant \frac{37}{60}$.
Moreover, the difference between successive sums is $\frac{1}{n}$, and this tends towards zero, so the difference between each "odd sum" and the next "even sum" tends to zero, so $s_{\text{even}} = s_{\text{odd}}$.

(b) The proof from part (a) carries over word for word, with "$\frac{1}{k}$" replaced at each stage by "a_k".

256. Let **P**(n) be the statement:

"f_k has at least k distinct prime factors".

- $f_1 = 2$ has exactly 1 prime factor, so **P**(1) is true.
- Suppose that **P**(k) is true for some $k \geqslant 1$.
  Then $f_{k+1} = f_k(f_k + 1)$. The first factor f_k has at least k distinct prime factors. And the second factor $f_k + 1 > f_k > 1$, so has at least one prime factor. Moreover $HCF(f_k, f_k + 1) = 1$, so the second bracket has no factor in common with f_k.
  Hence f_{k+1} has at least $k + 1$ distinct prime factors, so **P**(k + 1) is true.

Hence **P**(n) is true for all $n \geqslant 1$.

Note: This problem [suggested by Serkan Dogan] gives a different proof of the result (Problem **76**(d)) that the list of prime numbers goes on for ever.

257.

(a) Let **P**(n) be the statement: "n distinct points on a straight line divide the line into $n + 1$ intervals".

- 0 points leave the line in pristine condition – namely a single interval – so **P**(0) is true.

- Suppose that **P**(k) is true for some $k \geq 0$.
  Consider an arbitrary straight line divided by $k+1$ points $A_0, A_1, \ldots, A_k$.
  Then the k points $A_1, \ldots, A_k$ divide the line into $k+1$ intervals (by **P**(k)).
  The additional point A_0 is distinct from $A_1, \ldots, A_k$ and so must lie inside one of these $k+1$ intervals, and divides it in two – giving $(k+1)+1 = k+2$ intervals altogether.
  Hence **P**$(k+1)$ is true.

  Hence **P**(n) is true for all $n \geq 0$.

(b) (i) We want a function R satisfying

$$R_0 = 1, \ R_1 = 2, \ R_2 = 4, \ R_3 = 7.$$

If we notice that in part (a)

$$n + 1 = 1 + \binom{n}{1},$$

then we might guess that

$$R_n = 1 + \binom{n}{1} + \binom{n}{2}.$$

(ii) Let **P**(n) be the statement:

"n distinct straight lines in the plane divide the plane into at most

$$f(n) = 1 + \binom{n}{1} + \binom{n}{2}$$

regions".

- 0 lines leave the plane in pristine condition – namely a single region – so **P**(0) is true, provided that

$$1 + \binom{0}{1} + \binom{0}{2} = 1.$$

- Suppose that **P**(k) is true for some $k \geq 0$.
  Consider the plane divided by $k+1$ straight lines $m_0, m_1, \ldots, m_k$.
  Then the k lines $m_1, \ldots, m_k$ divide the plane into at most

$$R_k = 1 + \binom{k}{1} + \binom{k}{2}$$

regions (by **P**(k)).
The additional line m_0 is distinct from $m_1, \ldots, m_k$ and so meets each of these lines in at most a single point – giving at most k points on the line m_0. These points divide m_0 into at most $k+1$ intervals, and each of these intervals corresponds to a cut-line, where the line m_0 cuts one of the regions created

by the lines $m_1, m_2, \ldots, m_k$ into **two** pieces – giving at most

$$\begin{aligned}
R_k + (k+1) &= 1 + \binom{k}{1} + \binom{k}{2} + k + 1 \\
&= 1 + \binom{k+1}{1} + \binom{k+1}{2} \\
&= R_{k+1}
\end{aligned}$$

regions altogether.
Hence $\mathbf{P}(k+1)$ is true.

Hence $\mathbf{P}(n)$ is true for all $n \geqslant 0$.

(c) (i) We want a function S satisfying

$$S_0 = 1,\ S_1 = 2,\ S_2 = 4,\ S_3 = 8,\ S_4 = 15, \ldots$$

After thinking about the differences between successive terms in part (b), we might guess that

$$S_n = \binom{n}{0} + \binom{n}{1} + \binom{n}{2} + \binom{n}{3}.$$

(ii) Let $\mathbf{P}(n)$ be the statement:

"n distinct planes in 3-space divide space into at most

$$S_n = \binom{n}{0} + \binom{n}{1} + \binom{n}{2} + \binom{n}{3}$$

regions".

- 0 planes leave our 3D space in pristine condition – namely a single region – so $\mathbf{P}(0)$ is true – provided that

$$\binom{0}{0} + \binom{0}{1} + \binom{0}{2} + \binom{0}{3} = 1.$$

- Suppose that $\mathbf{P}(k)$ is true for some $k \geqslant 0$.
Consider 3D divided by $k+1$ planes $m_0, m_1, \ldots, m_k$.
Then the k planes $m_1, \ldots, m_k$ divide 3D into at most

$$S_k = \binom{k}{0} + \binom{k}{1} + \binom{k}{2} + \binom{k}{3}$$

regions (by $\mathbf{P}(k)$).
The additional plane m_0 is distinct from $m_1, \ldots, m_k$ and so meets each of these planes in (at most) a line – giving rise to at most k lines on the plane

m_0. This arrangement of lines on the plane m_0 divides m_0 into at most

$$R_k = 1 + \binom{k}{1} + \binom{k}{2}$$

regions, and each of these regions on the plane m_0 is the "cut" where the plane m_0 cuts an existing region into **two** pieces – giving rise to at most

$$\begin{aligned} S_k + R_k &= \left[\binom{k}{0} + \binom{k}{1} + \binom{k}{2} + \binom{k}{3}\right] + \left[1 + \binom{k}{1} + \binom{k}{2}\right] \\ &= \binom{k+1}{0} + \binom{k+1}{1} + \binom{k+1}{2} + \binom{k+1}{3} \\ &= S_{k+1} \end{aligned}$$

regions altogether. (There is no need for any algebra here: one only needs to use the Pascal triangle condition.)

Hence $\mathbf{P}(k+1)$ is true whenever $\mathbf{P}(k)$ is true.

Hence $\mathbf{P}(n)$ is true for all $n \geqslant 0$.

258. Notice that, given a dissection of a square into k squares, we can always cut one square into four quarters (by lines through the centre, and parallel to the sides), and so create a dissection with $k+3$ squares.

Let $\mathbf{P}(n)$ be the statement:

"Any given square can be cut into m (not necessarily congruent) smaller squares, for each m, $6 \leqslant m \leqslant n$".

- Let $n = 6$. Given any square of side s (say). We may cut a square of side $\frac{2s}{3}$ from one corner, leaving an L-shaped strip of width $\frac{s}{3}$, which we can then cut into 5 smaller squares, each of side $\frac{s}{3}$. Hence $\mathbf{P}(6)$ is true.

  Let $n = 7$. Given any square, we can divide the square first into four quarters; then divide one of these smaller squares into four quarters to obtain a dissection into 7 smaller squares. Hence $\mathbf{P}(7)$ is true.

  Let $n = 8$. Given a square of side s (say). We may cut a square of side $\frac{3s}{4}$ from one corner, leaving an L-shaped strip of width $\frac{s}{4}$, which we can then cut into 7 smaller squares, each of side $\frac{s}{4}$. Hence $\mathbf{P}(8)$ is true.

- Suppose that $\mathbf{P}(k)$ is true for some $k \geqslant 8$.

  Then $k - 2 \geqslant 6$, so any given square can be dissected into $k-2$ smaller squares (by $\mathbf{P}(k)$). Taking this dissection and dividing one of the smaller squares into four quarters gives a dissection of the initial square into $k-2+3$ squares. Hence $\mathbf{P}(k+1)$ is true.

Hence $\mathbf{P}(n)$ is true for all $n \geqslant 6$.

259. Let $\mathbf{P}(n)$ be the statement:

"Any tree with n vertices has exactly $n-1$ edges".

- A tree with 1 vertex is simply a vertex with 0 edges (since any edge would have to be a loop, and would then create a cycle). Hence $\mathbf{P}(1)$ is true.
- Suppose that $\mathbf{P}(k)$ is true for some $k \geqslant 1$.

Consider an arbitrary tree T with $k+1$ vertices.

[**Idea:** We need to find some way of reducing T to a tree T' with k vertices. This suggests "removing an end vertex". So we must first prove that "any tree T has an end vertex".]

Definition The number of edges incident with a given vertex v is called the *valency* of v.

Lemma Let S be a finite tree with $s > 1$ vertices. Then S has a vertex of valency 1 – that is an "end vertex".

Proof Choose any vertex v_0. Then v_0 must be connected to the rest of the tree, so v_0 has valency at least 1.

If v_0 is an "end vertex", then stop; if not, then choose a vertex v_1 which is adjacent to v_0.

If v_1 is an "end vertex", then stop; if not, choose a vertex $v_2 \neq v_0$ which is adjacent to v_1.

If v_2 is an "end vertex", then stop; if not, choose a vertex $v_3 \neq v_1$ which is adjacent to v_2.

Continue in this way.

All of the vertices $v_0, v_1, v_2, v_3, \ldots$ must be different (since any repeat $v_m = v_n$ with $m < n$ would define a cycle

$$v_m, v_{m+1}, v_{m+2}, \ldots, v_{n-1}, v_n = v_m$$

in the tree S). Since we know that the tree is finite (having precisely s vertices), the process must terminate at some stage. The final vertex v_e is then an "end vertex", of valency 1. QED

If we apply the Lemma to our arbitrary tree T with $k+1$ vertices, we can choose an "end vertex" v and remove both it and the edge e incident with it to obtain a tree T' having k vertices. By $\mathbf{P}(k)$ we know that T' has exactly $k-1$ edges, so when we reinstate the edge e, we see that T has exactly $(k-1)+1$ edges, so $\mathbf{P}(k+1)$ is true.

Hence $\mathbf{P}(n)$ is true for all $n \geqslant 1$.

260.

Note: All the polyhedra described in this solution are "spherical" by virtue of having their vertices located on the unit sphere.

6.10. *Chapter 6: Comments and solutions* 353

(a) (i) A regular tetrahedron.
  (ii) A square based pyramid with its apex at the North pole.
(b) If a spherical polyhedron has V vertices, E edges, and F faces, then
$$V - E + F = 2.$$
Now each edge has exactly two end vertices, so $2E$ counts the exact number of ordered pairs (v, e), where e is an edge, and v is a vertex "incident with e".

On the other hand, in a spherical polyhedron, each vertex v has valency at least 3; so each vertex v occurs in at least 3 pairs (v, e) of this kind. Hence $2E \geqslant 3V$.

In the same way, each edge e lies on the boundary of exactly 2 faces, so $2E$ counts the exact number of ordered pairs (f, e), where e is an edge of the face f.

On the other hand, in a spherical polyhedron, each face f has at least 3 edges; so each face f occurs in at least 3 pairs (f, e) of this kind. Hence $2E \geqslant 3F$.

If $E = 7$, then $14 \geqslant 3V$, and $14 \geqslant 3F$; now V and F are integers, so $V \leqslant 4$ and $F \leqslant 4$. Hence $V + F \leqslant 8$. However $V + F = E + 2 = 9$. This contradiction shows that no such polyhedron exists. QED

(c) We show by induction how to construct certain "spherical" polyhedra, with at most one non-triangular face. Let $\mathbf{P}(n)$ be the statement:

"There exists a spherical polyhedron with at most one non-triangular face, and with e edges for each e, $8 \leqslant e \leqslant n$".

- We know that there exists a such a spherical polyhedron with $n = 6$ edges – namely the regular tetrahedron (with four faces, which are **all** equilateral triangles).

We know there is no such polyhedron with $n = 7$ edges (by part (b)).

When $n = 8$, there is no spherical polyhedron with $n = 8$ edges and **all** faces triangular (since we would then have $16 = 2E = 3F$, as in part (b)). However, there exists a spherical polyhedron with $n = 8$ edges and just one non-triangular face – namely the square based pyramid with its apex at the North pole.

When $n = 9$, we can join three points on the equator to the North and South poles to produce a triangular bi-pyramid (the dual of a triangular prism), with **all** faces triangular, and with $n = 9$ edges.

When $n = 10$, there is no spherical polyhedron with $n = 10$ edges and with all faces triangles (since we would then have to have $20 = 2E = 3F$, as in part (b)); but there exists a spherical polyhedron with $n = 10$ edges and just one face which is not an equilateral triangle – namely the pentagonal based pyramid with its apex at the North pole.

This provides us with a starting point for the inductive construction. In particular $\mathbf{P}(8)$, $\mathbf{P}(9)$, and $\mathbf{P}(10)$ are all true.

- Suppose that $\mathbf{P}(k)$ is true for some $k \geqslant 10$. The only part of the statement $\mathbf{P}(k+1)$ that remains to be demonstrated is the existence of a suitable polyhedron with $k + 1$ edges.

Since $k \geqslant 10$, we know that $k - 2 \geqslant 8$, so (by $\mathbf{P}(k)$) there exists a polyhedron with all its vertices on the unit sphere, with at most one non-triangular face, and with $e = k - 2$ edges. Take this polyhedron and remove a triangular face ABC. Now add a new vertex X on the sphere, internal to the spherical triangle ABC, and add the edges XA, XB, XC and the three triangular faces XAB, XBC, XCA, to produce a spherical polyhedron with $e = (k-2)+3 = k+1$ edges, and with at most one non-triangular face. Hence $\mathbf{P}(k+1)$ is true.

Hence $\mathbf{P}(n)$ is true for all $n \geqslant 8$.

261. To prove a result that is given in the form of an "if and only if" statement, we have to prove two things: "if", and "only if".

We begin by proving the "only if" part:

> "a map can be properly coloured with two colours only if every vertex has even valency".

Let M be a map that can be properly coloured with two colours. Let v be any vertex of the map M.

The edges $e_1, e_2, e_3, \ldots$ incident with v form parts of the boundaries of the sequence of regions around the vertex v (with e_1, e_2 bordering one region; e_2, e_3 bordering the next; and so on). Since we are assuming that the regions of the map M can be "properly coloured" with two colours, the succession of regions around the vertex v can be properly coloured with just two colours. Hence the colours of the regions around the vertex v must alternate (say black-white-black- ...). And since the map is finite, this sequence must return to the start – so the number of such regions at the vertex v (and hence the number of edges incident with v – that is, the valency of v) must be even.

We now prove the "if" part:

> "a map can be properly coloured with two colours if every vertex has even valency".

Suppose that we have a map M in which each vertex has even valency. We must prove that any such map M can be properly coloured using just two colours.

Let $\mathbf{P}(n)$ be the statement:

> "any map with m edges, in which each vertex has even valency, can be properly coloured with two colours whenever m satisfies $1 \leqslant m \leqslant n$,".

- If $n = 1$, a map in which every vertex has even valency, and which has just one edge e, must consist of a single vertex v, with e as a loop from v to v (so v has valency 2, since the edge e is incident with v twice). This creates a map with two regions – the "island" inside the loop, and the "sea" outside; so we can colour the "island" black and the "sea" white. Hence $\mathbf{P}(1)$ is true.

- Suppose that $\mathbf{P}(k)$ is true for some $k \geqslant 1$.

  Most of the contents of the statement $\mathbf{P}(k+1)$ are already guaranteed by $\mathbf{P}(k)$. To prove that $\mathbf{P}(k+1)$ is true, all that remains to be proved is that

  > any map with exactly $k+1$ edges, in which every vertex has even valency, can be properly coloured using just two colours.

Consider an arbitrary map M with $k+1$ edges, in which each vertex has even valency.

[**Idea:** We need to find some way of reducing the map M to a map M' with $\leqslant k$ edges, in which every vertex still has even valency.]

Since $k \geqslant 1$, the map M has at least 2 edges. Choose any edge e of M, with (say) vertices u_1, u_2 as its endpoints, and with regions R, S on either side of e.

Suppose first that $u_1 = u_2$, so the boundary of the region R (say) consists only of the edge e. Hence e is a loop, and S is the only region neighbouring R. The edge e contributes 2 to the valency of u_1; so if we delete the edge e, we obtain a map M' in which every vertex again has even valency, in which the regions R and S have been amalgamated into a region S'. Since M' has just k edges, M' can be properly coloured with just two colours. If we now reinstate the edge e and the region R, we can give S the same colour as S' (in the proper colouring of M') and give R the opposite colour to S' to obtain a proper colouring of the map M with just two colours.

Hence we may assume that $u_1 \neq u_2$, so that e is not the complete boundary of R. We may then slowly shrink the edge e to a point – eventually fusing the old vertices u_1, u_2 together to form a new vertex u', where two new regions R', S' meet. The result is then a new map M', in which all other vertices are unchanged (and so have even valency), and in which

$$\text{valency}(u') = (\text{valency}(u_1) - 1) + (\text{valency}(u_2) - 1)$$

which is also even.

Hence every vertex of the new map M' has even valency. Moreover, M' has at most k edges, so (by $\mathbf{P}(k)$) we know that the map M' can be properly coloured with just two colours. And in this colouring of M', there are an odd number of colour changes as one goes from R' to S' through the other regions that meet around the old vertex u_1 of M, so S' receives the opposite colour to R'. The guaranteed proper two-colouring of M' therefore extends back to give a proper two-colouring of the original map M. Hence $\mathbf{P}(k+1)$ is true.

Hence $\mathbf{P}(n)$ is true for all $n \geqslant 1$.

262. Let $\mathbf{P}(n)$ be the statement:

"The 2^n sequences of length n consisting of 0s and 1s can be arranged in a cyclic list such that any two neighbouring sequences (including the last and the first) differ in exactly one coordinate position."

- When $n = 2$, the required cycle is obvious:

$$00 \to 10 \to 11 \to 01 \quad (\to 00).$$

So $\mathbf{P}(2)$ is true.

- The general construction is perhaps best illustrated by first showing how $\mathbf{P}(2)$ leads to $\mathbf{P}(3)$.

The above cycle for sequences of length 2 gives rise to two disjoint cycles for sequences of length 3:

– first by adding a third coordinate "0":

$$000 \to 100 \to 110 \to 010 \quad (\to 000)$$

– then by adding a third coordinate "1":

$$001 \to 101 \to 111 \to 011 \quad (\to 001).$$

Now eliminate the final join in each cycle ($010 \to 000$ and $011 \to 001$) and instead link the two cycles together by first reversing the order of the first cycle, and then inserting the joins $000 \to 001$ and $011 \to 010$ to form a single cycle.

In general, suppose that $\mathbf{P}(k)$ is true for some $k \geqslant 1$. Then we construct a single cycle for the 2^{k+1} sequences of length $k + 1$ as follows:

Take the cycle of the 2^k sequences of length k guaranteed by $\mathbf{P}(k)$, and form two disjoint cycles of length 2^k

- first by adding a final coordinate "0"
- then by adding a final coordinate "1".

Then link the two cycles into a single cycle of length 2^{k+1}, by eliminating the final step

$$v_1 v_2 \cdots v_k 0 \to 00 \cdots 00$$

in the first cycle, and

$$v_1 v_2 \cdots v_k 1 \to 00 \cdots 01$$

in the second cycle, reversing the first cycle, and inserting the joins

$$00 \cdots 00 \to 00 \cdots 01 \text{ and } v_1 v_2 \cdots v_k 1 \to v_1 v_2 \cdots v_k 0$$

to produce a single cycle of the required kind joining all 2^{k+1} sequences of length $k + 1$. Hence $\mathbf{P}(k + 1)$ is true.

Hence $\mathbf{P}(n)$ is true for all $n \geqslant 2$.

Note: The significance of what we call *Gray codes* was highlighted in a 1953 patent by the engineer Frank Gray (1887–1969) – where they were called *reflected binary codes* (since the crucial step in their construction above involves taking two copies of the previous cycles, reversing one of the cycles, and then producing half of the required cycle by traversing the first copy before returning backwards along the second copy). Their most basic use is to re-encode the usual binary counting sequence

$$1 \to 10 \to 11 \to 100 \to 101 \to 110 \to 111 \to 1000 \to 1001 \to 1010 \to \cdots,$$

where a single step can lead to the need to change arbitrarily many binary digits (e.g. the step from 3 = "11" to 4 = "100" changes 3 digits, and the step from 7 = "111" to 8 = "1000" changes 4 digits, etc.) – a requirement that is inefficient in terms of electronic "switching", and which increases the probability of errors. In contrast, the *Gray code* sequence changes a single binary digit at each step. However, the physical energy which is saved through reducing the amount of "switching" in the circuitry corresponds to an increase in the need for unfamiliar mathematical formulae, which re-interpret each vector in the *Gray code* as the ordinary integer in the counting sequence to which it corresponds.

263.

(a) The whole construction is inductive (each label derives from an earlier label). So let $\mathbf{P}(n)$ be the statement:

"if $HCF(r,s) = 1$ and $2 \leqslant r+s \leqslant n$, then the positive rational $\frac{r}{s}$ occurs once and only once as a label, and it occurs in its lowest terms".

- By construction the root is given the label $\frac{1}{1}$, so $\frac{1}{1}$ occurs. And it cannot occur again, since the numerator and denominator of each parent vertex are both positive, neither i nor j can ever be 0. Hence $\mathbf{P}(2)$ is true.

Notice that the basic construction:

"if $\frac{i}{j}$ is a 'parent' vertex, then we label its 'left descendant' as $\frac{i}{i+j}$, and its 'right descendant' $\frac{i+j}{j}$"

guarantees that, since we start by labelling the root with the positive rational $\frac{1}{1}$, **all** subsequent 'descendants' are **positive**.

Moreover, if any 'descendant' were suddenly to appear not "in lowest terms", then either

- $HCF(i, i+j) > 1$, in which case $HCF(i, i+j) = HCF(i, j)$, so $HCF(i, j) > 1$ at *the previous stage*; or
- $HCF(i+j, j) > 1$, in which case $HCF(i+j, j) = HCF(i, j)$, so $HCF(i, j) > 1$ at *the previous stage*.

Since we begin by labelling the root $\frac{1}{1}$, where $HCF(1,1) = 1$, it follows that **all** subsequent labels are positive rationals in **lowest terms**.

- Suppose that $\mathbf{P}(k)$ is true from some $k \geqslant 2$.
  Most parts of the statement $\mathbf{P}(k+1)$ are guaranteed by $\mathbf{P}(k)$. To show that $\mathbf{P}(k+1)$ is true, it remains to consider cases where $HCF(r,s) = 1$ and $r + s = k + 1$ ($\geqslant 3$). Either (i) $r > s$, or (ii) $s > r$.

(i) Suppose that $r > s$. Then $\frac{r}{s}$ arises in this (fully cancelled) form only as a direct (right) descendant of $\frac{r-s}{s}$. So $\frac{r}{s}$ occurs. Moreover, every label occurs in its lowest terms, so $\frac{r}{s}$ cannot occur again.

(ii) Suppose that $s > r$. Then $\frac{r}{s}$ arises in this (fully cancelled) form only as a direct (left) descendant of $\frac{r}{s-r}$. So $\frac{r}{s}$ occurs. Moreover, every label occurs in its lowest terms, so $\frac{r}{s}$ cannot occur again.
Hence $\mathbf{P}(k+1)$ is true.

Hence $\mathbf{P}(n)$ is true for all $n \geqslant 2$.

(b) The fact that the labels are left-right symmetric is also an inductive phenomenon. We note that the one fully "left-right symmetric" label, namely $\frac{1}{1}$, occurs in the only fully "left-right symmetric" position – namely at the root.

All other labels occur in reciprocal pairs: $\frac{r}{s}$ and $\frac{s}{r}$, where we may assume that $r > s$. The fact that these occur as labels of "left-right symmetric" vertices derives from the fact that

$\frac{r}{s}$ is the '**right** descendant' of $\frac{r-s}{s}$ and

$\frac{s}{r}$ is the '**left** descendant' of $\frac{s}{r-s}$.

So if we know that the earlier reciprocal pair reciprocal pair $\frac{r-s}{s}$ and $\frac{s}{r-s}$ occur as labels of symmetrically positioned vertices, then it follows that the same is true of the descendant reciprocal pair $\frac{r}{s}$ and $\frac{s}{r}$. We leave the reader to write out the proof by induction – for example, using the statement

> "$\mathbf{P}(n)$: if $r, s > 0$, and $2 \leq r + s \leq n$, then the reciprocal pair $\frac{r}{s}$, $\frac{s}{r}$ occur as labels of vertices at the same level below the root, with the two labelled vertices being mirror images of each other about the vertical mirror through the root vertex."

264. The intervals in this problem may be of any kind (including finite or infinite). Each interval has two "endpoints", which are either ordinary real numbers, or $\pm\infty$ (signifying that the interval goes off to infinity in one or both directions).

Let $\mathbf{P}(n)$ be the statement:

> "if a collection of n intervals on the x-axis has the property that any two intervals overlap in an interval (of possibly zero length – i.e. a point), then the intersection of all intervals in the collection is a non-empty interval".

When $n = 2$, the hypothesis of $\mathbf{P}(2)$ is the same as the conclusion. So $\mathbf{P}(2)$ is true.

Suppose that $\mathbf{P}(k)$ is true for some $k \geq 2$. We seek to prove that $\mathbf{P}(k+1)$ is true.

So consider a collection of $k+1$ intervals with the property that any two intervals in the collection intersect in a non-empty interval. If this collection includes one interval that is listed more than once, then the required conclusion follows from $\mathbf{P}(k)$. So we may assume that the intervals in our collection are all different.

Among the $k+1$ intervals, consider first those with the largest right hand endpoint. If there is only one such interval, denote it by I_0; if there is more than one interval with the same largest right hand endpoint, let I_0 be the interval among those with the largest right hand endpoint that has the largest left hand endpoint. In either case, put I_0 aside for the moment, leaving a collection S of k intervals with the required property.

By $\mathbf{P}(k)$ we know that the intervals in the collection S intersect in a non-empty interval I, with left hand endpoint a and right hand endpoint b (say).

We have to show that the intersection $I \cap I_0$ is non-empty.

The proof that follows works if the endpoint b is included in the interval I. The slight adjustment needed if b is not included in the interval I is left to the reader.

Since the right hand endpoint of I_0 is the largest possible, and since points between a and b belong to all the intervals of S, we can be sure that the right hand endpoint of I_0 is $\geq b$.

6.10. Chapter 6: Comments and solutions 359

Moreover, for each point $x > b$, we know that there must be some interval I_x in the collection S which does not stretch as far to the right as x. Since, by hypothesis, the intersection $I_0 \cap I_x$ is non-empty, the left hand endpoint of I_0 lies to the left of every such point x, so I_0 must overlap the interval I, whence it follows that $I \cap I_0$ is a non-empty interval as required.

Hence $\mathbf{P}(k+1)$ is true.

Hence $\mathbf{P}(n)$ is true for all $n \geqslant 2$.

265. If one experiments a little, it should become clear

- that if tank T_2 contains more than tank T_1, then linking tank T to T_2 leads to a better immediate outcome (i.e. a larger amount in tank T) than linking T to T_1;
- that if, at some stage in the linking sequence, tank T contains an interim amount of a litres, and is about to link successively to a tank containing b litres, and then to the tank containing c litres, this ordered pair of changes alters the amount in the tank T to $\frac{a+b+2c}{4}$ litres; so for a better final outcome we should always choose the sequence so that $b < c$;
- once the tap linking tank T to another tank has been opened, so that the two levels become equal, there is no benefit from opening the tap linking these two tanks ever again, so tank T should be linked with each other tank at most once.

These three observations essentially determine the answer – namely that tank T should be joined to the other tanks *in increasing order of their initial contents*.

For a proof by induction, let $\mathbf{P}(n)$ be the statement:

"given n tanks containing $a_1, a_2, a_3, \ldots, a_n$ litres respectively, where

$$a_1 < a_2 < a_3 < \cdots < a_n,$$

if T is the tank containing the smallest amount a_1 litres, then the optimal sequence for linking the other $n - 1$ tanks to tank T (optimal in the sense that it transfers the maximum amount of water to tank T) is the sequence that links T successively to the other tanks in increasing order of their initial contents".

- When $n = 2$, there is only one possible sequence, which is the one described, so $\mathbf{P}(2)$ is true.
- Suppose that $\mathbf{P}(k)$ is true for some $k \geqslant 2$, and consider an unknown collection of $k + 1$ tanks containing $a_1, a_2, a_3, \ldots, a_{k+1}$ litres respectively, where

$$a_1 < a_2 < a_3 < \cdots < a_{k+1},$$

and where T is the tank which initially contains a_1 litres.

Suppose that in the optimal sequence of k successive joins to the other k tanks (that is, that transfers the largest possible amount of water to tank T), the succession of joins is to join T first to tank T_2, then to tank T_3, and so on up to tank T_{k+1} (where tank T_m is not necessarily the tank containing a_m litres). There are now two possibilities: either

(i) T_{k+1} is the tank containing a_{k+1} litres, or
(ii) T_{k+1} contains less than a_{k+1} litres.

(i) Suppose tank T_{k+1} is the tank containing a_{k+1} litres. Then the first $k-1$ joins involve the k tanks containing $a_1, a_2, a_3, \ldots, a_k$ litres. And we know (by the very first bullet point above) that, in order to maximize the final amount in tank T, the amount in tank T after linking it to tank T_k must be "as large as it could possibly be". Hence, by $\mathbf{P}(k)$, the first $k-1$ joins link T successively to the other tanks in increasing order of their contents – before finally linking to the tank containing a_{k+1} litres. Hence the conclusion of $\mathbf{P}(k+1)$ holds.

(ii) Now suppose that tank T_{k+1} contains $a_m < a_{k+1}$ litres.

By the very first bullet point, in order to guarantee the optimal overall outcome of the final linking with tank T_{k+1} the amount in tank T after it has been linked to tank T_k must be "as large as it can possibly be" (given the amounts in the tanks $T, T_2, T_3, \ldots, T_k$). Hence statement $\mathbf{P}(k)$ applies to the initial sequence of $k-1$ joins (of T to T_2, then to T_3, and so on up to T_k), and guarantees that these tanks must be in increasing order of their initial contents. In particular, the last tank in this sequence, T_k, must be the one containing a_{k+1} litres. But if we denote by a litres the amount in tank T just before it links with tank T_k (containing a_{k+1} litres), then the last two linkings, with $b = a_{k+1}$ and $c = a_m$ contradict the second bullet point at the start of this solution. Hence case (ii) cannot occur.

Hence $\mathbf{P}(k+1)$ is true.

Hence $\mathbf{P}(n)$ is true for all $n \geqslant 2$.

266.

Note: Like all practical problems, this one requires an element of initial "modelling" in order to make the situation amenable to mathematical analysis.

'Residue' clings to surfaces; so the total amount of 'residue' will depend on

(a) the viscosity of the chemical (how 'thick', or 'sticky' it is), and
(b) the total surface area of the inside of the flask.

Since we are given no information about quantities, we may fix the amount of residue remaining in the 'empty' flask at "1 unit", and the amount of solvent in the other flask as "s units".

If we add the solvent, we get a combined amount of $1 + s$ units of solution – which we may assume (after suitable shaking) to be homogeneous, with the chemical concentration reduced to "1 part in $1 + s$".

The first modelling challenge arises when we try to make mathematical sense of what remains at each stage after we empty the flask. The internal surface area of the flask, to which any diluted residue may adhere, is fixed. If we make the mistake of thinking of the original chemical as "thick and sticky" and the solvent as "thin", then the viscosity of the diluted residue will change relative to the original, and will do so in ways that we cannot possibly know. Hence the only reasonable assumption (which may or may not be valid in a particular instance) is to *assume* that the viscosity of the original chemical is roughly the same as the viscosity of the chemical-solvent mixture. This then suggests that, on emptying the diluted mixture, roughly the same amount (1 unit) of diluted mixture will remain adhering to the walls of the flask. So we will be left with 1 unit of residue, with a concentration of $\frac{1}{1+s}$. In particular, if $s=1$, using all the solvent at once reduces the amount of toxic chemical residue to $\frac{1}{2}$ unit (with the other $\frac{1}{2}$ unit consisting of solvent).

But what if we use only half of the solvent first, empty the flask, and then use the other half? Adding $\frac{s}{2}$ units of solvent (and shaking thoroughly) produces $1+\frac{s}{2}$ units of homogeneous mixture, with a concentration of 1 part per $1+\frac{s}{2}$. When we empty the flask, we expect roughly 1 unit of residue with this concentration — so just $\frac{2}{2+s}$ units of the chemical, with $\frac{s}{2+s}$ units of solvent.

If we then add the other $\frac{s}{2}$ units of solvent, this produces $1+\frac{s}{2}$ units of mixture with a concentration of 1 part per $\left(1+\frac{s}{2}\right)^2$. When we empty the flask, we expect roughly 1 unit of residue with concentration 1 part per $\left(1+\frac{s}{2}\right)^2$. In particular, if $s=1$, this strategy reduces the amount of toxic chemical in the 1 unit of residue to $\frac{4}{9}$ units. Since $\frac{4}{9}<\frac{1}{2}$ this two-stage strategy seems more effective than the previous "all at once" strategy.

Suppose that we use a four-stage strategy – using first one quarter of the solvent, then another quarter, and so on. We then land up with roughly 1 unit of residue with concentration 1 part per $\left(1+\frac{s}{4}\right)^4$. In particular, if $s=1$, we land up with the amount of toxic chemical in the 1 unit of residue equal to $\frac{256}{625}$ units, and $\frac{256}{625}<\frac{4}{9}$.

More generally, if we use $\left(\frac{1}{n}\right)$th of the solvent, n times, the final amount of toxic chemical in the 1 unit of residue is equal to $\left(1+\frac{s}{n}\right)^{-n}$. And as n gets larger and larger, this expression gets closer and closer to e^{-s}. In particular, when $s=1$, this strategy leaves a final amount of chemical in the 1 unit of residue approximately equal to $\frac{1}{e}=0.367879\cdots$.

Note: The situation here is similar to that faced by a washing machine designer, who wishes to remove traces of detergent from items that have been washed, without using unlimited amounts of water. The idea of having a "fixed amount of solvent" corresponds to the goal of "water efficient" rinsing. However, the washing machine cycle, or programme, clearly cannot repeat the rinsing indefinitely (as would be required in the limiting case above).

267.

(i) If $\sqrt{2}=\frac{m}{n}$, then
$$2n^2=m^2 \quad (*)$$

Hence m^2 is even.

It follows that m must be even.

Note: It is a fact that, if $m=2k$ is even, then $m^2=4k^2$ is also even. But this is **completely irrelevant** here. In order to conclude that "m must be even", we have to prove:

Claim m cannot be odd.
Proof Suppose m is odd.
$\therefore m = 2k + 1$ for some integer k.
But then
$$m^2 = (2k+1)^2 = 4k^2 + 4k + 1$$
would be odd, contrary to "m^2 must be even".
Hence m cannot be odd. QED

(ii) Since m is even, we may write $m = 2m'$ for some integer m'.
Equation (∗) in (i) above then becomes $n^2 = 2(m')^2$, so n^2 is even. Hence, as in the **Note** above, n must be even, so we can write $n = 2n'$ for some integer n'.

(iii) If $\sqrt{2} = \frac{m}{n}$, then $m = 2m'$, and $n = 2n'$ are both even.
$\therefore \sqrt{2} = \frac{m}{n} = \frac{2m'}{2n'} = \frac{m'}{n'}$.
In the same way, it follows that m' and n' are both even, so we may write $m' = 2m''$, $n' = 2n''$ for some integers m'', n''.
Continuing in this way then produces an endless decreasing sequence of positive denominators
$$n > n' > n'' > \cdots > 0.$$
contrary to the fact that such a sequence can have length at most $n - 1$ (or in fact, at most $1 + \log_2 n$).

268.

(i) If $a < b$ and $c > 0$, then $ac < bc$.
If $0 < \sqrt{2} \leqslant 1$, then (multiplying by $\sqrt{2}$) it follows that $2 \leqslant \sqrt{2} \leqslant 1$, which is false. Hence $\sqrt{2} > 1$.
We now know that $1 < \sqrt{2}$, so multiplying by $\sqrt{2}$ gives $\sqrt{2} < 2$. Hence $1 < \sqrt{2} < 2$. In particular, $\sqrt{2}$ cannot be written as a fraction with denominator 1, so $\mathbf{P}(1)$ is true.

(ii) Suppose $\mathbf{P}(k)$ is true for some $k \geqslant 1$. Most of the statement $\mathbf{P}(k+1)$ is implied by $\mathbf{P}(k)$: all that remains to be proved is that $\sqrt{2}$ cannot be written as a fraction with denominator $n = k + 1$.
Suppose $\sqrt{2} = \frac{m}{n}$, where $n = k + 1$ and m are positive integers.
Then $m = 2m'$ and $n = 2n'$ are both even (as in Problem **267**).
So $\sqrt{2} = \frac{m'}{n'}$ with $n' \leqslant k$, contrary to $\mathbf{P}(k)$. Hence $\mathbf{P}(k+1)$ holds.
$\therefore \mathbf{P}(n)$ is true for all $n \geqslant 1$.

269.

(a) Suppose that S is not empty. Then by the *Least Element Principle* the set S must contain a least element k: that is, a smallest integer k which is not in the set T. Then $k \neq 1$ (since we are told that T contains the integer 1). Hence $k > 1$.

Therefore $k-1$ is a positive integer which is smaller than k. So $k-1$ is not an element of S, and hence must be an element of T. But if $k-1$ is an element of T, then we are told that $(k-1)+1$ must also be a member of T. This contradiction shows that S must be empty, so T contains all positive integers.

(b) Suppose that T does not have a smallest element. Clearly 1 does not belong to the set T (or it would be the smallest element of T). Hence 1 must be an element of the set S.

Now suppose that, for some $k \geqslant 1$, all positive integers $1, 2, 3, \ldots, k$ are elements of S, but $k+1$ is not an element of S. Then $k+1$ would be an element of T, and would be the smallest element of T, which is not possible. Hence S has the property that

> "whenever $k \geqslant 1$ is an element of S, we can be sure that $k+1$ is also an element of S."

The *Principle of Mathematical Induction* then guarantees that these two observations (that 1 is an element of S, and that whenever k is an element of S, so is $k+1$) imply that S contains all the positive integers, so that the set T must be empty – contrary to assumption.

Hence T must have a smallest element.

270.

(i) Triangle $OP'Q'$ is a right angled triangle with $\angle P'OQ' = 45°$. Hence the two base angles (at O and at Q') are equal, so the triangle is isosceles: $P'Q' = P'O$.

(ii) Triangles $QQ'P'$ and $QQ'R$ are congruent by RHS (since they share the hypotenuse QQ', and have equal sides ($QP' = QP = QR$). Hence $Q'P' = Q'R$.

(iii) If $OQ : OP = m : n$, we may choose a unit so that OQ has length m units and OP has length n units. Then

$$OP' = OQ - QP' = OQ - QP = m - n,$$

and

$$OQ' = OR - Q'R = OR - Q'P' = OR - OP' = n - (m - n).$$

(iv) $OP < OQ < OP + PQ$, so $n < m < n + n$. Hence $0 < m - n < n$, and $0 < 2n - m$.

(v) In the square $OP'Q'R'$, the ratio "diagonal : side" $= OQ' : OP' = \sqrt{2} : 1$.
If the ratio $OQ : OP = m : n$, with m, n positive integers, then the construction here replaces the positive integers m, n by smaller positive integers $2n - m$ and $m - n$, with $m - n < n$. And the process can be repeated indefinitely to generate an endless sequence of decreasing positive integers

$$n > m - n > (2n - m) - (m - n) = 3n - 2m > \cdots > 0.$$

> *Zarathustra's last, most vital lesson:*
> *"Now do without me."*
> George Steiner (1929–)

Index

10-gon
    regular, 210
n-gon, 39, 195, 196, 200, 205, 292
    regular, 15, 28, 47, 131, 196, 199, 201, 205, 211–214, 216, 250, 251, 265, 269
    regular circumscribed, 213–217, 270
    regular inscribed, 211, 213–217, 254, 270
x-axis, 15
    positive, 15
Ars Magna, 129, 132, 156
Astronomia Nova, 199
De Divina Proportione, 199
De revolutionibus, 132
Disquisitiones arithmeticae, 20, 201
Elements, 13, 14, 181, 199, 207, 210, 230
Liber Abaci, 56
The Sand Reckoner, 283
Trattato d'aritmetica, 109
10-gon
    regular, 264

abacists, 111
acceleration, 96, 106
    constant, 96, 106, 107
    uniform, 105, 106
addition, 8, 10, 11, 35
Al-Khwarizmi, Muhammad ibn Musa (c.780–c.850), 111
algebra, viii, 2, 10, 51, 111, 112, 145, 156, 171, 172, 197, 198
    abstract, 173
    Lie, vii, 109
    linear, 172, 173
    mental, 2
    symbolical, 131
algorithm

division, 134, 166
    Euclidean, 6, 9, 133, 134, 166
    standard written, 172
analysis, 173
    dimensional, vi, 109
    method of, 197, 205
Anaximenes (c. 586–c. 526 BC), 194
angle, 4, 14–16, 23–25, 48, 130, 171, 174, 219
    acute, 189, 262
    angle sum, 249, 256, 258–260, 319, 320
    apex, 279
    dihedral, 209, 263, 265
    exterior, 180, 184, 185, 234–236
    obtuse, 185, 262
    reflex, 235
    right, 14, 17, 23, 24, 40, 44, 176, 185, 218, 220, 231, 234, 237, 239, 241, 254, 257, 272–275
    sector, 269
    solid, 201
    straight, 176, 179–185, 189, 201, 231, 232, 234–236, 238, 241, 242, 244
    subtended at the centre of the circle, 267
    subtended by a chord, 44, 184, 235, 243
    subtended by a diameter, 44, 194, 254
    subtended by an arc, 269, 273
angles
    alternate, 183, 237–239, 319
    base, 47, 178, 250, 260, 263
    congruent, 28
    corresponding, 183
    internal, 182
    measured in degrees, 23

opposite, 17
special, 16
supplementary, 184, 185, 262
vertically opposite, 238, 241–243, 248, 250
anthyphairesis, 9
antiprism, 205
n-gonal, 258
apex, 224
of an isosceles triangle, 177
of a pyramid, 193
arc, 24, 25, 184, 211, 214, 218–220, 235, 236, 243, 264, 268, 269, 272
circular, 24
length of, 274
of a great circle, 272
arc length, 24
Archimedes (c. 287–212 BC), 198, 199, 283
area, 211
of a circle, 214, 215, 217
of a circumscribed regular n-gon, 217
of a parallelogram, 97, 186, 188, 237, 243, 244, 272
of a quarter circle, 216
of a rectangle, 188, 237
of a regular n-gon, 214, 215
of a regular polygon, 211
of a sector, 24
of a sector of a circle, 216, 267, 268
of a semicircle, 216
of a spherical triangle, 27, 45
of a trapezium, 193
of a triangle, 15, 97, 106, 191, 206, 237, 244
of an ellipse, 223
of an inscribed regular n-gon, 216
surface area of a pyramid, 211
surface area of a right circular cone, 216, 269
surface area of a right cylinder, 215
surface area of a right prism, 211, 216
surface area of a right pyramid, 216
surface area of a sphere, 26
under the graph, 104, 106, 107
argument, 130

arithmetic, 11, 111, 171
inverse, 156
mental, 2, 3, 5, 8, 20, 172
modular, 101
of complex numbers, 42, 129
of norms, 20
of surds, 9
structural, 12, 34
written, 2
Arnold, Vladimir (1937–2010), vi, 109
average, 93, 98
averaging, 104

Babylonians, 14, 23
base, 10, 237
of an isosceles triangle, 177
of a cone, 244
of a pyramid, 193, 218, 244, 245
of a rectangle, 106, 188
of a right cylinder, 215
of a right prism, 216
Bernoulli, Jacob (1654–1705), 302
bi-pyramid, 353
bisector, 171, 186
of angle, 186, 237
perpendicular, 22, 44, 47, 179, 221, 223, 224, 232, 240, 254–256, 260, 275–277, 281
Bolyai, János (1802–1860), 182
bracket, 10, 11
Bragg, Sir William Lawrence (1890–1971), xi
Bronowski, Jacob (1908–1974), 170
Bruno, Giordano (1548–1600), 284

calculation, 1
direct, 11
exact mental, 14
mental, xiii
calculus, 11, 173
cancellation, 34
Cantor, Georg (1845–1918), 142, 284
Cardano, Girolamo (1501–1576), 129, 132, 156
Catalan, Eugène Charles (1814–1894), 77
caustic, vii, 109
centroid, 189
chord, 44, 184, 235, 242, 243, 320

Christ, Jesus, 14
circle, 9, 15, 24, 44, 174
    area of, 214–217
    great, 25, 26, 218
    longitude, 275
    of Apollonius, 223
    of longitude, 273
    perimeter of, 213, 216, 268
    unit, 15, 16
circumcentre, 179, 180, 184, 196, 199, 200, 204, 205, 210, 236, 240, 254–256, 263, 270
circumcircle, 22, 184, 185, 210, 236, 241, 254–256, 264, 265
circumference, 25, 44, 179, 195, 269
circumradius, 180, 199, 210, 213, 264, 265
code
    Gray, 282, 312
coefficient, 124
concave, 107
concavity, 107
concept
    abstract, xiii
cone, 244
    apex of, 224
    generator of, 224
    right, 216
congruence, 171, 174
    ASA-congruence, 177, 178, 237–239, 250
    RHS-congruence, 136, 185, 186, 236, 237, 239, 257
    SAS-congruence, 14, 15, 48, 177, 178, 185, 190, 232, 233, 238, 240, 242, 249
    SSS-congruence, 47, 177, 178, 186, 231, 232, 238, 257
conic sections, 225, 227
constant
    of proportionality, 101
construction
    geometrical, 199, 206
    of a cube, 227
    of a parabola, 221
    of a regular pentagon, 255
    of an equilateral triangle, 175
    of regular polygons, 175, 199–201
    of regular polyhedra, 199
    ruler and compasses, 9, 174, 175, 179, 181, 183, 193, 199–201, 230
converse, 14, 16, 17
    statement, 16
convexity, 107
coordinate, 15
coordinates
    Cartesian, 129
Copernicus, Nicolaus (1473–1543), 132
counterexample, 314
Courant, Richard (1888–1972), v
cow
    spherical, 91
Coxeter, Harold Scott MacDonald (1907–2003), vii, 109
Cramer, Gabriel (1704–1752), 114
criteria
    congruence, 174, 178, 181, 185, 191, 206
    parallel, 174, 179, 181, 182, 206
    similarity, 174, 191–193, 206
cube, 28, 29, 48, 49, 201, 203–205, 209, 210, 227, 244, 260
    cross-section of, 28
    truncated, 259
    unit, 209
cubes, 2, 3, 56, 61, 119
cuboctahedron, 258
curriculum, xiii
curve, vii, 109
    quadratic, 227
cylinder
    right, 215

d'Alembert, Jean le Rond (1717–1783), 283, 286
Dürer, Albrecht (1471–1528), 280
da Vinci, Leonardo (1452–1519), 199
Dandelin, Germinal Pierre (1794–1847), 224
de la Vallée Poussin, Charles-Jean (1866–1962), 68
de Moivre, Abraham (1667–1754), 130
decimal, 172
degree, 23
del Ferro, Scipione (1465–1526), 132
del Fiore, Antonio (1506–??), 132

dell'Abbaco, Paolo (1282–1374), xi, 97, 109, 111
denominator, 5, 34, 65, 84, 105
derivative, 24
Descartes, René (1596–1650), 3, 112, 156, 193, 197, 198
diagonal, 4, 15, 16, 48, 49, 187, 188, 196, 206, 238, 239, 249, 250, 255, 262, 263, 316, 363
difference, 6
digit, 2
    leading, 56, 288
    units, 56, 288
digit-sum, 55
dimension, 2, 35
directrix, 221, 224
    of ellipse, 223
    of hyperbola, 224
    of parabola, 221, 280
disc
    circular, 269
discount, 8, 35
discriminant, 127
division, 10, 11, 40
    with reminder, 134
divisor, 6, 134
dodecagon
    regular, 16, 195
dodecahedron
    regular, 205, 260
    snub, 258
    truncated, 259
Dogan, Serkan, 348

Earth, vii
    circumnavigation of, vii
eccentricity, 221
    of a hyperbola, 226
    of an ellipse, 226
    of parabola, 221
edge, 312
Edwards, Harold M. (1936–), 1
Ehrlich, Paul R. (1932–), 111
ellipse, 223, 224, 226, 280
    area of, 223
encryption, 6
equation, 4
    cubic, 127–129, 131–133, 156, 165, 166

differential, 11
    linear, 201
    Pell, 76, 290
    quadratic, 4, 9, 123, 125, 129, 133, 156, 201, 226, 227
    quartic, 127, 129, 132, 156, 158, 159
equator, 25, 273, 274
equidistant, 28, 44, 179, 180, 186, 221, 225, 232, 237, 254
Eratosthenes, (c. 276 – 194 BC), 31
Euclid, (flourished c. 300 BC), 13, 14, 175, 181, 199, 207, 210, 230, 231
Euler, Leonhard (1707–1783), 19, 20, 42, 112, 122, 123, 153, 302, 310, 312
expansion
    base 2, 115
    base 3, 115
    decimal, 115
exponent, 30

face, 311
factor, 6, 29, 67, 74, 76, 77, 84, 87, 90, 119–121, 153, 154
    common, 5, 6, 18, 76, 148, 167
    enlargement, 8
    highest common, 33
    linear, 119, 120, 131, 165
    linear conjugate, 165
    prime, 31, 67, 71, 90, 311, 317, 348
    quadratic, 165
    scale, 18, 191
factorial, 7, 326
factorisation, 6, 11, 69, 71, 76, 78, 120, 121, 152, 164
    algebraic, 72, 73
    of integers, 6, 32
    of polynomials, 12
    prime, 5, 31, 72
    prime power, 6
Fermat, Pierre de (1601–1665), 112, 123, 154, 197, 201, 289
Ferrari, Lodovico (1522–1565), 132
Fibonacci, Leonardo Pisano (c. 1170 – c. 1250), 56
focus, 224
    of ellipse, 223, 224

of hyperbola, 224
of parabola, 221
formula
    Euler's, 312
fraction, viii, 5, 7
    continued, 9
    decimal, 85
    unit, 65, 172
fractions
    decimal, 172
frustum, 193, 245
function, 3, 4, 9, 11, 129
    constant, 104
    trigonometric, 16, 24, 273

Galen (c. 130 – c. 210), vii
Galileo, Galilei (1564–1642), 174
Gauss, Carl Friedrich (1777–1855), 20, 68, 69, 173, 182, 201
geometry, 9, 173
    algebraic, vii, 109
    analytic, 173
    Cartesian, 173
    coordinate, 172, 173
    dynamic, 172, 173
    elementary, 172
    Euclidean, 170, 172, 173
    of surfaces, 173
gnomon, 293
Gog, Julia, 48
Goldbach, Christian (1690–1764), 122
Golden Ratio, 30, 31, 37, 199
Golding, William (1911–1993), xiii
Grassmann, Hermann Günther (1809–1877), 173
Gray, Frank (1887–1969), 356
group, vii, 109, 173
    Coxeter, vii, 109
    of transformations, 173

Hadamard, Jacques (1865–1963), 68
Harte, John, 91
HCF, 5–7, 10
height, 14, 237
    of a rectangle, 188
    of a parallelogram, 188
    of a pyramid, 244, 245
    of a right cylinder, 215
    of a right prism, 216

of a triangle, 97, 106
hemisphere, 274
hexagon, 49, 252
    regular, 16, 28, 29, 49, 196, 210, 231, 247, 251, 253, 264, 265, 281
Hilbert, David (1862–1943), 174
Hildebrand, Joel Henry (1881–1983), 51
Hipparchus, (died c. 125 BC), 208
hyperbola, 224, 226, 281
hypotenuse, 14, 15, 39, 43

icosahedron
    regular, 204, 205, 209, 210, 257
    truncated, 260
icosidodecahedron, 258
identity
    trigonometric, 5, 27
incentre, 186, 187
incircle, 9, 113, 181, 187
induction
    mathematical, 57, 78, 88, 118, 145, 147, 162, 270, 271, 282, 285, 288, 290, 292–294, 300, 314, 315, 324
inequality
    triangle, 180, 270, 271, 277
infinite descent, 314–316
inradius, 187, 213, 216
integer part, 54
integers, 6, 10
    consecutive, 6
    coprime, 5
    cubes, 3
    Gaussian, 134, 166–168
    positive, 3, 5
    powers of 2, 3
    relatively prime, 5, 6, 32, 33, 41
    squares, 3
integral, 104, 310
integration, 11
interval, 115

Kapitsa, Peter (1894–1984), ix
Kepler, Johannes (1571–1630), 199, 224, 263
kinematics, 105
Klein, Felix (1849–1925), 173

Lagrange, Joseph-Louis (1736–1813), 20
law
    associative, 60, 82
    commutative, 60, 82
    distributive, 60, 82
    index laws for powers, 2
LCM, 5
Legendre, Adrien-Marie (1752–1833), 20, 68, 248
Leibniz, Gottfried Wilhelm (1646–1716), 112, 304, 306, 310
length, 9, 211
line, 14, 25, 171, 173, 174
    parallel, 16, 30, 44
    segment, 25, 49, 174, 175
Livio, Mario, 31
Lobachevski, Nikolai (1792–1856), 182
locus, 179, 180, 186, 198, 220–223, 254, 275–278
logarithm
    natural, 310
longitude, 25
lune, 26, 27

map, 312
mapping
    one-to-one, 10
    two-to-one, 10
mathematics
    elementary, ix, xi, 4, 28
mean
    arithmetic, 93, 96, 105, 118, 149, 305
    geometric, 97, 118, 149, 305
    harmonic, 96, 105, 108, 118, 149, 305
    quadratic, 118, 149
mechanics, 107
median, 189, 241, 261, 262
Menelaus, (c. 70–130 AD), 208
Mersenne, Marin (1588–1648), 123, 289
midpoint, 15, 21, 99, 113, 114, 136, 137, 178–180, 187–189, 193, 203, 231–234, 239, 240, 242, 243, 245, 246, 255–257, 261, 263–265, 267, 268, 270, 308
modulus, 42, 115, 130
Moise, Edwin (1918–1998), 174

Morotzkin, I.V., vi, 109
multiple, 5, 10, 30
    common, 5
multiplication, 10, 11, 35
    long, 64, 72
    of complex numbers, 129
multiplier, 10

net, 48
Newton, Issac (1642–1727), 112, 125, 224
norm, 42, 167
notation
    algebraic, 2, 156
number, 2, 9, 171
    complex, 1, 20, 27, 42, 76, 77, 126–131, 134, 164
    complex conjugate, 128, 129, 164
    complex in modulus form, 27
    complex in polar form, 129, 130
    Fermat, 123, 289
    Fibonacci, 30, 56, 58, 285, 294
    irrational, 63, 126
    positive, 3, 4
    prime, 5, 76, 90, 119
    rational, 63
    real, 3
    real positive, 8, 28
    square triangular, 76
    triangular, 56, 57, 61, 75–77, 284, 293
numeral system
    binary, 54
    Hindu-Arabic, 56
numerator, 34, 105
numerosity, 171

octagon, 39
    regular, 9, 16, 39
octahedron, 203, 257
    regular, 203, 205, 208–210, 257, 258, 262, 263
    stellated, 263
    truncated, 260
octant, 209
operation, 2, 3
    arithmetic, 11
    direct, 3, 10
    inverse, 3

origin, 15
orthocentre, 188

Pólya, George (1887–1985), xi, 29
Pacioli, Luca (c. 1445–1509), 199
Pappus, c. 290–c. 350 AD), 198
parabola, 221, 224–226, 280
parallelogram, 17, 30, 149, 187, 188, 237, 239, 240, 242–244, 246, 254, 272
    Fibonacci, 59
parallels, 171, 174
parameter, 108, 112
    hidden, 108
    hidden intermediate, 108
parity, 18, 41
Pascal triangle, 351
pattern, 2
Pell, John (1611–1685), 290
Penrose, Roger (1930 –), 169
pentagon, 29, 113, 114, 264
    convex, 195
    regular, 4, 9, 28, 30, 49, 124, 195, 196, 205, 209, 210, 249, 250, 252, 255, 263
pentagram, 195
percentage, 7, 35, 60
perimeter, 21, 213, 270–272
    of a circle, 213, 216, 217, 268
    of a convex polygon, 217, 270, 271
    of a regular polygon, 211–213, 216, 217
    of a sector, 216
    of a triangle, 43, 189, 240
    of an ellipse, 223
perpendicular, 14, 40, 45, 176, 178–181, 183, 186, 188, 206, 210, 221, 224, 232, 234, 237, 239, 240, 244–246, 254–256, 261, 262
Petrie, John Flinders (1907–1972), 49
Petrie, Sir William Matthew Flinders (1853–1942), 49
Pigeon Hole Principle, 33
place value, 2, 51, 171, 172
plane, vii, 109, 174
    complex, 129
    real, vii, 109
Plato, (c. 428–347 BC), 205
Poincaré, Henri (1854–1912), 283

point, 171, 173, 174
    decimal, 115, 172
polygon, 24, 28, 49, 292
    angles of, 24
    convex, 217
    cyclic, 28
    Petrie, 49
    regular, 28, 199, 201, 205, 211, 256
    regular inscribed, 202
polyhedron, 49, 198
    convex, 201
    dual, 205
    regular, 28, 199, 202, 203, 205, 210, 265
    semi-regular, 205, 206, 259
    spherical, 311, 312, 352–354
polynomial, 126
    anti-symmetric, 125
    cubic, 127
    long division of, 126
    quadratic, 112, 119, 122, 124–126, 131, 158
    quartic, 158, 159
    symmetric, 124
postulate
    parallel, 181
powers, 2, 10
    algebraic, 3
    numerical, 3
    of 10, 2, 62, 71, 84, 85
    of 2, 2, 3, 56–58, 64, 65, 76, 88
    of 3, 88
    of 4, 12, 38, 56
    proper, 77
    sums of, 20
prime numbers, 5, 6, 12, 18–20, 31, 32, 42, 43, 53, 119, 120, 122, 151, 153, 154, 288, 289, 297, 335, 348
    distribution of, 68
    Fermat, 123, 154, 201, 290
    Mersenne, 123
prism, 205, 211
    3-gonal, 259
    n-gonal, 260
    right, 216
problems, xi
    inverse, 10, 11

multi-step, viii
one-step, viii
word, 91, 92, 98, 172
procedure
    deterministic, vii
    direct, 10, 11
    inverse, 10, 11
progression
    arithmetic, 19, 41
projectile, 221
projection, 49
proof, 1
proportion, viii, 100, 101, 109
Ptolemy, (died 168 AD), 208
pyramid, 193, 211, 218, 244, 245
    right, 216
Pythagoras, 13
Pythagorean triple, 18
Pythagorean triples, 14, 17
    primitive, 18, 19, 197
Pythagoreans, 13

quadrant, 15, 16
quadrilateral, 17, 113, 114, 137, 174, 184, 238, 242
    cyclic, 27, 184, 185, 236, 241, 257
    planar, 257
quantifier, 317
quantity, 91, 171
    averaged, 104
    continuous, 97, 98, 171
    discrete, 104
    unknown, 98, 108
quotient, 126, 160, 166

radians, 24, 25, 196, 205, 214, 216, 248, 249, 262, 273, 292
radius, 15, 24, 174, 211
ratio, 5, 7, 30, 100
recognising structure, 91, 92, 99
rectangle, 14, 21, 28, 35, 44, 48, 106, 187, 188, 211, 217, 223, 237, 239, 246, 254, 257, 269
    Golden, 257
    similar rectangles, 7
rectangular, 28
recursion, 54, 284
relation
    Fibonacci recurrence, 56

recurrence, 58
remainder, 40, 54, 55, 90, 126, 134, 160, 161, 166–168
rhombicosidodecahedron
    great, 260
    small, 259
rhombicuboctahedron, 259
    great, 260
rhombus, 4, 124, 187, 229, 250, 257
Riemann, Bernhard (1826–1866), 173
Robbins, Herbert (1915–2001), v
root, 3, 4, 10, 124–129
    complex, 131
    conjugate, 158, 161
    cube, 133, 166
    cube complex, 251
    of a polynomial, 125
    of a quadratic equation, 37, 131, 149
    of a quadric, 124, 125
    of unity, 130, 131
    real, 127, 128
    repeated, 132, 154
    square, 4, 11, 66, 124, 128, 129, 151, 155, 201
Rule
    Cosine, 5, 17, 170, 206–208, 219, 261, 262, 264, 273, 274
    Cramer's, 139
    Sine, 5, 22, 23, 45, 47, 206–208, 219, 220, 261
ruler and compasses, 9, 174, 175, 179, 181, 183, 193, 199–201, 230

Sarton, George (1884–1956), 169
scales, 112
scaling, vi, 23, 109
scaling argument, 109
secant, 242
sector, 267, 269
semicircle, 216
sequence, 12, 13, 52, 53, 56, 59
    constant, 57
    Farey, 117
    Fibonacci, 56–58, 80, 294
    of cubes, 56
    of differences, 57
    of natural numbers, 56, 57
    of partial sums, 58

of powers, 58
of prime numbers, 56
of second differences, 57
of squares, 56
of triangular numbers, 56, 57
recurrent, 13
series
    alternating harmonic, 310
    alternating, 310
    divergent, 306
    Farey, 117, 118, 145–149
    geometric, 121, 122, 141, 142, 297–300, 302, 303, 305, 336
    harmonic, 305, 308, 310
    power, 310
Shakespeare, William (1564–1616), ix
sharing, 92, 172
Sieve of Eratosthenes, 31
similarity, 171, 174
    AAA-similarity, 191, 192, 242–245
    of parabolas, 221
    SAS-similarity, 190, 191, 240
    SSS-similarity, 191
simplifying, 91, 92
singularity, vii, 109
speed, 94, 96
    as a function of time, 104, 107
    average, 95, 96, 103–107
    constant, 101, 110
    negative, 101
    relative, 108
    varying, 104
sphere, 25
    surface area of, 26
    unit, 26, 218, 220
square, 3–5, 9, 14–16, 21, 28, 29, 44, 187
    unit, 227
square lattice, 20
Square Root Test, 31, 71
squares, 2, 3, 56, 119
squaring, 3, 11
star
    pentagonal, 195, 248
statement
    converse, 16, 17
Steiner, George (1929–), ix, 91, 363
stella octangula, 263
Stevin, Simon (1548–1620), 112

structure, 2, 31, 35, 51, 57, 60, 73, 92, 93, 98
    recognising, 91, 92, 99
subtraction, 10, 11, 35
sum, 6
    alternating, 87, 309, 310
    of four squares, 20
    of two squares, 19
    partial, 58
surds, 9, 14, 37, 125, 131, 208, 212, 214
Swetz, Frank, 14
symbol, 5, 9
system
    numeral, 10
Szegő, Gábor (1895–1985), 29

tables
    multiplication, 2, 13
    times, 2
tablet
    clay, 14
Tartaglia, Niccolò (1499–1557), 132
Test
    Square Root, 31, 71
test
    divisibility, 54
tetrahedron, 48, 210, 220, 263
    regular, 28, 48, 203, 208–210, 258, 261–263
    truncated, 259
Thales (c. 620–c. 546 BC), 193, 194
Theorem
    de Moivre's, 130
    Binomial, 336
    Fermat's Last, 1, 145
    Gougu, 14
    Midpoint, 113, 189, 190, 192
    Prime Number, 68
    Ptolemy's, 27
    Pythagoras', 9, 13–17, 39, 40, 170, 186, 188, 195, 206, 207, 217–219, 248, 264, 272, 273
    Pythagoras' converse, 17, 40, 264
    Pythagoras', spherical version of, 273
    Remainder, 332
    Thales', 193
theory
    elementary number, 17

tiling, 199, 253
    Archimedean, 198
    regular, 196, 197, 210
    semi-regular, 198, 205, 252, 253, 258
    semi-regular tilings, 199
time scales, 171
times tables, 2
TIMSS, 99
topology, vii, 109, 172, 173
    four-dimensional, vii, 109
transversal, 182, 183
trapezium, 44, 239
    area of, 193
    isosceles, 21, 263
tree
    binary, 313
    Calkin-Wilf, 313
triangles, 17, 22, 171, 173, 174, 176, 177
    acute angled, 189, 207
    congruent, 45, 136, 176, 177, 181, 185, 191
    equiangular, 253
    equilateral, 15, 21, 24, 28, 29, 43, 48, 175, 196, 199, 204, 205, 208, 230, 231, 247, 251, 253, 254, 257, 260, 262, 281
    isosceles, 30, 39, 47, 177, 178, 185, 231, 233–236, 240, 247, 249, 250, 253–256, 260–263, 267, 268
    isosceles right angled, 48
    obtuse angled, 207
    on a sphere, 26
    orthic, 188, 189, 240, 241
    regular, 16
    right angled, 14, 17, 23, 39, 40, 43, 48, 135, 136, 148, 186, 192, 210, 217–220, 239, 247, 254, 257, 261, 264, 265, 272, 274
    similar, 4, 190–193, 240
    spherical, 23, 25, 26, 45, 218–220
    spherical, area of, 25
trigonometry, 14, 192, 195, 206–208
triple
    Pythagorean, 14, 19, 197

Ulam's spiral, 153
Ulam, Stanisław (1909–1984), 153

unit, 100, 106, 108, 172
    abstract, 108
    dimensionless, 109
    of distance, 106, 107
    of measure, 23
    of measurement, 109
    of speed, 106
    repeated, 171
unitary method, 102

valency, 352
value
    approximate, 30
variable, 112
variety
    toric, vi, 109
VAT, 8
vector, 172
    position, 114
    product, 272
vertex, 312
vertex figure, 196–198, 201–206, 259
Vesalius, Andreas (1514–1564), vii
Viète, François (1540–1603), 112
visualisation, xiii
volume
    of a cone, 244, 245
    of a pyramid, 244
von Neumann, John (1903–1957), xi, 299

Wagner, Roy, 109
Waring, Edward (1736–1798), 20
wave, vii, 109
Whitehead, Alfred North (1861–1947), 134
word, 5

Yevdokimov, Oleksiy, 150

Zarathustra, 363
Zeno of Elea (c. 495 BC – c. 430 BC), 284

This book need not end here...

At Open Book Publishers, we are changing the nature of the traditional academic book. The title you have just read will not be left on a library shelf, but will be accessed online by hundreds of readers each month across the globe. We make all our books free to read online so that students, researchers and members of the public who cant afford a printed edition can still have access to the same ideas as you.

Our digital publishing model also allows us to produce online supplementary material, including extra chapters, reviews, links and other digital resources. Find *The Essence of Mathematics Through Elementary Problems* on our website to access its online extras. Please check this page regularly for ongoing updates, and join the conversation by leaving your own comments:

http://www.openbookpublishers.com/isbn/9781783769996

If you enjoyed this book, and feel that research like this should be available to all readers, regardless of their income, please think about donating to us. Our company is run entirely by academics, and our publishing decisions are based on intellectual merit and public value rather than on commercial viability. We do not operate for profit and all donations, as with all other revenue we generate, will be used to finance new Open Access publications.

For further information about what we do, how to donate to OBP, additional digital material related to our titles or to order our books, please visit our website: http://www.openbookpublishers.com

You may also be interested in...

http://www.openbookpublishers.com/product/340

Teaching Mathematics is nothing less than a mathematical manifesto. Arising in response to a limited National Curriculum, and engaged with secondary schooling for those aged 11–14 (Key Stage 3) in particular, this handbook for teachers will help them broaden and enrich their students' mathematical education. It avoids specifying how to teach, and focuses instead on the central principles and concepts that need to be borne in mind by all teachers and textbook authors—but which are little appreciated in the UK at present. This study is aimed at anyone who would like to think more deeply about the discipline of 'elementary mathematics', in England and Wales and anywhere else. By analysing and supplementing the current curriculum, *Teaching Mathematics* provides food for thought for all those involved in school mathematics, whether as aspiring teachers or as experienced professionals. It challenges us all to reflect upon what it is that makes secondary school mathematics educationally, culturally, and socially important.

www.ingramcontent.com/pod-product-compliance
Lightning Source LLC
Chambersburg PA
CBHW051625230426
43669CB00013B/2180